CONSTRUCTED WETLANDS FOR POLLUTION CONTROL

PROCESSES, PERFORMANCE, DESIGN AND OPERATION

by

IWA Specialist Group on
Use of Macrophytes in Water Pollution Control

AF729551

Published by IWA Publishing in its Scientific and Technical Report series.

Published by IWA Publishing, Alliance House, 12 Caxton Street, London SW1H 0QS, UK
Telephone: +44 (0) 20 7654 5500; Fax: +44 (0) 20 7654 5555; Email: publications@iwap.co.uk
Web: www.iwapublishing.com

First published 2000
Reprinted 2001, 2005 and 2006
© 2000 IWA Publishing

Typeset in 10.5/12pt New Caledonia by Perfect Page Publishing Services, London, UK
Index prepared by June Morrison, Helensburgh, Scotland, UK
Cover design by Hilary Basing at Cato Street Design, London, UK

Apart from any fair dealing for the purposes of research or private study, or criticism or review, as permitted under the UK Copyright, Designs and Patents Act (1998), no part of this publication may be reproduced, stored or transmitted in any form or by an means, without the prior permission in writing of the publisher, or, in the case of photographic reproduction, in accordance with the terms of licences issued by the Copyright Licensing Agency in the UK, or in accordance with the terms of licenses issued by the appropriate reproduction rights organization outside the UK. Enquiries concerning reproduction outside the terms stated here should be sent to IWA Publishing at the address printed above.

The publisher makes no representation, express or implied, with regard to the accuracy of the information contained in this book and cannot accept any legal responsibility or liability for errors or omissions that may be made.

British Library Cataloguing in Publication Data
A CIP catalogue record for this book is available from the British Library

ISBN 978-1-900222-05-1
ISSN 1025-0913

Preface

This Scientific and Technical Report is not an original work. The authors have borrowed freely from a variety of published sources, and have tried to acknowledge those sources throughout the document. Three publications, listed at the end of this preface, must be given special credit, because a large amount of material has been extracted from them. The publishers' permission has been secured for this purpose.

It is the purpose of this report to provide the broad outline of the state of constructed wetland technology. Much more detail is available in the large literature on this subject. The reader is therefore cautioned not to use this report as a design manual, but rather to seek out the relevant publications that deal directly and comprehensively with the specific case under consideration. General and specific references to this literature are provided in Chapter 11.

Constructed wetland technology has grown enormously over the past three decades. First, and still foremost, there has been an exponential growth in their application to the treatment of domestic wastewater. Individual homes, small communities and even rather large cities have used various forms of wetland treatment. Primary, secondary, tertiary and higher treatment levels have been the goals, and both subsurface-flow reed beds and free water surface wetlands have been implemented. There has also been a second type of growth of the use of macrophyte systems: many new application areas have emerged and developed, primarily within the past ten years. These include the treatment of industrial effluents, urban and agricultural stormwater runoff, animal wastewaters, leachates and sludges.

It is clear that aquatic macrophyte assemblages are an important component of the suite of natural systems, which form the foundation of the appropriate technology for wastewater treatment in rural and developing settings. The capital costs of the systems can be small or large, depending on site factors, but the operation is simple and inexpensive. Operators do not require extensive technical skills. Thus the continued growth and expansion of the technology seems assured.

Robert H. Kadlec
Robert L. Knight
Jan Vymazal
Hans Brix
Paul Cooper
Raimund Haberl

References

Constructed wetlands for wastewater treatment in Europe(ed. J. Vymazal, H. Brix, P.F. Cooper, M.B. Green and R. Haberl). Backhuys Publishers, Leiden, The Netherlands, 1998. (366 pages.)

Reed beds and constructed wetlands for wastewater treatment by P.F. Cooper, J.D. Job, M.B. Green and R.B.E. Shutes. WRc Swindon, Wiltshire, 1996. (Book, bibliography and UK database diskette.)

Treatment wetlands by R.H. Kadlec and R.L. Knight. CRC Press, Boca Raton, FL, USA, 1996. (893 pages.)

Contents

Preface iii
List of authors vii
Acknowledgements viii
Glossary of terms ix

1. Biological methods for the treatment of wastewaters 1
- 1.1 Introduction 1
- 1.2 Constructed wetlands 9
- 1.3 Companion natural technologies 12
- 1.4 Integrated natural systems 14

2. Types of constructed wetland 17
- 2.1 Free water surface treatment wetlands 17
- 2.2 Subsurface flow treatment wetlands 20

3. Applications of the technology 23
- 3.1 Domestic and municipal wastewaters 23
- 3.2 Combined sewer overflows 28
- 3.3 Urban stormwater 29
- 3.4 Agricultural 30
- 3.5 Industrial 32
- 3.6 Landfill leachate 37
- 3.7 Sludge consolidation 38

4. Framework for interpreting and predicting water quality improvement 41
- 4.1 Hydraulics 41
- 4.2 Internal flow patterns 47
- 4.3 Thermal effects 48
- 4.4 Performance considerations 50
- 4.5 Stochastic variability 54

5. Mechanisms and results for water quality improvement 55
- 5.1 Suspended solids 55
- 5.2 Biochemical oxygen demand 58
- 5.3 Nitrogen 61
- 5.4 Phosphorus 69
- 5.5 Pathogens 72
- 5.6 Metals 74
- 5.7 Ancillary water chemistry 77

6. Design 79
- 6.1 Scaling the system 79
- 6.2 System layout 85
- 6.3 Compartmentation 85
- 6.4 Pond zones 86
- 6.5 Sealing the basin 86
- 6.6 Substrate selection 87
- 6.7 Inlet and outlet structures 88

7. Plants and planting 91
- 7.1 Varieties of vegetation 91
- 7.2 Role of macrophytes in the treatment of wetlands 93
- 7.3 Establishment 97
- 7.4 Inspection and maintenance 100
- 7.5 Weeds 101

8. System start-up **103**

8.1 Basin testing 103
8.2 Antecedent conditions 103
8.3 Ecological transitions 103
8.4 Performance transitions 106

9. Economics **109**

9.1 Capital costs 109
9.2 Operating costs 113
9.3 Present worth (net present value) 115

10. Case studies **117**

10.1 Free water surface constructed wetlands 117
10.2 Soil-based reed beds 127
10.3 Horizontal subsurface flow 128
10.4 Vertical subsurface flow 132
10.5 Integrated natural treatment system 137
10.6 Floating aquatic plant system 138

11. Bibliography **141**

Index 151

List of authors

Robert H. Kadlec (co-editor)
Wetland Management Services
6995 Westbourne Drive
Chelsea
MI 48118-9527
USA
Telephone: +1 734 475 7256
Fax: +1 734 475 3516
Email: rhkadlec@ic.net

Robert L. Knight (co-editor)
2809 N.W. 161 Court
Gainesville
FL 32609
USA
Telephone:+1 904 462 1003
Fax: +1 904 462 3196
Email: bknight@fdt.net

Jan Vymazal
Ricanova 40
169 00 Praha 6
Czech Republic
Telephone:+42 2 350 761
Fax: +42 2 350 762
Email: vymazal@yahoo.com

Hans Brix
Aarhus University
Department of Plant Ecology
Nordlandsvej 68
DK-8240 Risskov
Denmark
Telephone:+45 8942 4714
Fax: +45 8942 4747
Email: hans.brix@biology.aau.dk

Paul Cooper
The Ladder House
Cheap Street
Chedworth
near Cheltenham
Gloucestershire GL54 4AB
UK
Telephone/fax: +44 1285 720681
Email: paul.cooper@ladderhouse.demon.co.uk

Raimund Haberl
Institute for Water Provision
Water Ecology and Water Management
Department for Sanitary Engineering and Water Pollution Control
Universität für Bodenkultur Wien
Muthgasse 18
A-1190 Wien
Austria
Telephone:+43 1 36 006 5800
Fax: +43 1 36 89 949
Email: haberl@iwgf-sig.boku.ac.at

Acknowledgements

General editing was provided by Christine Waresak. Funding was provided by the International Association on Water Quality (IAWQ). This document was sponsored by the Specialist Group on Use of Macrophytes in Water Pollution Control. Final editing and document production was led by Chris Purdon of IWA Publishing.

Glossary of terms

ABS Alkyl benzene sulphonate plastic used for making pipes.
activated sludge Material consisting largely of naturally occurring bacteria and protozoa, used in and produced by one method of sewage disposal. Sewage is mixed with some activated sludge and agitated with air; organisms of the sludge multiply and purify the sewage. When allowed to settle, they separate out as a greatly increased amount of activated sludge. Part of this is added to new sewage and part is disposed of.
adsorption The adherence of a gas, liquid, or dissolved chemical to the surface of a solid.
advanced wastewater treatment (AWT) Treatment of wastewater beyond the secondary treatment level. In some areas AWT represents treatment to less than 5 milligrams per litre (mg l^{-1}) of 5-day biochemical oxygen demand (BOD_5), 5 mg l^{-1} of total suspended solids (TSS), 3 mg l^{-1} of total nitrogen (TN), and 1 mg l^{-1} of total phosphorus (TP). See also **Tertiary treatment**.
aeration The addition of air to water, usually for the purpose of providing higher oxygen concentrations for chemical and microbial treatment processes.
aerobic Pertaining to the presence of elemental oxygen.
algae A group of autotrophic plants that are unicellular or multicellular and typically grow in water or humid environments.
alkalinity A measure of the capacity of water to neutralize acids because of the presence of one or more of the following bases in the water: carbonates, bicarbonates, hydroxides, borates, silicates or phosphates.
allochthonous External input of organic material into a stream or wetland.
ammonification Bacterial decomposition of organic nitrogen to ammonia.
anaerobic Pertaining to the absence of all oxygen (both free oxygen and chemically bound oxygen).
annual Occurring over a 12-month period.
anoxic Pertaining to the absence of free oxygen but with nitrate, nitrite or sulphate present.
aquaculture Propagation and maintenance of plants or animals by humans in aquatic and wetland environments.
aquatic Pertaining to flooded environments. Over a hydrological gradient, the aquatic environment is the area waterward from emergent wetlands and is characterized by the growth of floating or submerged plant species.
aerenchyma Porous tissues in vascular plants that have large air-filled spaces and thin cell walls. Aerenchymous tissues allow gaseous diffusion between above-ground and below-ground plant structures, thus permitting plants to grow in flooded conditions.
aspect ratio Ratio of wetland cell length to width.
autochthonous Pertaining to substances (usually organic carbon) produced internally in an aquatic or wetland ecosystem.
autotrophic An organism (process) that derives nutrition from inorganic compounds. Photosynthesis is an example of an autotrophic process.
bacteria Microscopic, unicellular organisms lacking chlorophyll. Most bacteria are heterotrophic (some are chemoautotrophs), and many species perform chemical transformations that are important in nutrient cycling and wastewater treatment.
benthic Pertaining to occurrence on or in the bottom sediments of wetland and aquatic ecosystems.
bioassay Biological experiment where plants or animals are used as test organisms.
biomass The mass of living tissues (plant and animal).
BOD (biochemical oxygen demand) Amount of dissolved oxygen that disappears from a water sample in a given time at a certain temperature, through decomposition of organic matter by microorganisms
BOD_5 Five-day biochemical oxygen demand.
bog An acidic, freshwater wetland, dominated by mosses, which typically accumulates peat.
bottomland Floodplain wetlands typically dominated by wetland tree species.
bulk density A measurement of the mass of soil occupying a given volume.
carbonate An inorganic chemical compound containing one carbon atom and three oxygen atoms (CO_3^{2-}).
$cBOD_5$ Carbonaceous BOD_5.
CEC (cation exchange capacity) A measure of the ability of a soil or other substance to bind positively charged ions.
channel A deeper portion of a water flow-way that has faster current and water flow.
channelization The creation of a channel or channels resulting in faster water flow, a decrease in hydraulic residence time, and less contact between waters and solid surfaces within the water body.
clarifier A circular or rectangular sedimentation tank used to remove settled solids from water or wastewater.
COD (chemical oxygen demand) A measure of the oxygen equivalent of the organic matter in water based on reaction with a strong chemical oxidant.
constructed wetland A wetland that is purposely constructed by humans in a non-wetland area.
CSO Combined sewer outflow.
CW Constructed wetland.
denitrification The microbial transformation of nitrate to nitrogen gas.
detritus Dead plant material that is in the process of microbial decomposition.
diffusion The transfer of mass through a gas or liquid from a region of high concentration to a region of lower concentration.
disinfection The use of chemical compounds and physical processes to kill microorganisms.
dispersion Scattering and mixing within a water or gas volume.
diurnal Occurring on a daily basis or during the daylight period.

diversity In ecology, diversity refers to the number of species of plants and animals within a defined area. Diversity is measured by a variety of indices that consider the number of species and, in some cases, the distribution of individuals among species.
down-flow system See VF (vertical-flow system).
ds Digested solids content for sludges.
EA (UK) Environmental Agency.
EC European Community.
ecology The study of the interactions of organisms with their physical environment and with each other and of the results of such interactions.
ecosystem All organisms and the associated non-living environmental factors with which they interact.
EFF Percentage concentration decrease efficiency.
effluent A liquid or gas that flows out of a process or treatment system. Effluent can be synonymous with wastewater after any level of treatment.
E_h Redox potential.
emergent plant A rooted, vascular plant that grows in periodically or permanently flooded areas and has portions of the plant (stems and leaves) extending through and above the water column.
EPA (US) Environmental Protection Agency.
ET Evapotranspiration.
EU European Union (formerly EC)
eutrophic Water with an excess of plant growth nutrients that typically results in algal blooms and extreme (high and low) dissolved oxygen concentrations.
evaporation The process by which water in a lake, river, wetland or other water body becomes a gas.
evapotranspiration The combined processes of evaporation from the water or soil surface and transpiration of water by plants.
exotic species A plant or animal species that has been intentionally or accidentally introduced and that does not naturally occur in a region.
facultative Having the ability to live under different conditions (for example, with or without free oxygen).
faecal Pertaining to faeces (feces in USA).
faecal coliform Aerobic and facultative, Gram-negative, non-spore-forming, rod-shaped bacteria capable of growth at 44 °C (112 °F), and associated with faecal matter of warm-blooded animals.
floating aquatic plant (FAP) A rooted or non-rooted (free-floating) vascular plant that is adapted to have some plant organs (generally the chlorophyll-bearing leaves) floating on the surface of the water in wetlands, lakes and rivers.
fresh water Water with a total dissolved solids content less than 500 mg l^{-1} (0.5 parts per thousand salts).
fungi Microscopic or small non-chlorophyll-bearing, heterotrophic, plant-like organisms that lack roots, stems or leaves, and typically grow in dark and moist environments.
FWS Free water surface. A treatment wetland category that is designed to have a free water surface, above the ground level.
groundwater Water that is located below the ground surface.
ha hectare = 10,000 m^2.
habitat The environment occupied by individuals of a particular species, population or community.
HDPE High-density polyethylene. Used to seal some reed beds.
heavy metals Metallic elements that have an atomic mass of more than 21 in the Periodic Table.
herbaceous Plant parts that contain chlorophyll and are non-woody.
herbivore An animal that feeds primarily on plant tissues.
heterotrophic An organism that derives nutrition from organic carbon compounds.
HF Horizontal-flow constructed treatment wetland.
hybrid system System containing a number of stages comprising horizontal-flow and vertical-flow systems.
hydraulic conductivity, k_f Ability of medium to allow water transmission.
hydraulic loading rate (HLR) A measure of the application of a volume of water to a land area with units of volume per area per time or simply reduced to applied water depth per time (for example, $m^3\ m^{-2}\ d^{-1}$, or cm d^{-1}).
hydraulic residence time (HRT) A measure of the average duration for which water occupies a given volume, with units of time. The theoretical HRT is calculated as the volume divided by the flow (for example, $m^3\ m^{-3}\ d^{-1}$ or d^{-1}). The actual HRT is estimated from tracer studies with conservative tracers such as lithium or dyes.
hydric soil A soil that is saturated, flooded or ponded long enough during the growing season to develop anaerobic conditions. Hydric soils that occur in areas having indicators of hydrophytic vegetation and wetland hydrology are wetland soils.
hydrograph A record of the rise and fall of water levels during a given time period.
hydrology A science dealing with the properties, distribution and circulation of water on the land surface and in the soil, underlying rocks and atmosphere.
hydroperiod The period of wetland soil saturation or flooding. Hydroperiod is often expressed as a number of days or a percentage of time flooded during an annual period (for example, 25 days or 7%).
hydrophytic Plant species that tolerate and typically grow in areas with periodic or continuous flooding.
influent Water, wastewater or other liquid flowing into a water body or treatment unit.
inorganic All chemicals that do not contain organic carbon.
invertebrate All animals that do not have backbones.
I/O Input/output.
k_f Hydraulic conductivity, m s^{-1} or m d^{-1}.
kinetics Pertaining to the rates at which changes occur in chemical, physical and biological processes.
lagoon Any large holding or detention pond, usually with earthen dikes, used to hold wastewater for sedimentation or biological oxidation.
leachate Liquid that has percolated through permeable solid waste and has extracted soluble dissolved or suspended materials from it.
LDPE Low-density polyethylene. Used in Europe as a synthetic liner material.
macroscopic Visible to the unassisted eye.
marsh A wetland dominated by herbaceous, emergent plants.
mass loading The total amount, on a mass or mass per area basis, of a constituent entering a system.
metabolism The chemical oxidation of organic compounds resulting in the release of energy for maintenance and growth of living organisms.
MGD Million US gallons per day (3785 $m^3\ d^{-1}$).
micronutrient A chemical substance that is required for biological growth in relatively low quantities and in small proportion to the major growth nutrients. Some typical micronutrients include molybdenum, copper, boron, cobalt, iron and iodine.
microorganism An animal or plant that can be viewed only with the aid of a microscope.

MMP Meadow/marsh/pond.

MPIP Max Planck Institute Process.

natural wetland A wetland ecosystem that occurs without the aid of humans.

NA Naphthenic acid.

NADB North American Treatment Wetland Database.

NH_4-N (ammonia nitrogen) A reduced form of nitrogen produced as a by-product of organic matter decomposition and synthesized from oxidized nitrogen by biological and physical processes.

nitrification Biological transformation (oxidation) of ammonia nitrogen to nitrite and nitrate forms.

nitrogen fixation A microbial process in which atmospheric nitrogen gas is incorporated into the synthesis of organic nitrogen.

NO_3 + NO_2-N (nitrate plus nitrite nitrogen) Oxidized nitrogen.

NOD Nitrogenous oxygen demand,

NPDES National Pollutant Discharge Elimination System.

NPS Non-point source.

NRCS Natural Resources Conservation Service.

NSCS Nutrient–sediment control system.

nutrient A chemical substance that provides a raw material necessary for the growth of a plant or animal.

NVSS Non-volatile component of TSS.

O&M Operating and maintenance.

OD Oxygen demand.

oligotrophic Water quality characterized by a deficiency of plant growth nutrients.

omnivore An animal that feeds on a mix of plant and animal foods.

organic Pertaining to chemical compounds that contain reduced carbon bonded with hydrogen, oxygen and a variety of other elements. Organic compounds are typically volatile, combustible or biodegradable, and include proteins, carbohydrates, fats and oils.

Org-N (organic nitrogen) Nitrogen that is bound in organic compounds.

OTR Oxygen transfer rate.

oxidation A chemical reaction in which the oxidation number (valence) of an element increases because of the loss of one or more electrons. Oxidation of an element is accompanied by the reduction of the other reactant and, in many cases, by the addition of oxygen to the compound.

PAH Polycyclic aromatic hydrocarbon.

PALD Passive anoxic limestone drain.

parasite An organism that lives within or on another organism and derives its sustenance from that organism without providing a useful return to its host.

PE Population (or person) equivalent. The EU UWWTD defines PE as 65 g BOD_5 d^{-1} per PE A value of 60 g BOD_5 d^{-1} per PE has been widely used in the UK. In flow terms, the UK guide value is *ca.* 200 l d^{-1} per PE. This is an approximate personal contribution of 150 l d^{-1} per PE plus 50 l d^{-1} per PE for infiltration. Guide values for nutrients are 12 g NH_4-N d^{-1} per PE and 2 g TP d^{-1} per PE.

peat Partly decomposed but relatively stable organic matter formed from dead plants in flooded environments.

peatland An area where the soil is predominantly peat.

periphyton The community of microscopic plants and animals that grows on the surface of submergent subjects in water bodies.

perennial Persisting for more than one year. Perennial plant species persist as woody vegetation from year to year or re-sprout from their rootstock on an annual basis.

photosynthesis The biological synthesis of organic matter from inorganic matter in the presence of sunlight and chlorophyll.

phytoplankton Microscopic algae that are suspended in the water column and are not attached to surfaces.

plant community All of the plant species and individuals occurring in a shared habitat or environment.

plug flow Linear flow along the length of a wetland cell.

pretreatment The initial treatments of wastewater to remove substances that might harm downstream treatment processes or to prepare wastewater for subsequent treatment (preliminary plus primary treatments).

primary production The production of organic carbon compounds from inorganic nutrients. The energy source for this production is generally sunlight for chlorophyll-containing plants (photoautotrophs), but in some cases can be derived from reduced chemicals (chemoautotrophs).

primary treatment The first step in the treatment of wastewaters. Primary treatment usually consists of screening and sedimentation of particulate solids.

protozoa Small, one-celled animals including amoebae, ciliates and flagellates.

PVC Poly(vinyl chloride).

RBTS Reed bed treatment system. General term used widely in the UK for constructed reed beds for wastewater treatment. See also constructed wetlands (CW).

receiving water A water body into which wastewater or treated effluent is discharged.

reclaimed wastewater Wastewater that has received treatment sufficient to allow beneficial reuse.

RED Load decrease (percentage mass removal efficiency).

redox potential (E_h) A measure of the electron pressure or availability in a solution; it is often used to quantify further the degree of electrochemical reduction in wetland soils.

reduction A chemical reaction in which the oxidation state (valence) of a chemical is lowered by the addition of electrons. The reduction of a chemical is simultaneous with the oxidation of another chemical and frequently involves the loss of oxygen.

respiration The intake of oxygen and the release of carbon dioxide as a result of metabolism (biological oxidation of organic carbon).

rhizosphere Zone of soil immediately surrounding root and rhizomes and modified by them. Characterized by enhanced microbial activity and by changes in the ratios of organisms compared with surrounding soil. More specifically, a wetland rhizosphere is the chemical sphere of influence of plant roots growing in flooded soils. Depending on the overall oxygen balance (availability and consumption), the rhizosphere can be oxidized, resulting in the presence of aerobic soil properties in an otherwise anaerobic soil environment.

riparian Pertaining to a stream or river. Plant communities occurring in association with any spring, lake, river, stream, creek, wash, arroyo or other body of water or channel having banks and a bed through which waters flow at least periodically.

RZM Root-zone method. Horizontal-flow system built to the Kickuth design (Germany).

salinity A measure of the total salt content of water. Salinity is usually reported as parts per thousand (ppt). The salinity of normal seawater is *ca.* 35 ppt.

saturated soil Soil in which the pore space is filled with water.

SDRB Sludge drying reed bed. (UK)

SE Standard error.
secondary production The production of biomass by consumer organisms by feeding on primary producers or lower trophic level consumers.
secondary treatment Generally refers to wastewater treatment beyond initial sedimentation. Secondary treatment typically includes biological reduction in concentrations of particulate and dissolved concentrations of oxygen-demanding pollutants.
sediment Mineral and organic particulate material that has settled from suspension in a liquld.
seed bank The accumulation of viable plant seeds occurring in soils and available for germination under favourable environmental conditions.
SF Surface flow (*q.v.*).
sheet flow Water flow with a relatively thin and uniform depth.
short-circuit A faster, channelized water flow route that results in a lower actual hydraulic residence time than the theoretical hydraulic residence time.
sludge The accumulated solids separated from liquids, such as water or wastewater, during the treatment process.
soil The upper layer of the earth that can be dug or ploughed and in which plants grow.
stabilization pond A type of treatment pond in which the biological oxidation of organic matter results from the natural or artificially enhanced transfer of oxygen from the atmosphere to the water.
SS Suspended solids.
SSF Subsurface flow (*q.v.*).
submerged plants Plants that have their photosynthetic tissue entirely submerged.
substrate Substances used by organisms for growth in a liquid medium.
substratum Surface area of solids or soils used by organisms for attachment.
subsurface flow (SSF) Flow of water or wastewater through a porous medium such as soil, sand or gravel.
surface flow (SF) Flow of water or wastewater over the surface of the ground.
swamp A wetland dominated by woody plant species including trees and shrubs.
terrestrial Living or growing on land that is not normally flooded or saturated.
tertiary treatment Wastewater treatment beyond secondary and often implying the removal of nutrients.
TKN (total Kjeldahl nitrogen) A measure of reduced nitrogen equal to the sum of Org-N and NH_4-N.
TN (total nitrogen) A measure of all organic and inorganic nitrogen forms in a water sample. Functionally, TN is equal to the sum of TKN and NO_3 + NO_2-N.
TOC (total organic carbon) A measure of the total reduced carbon in a water sample.
toxicity The adverse effect of a substance on the growth or reproduction of living organisms.
TON Total oxidized nitrogen. This is the sum of NO_2 N and NO_3-N.
TP (total phosphorus) A measure of the total phosphorus in a water sample, including organic and inorganic phosphorus in particulate and soluble forms.
transpiration The transport of water vapour from the soil to the atmosphere through actively growing plants.
trickling filter A filter with coarse substrate or medium to provide secondary treatment of wastewater. Microorganisms attached to the filter mediun use and decrease concentrations of soluble and particulate organic substances in the wastewater.
trophic level A level of biological organization characterized by a consistent feeding strategy (for example, all primary consumers are in the same trophic level in an ecosystem).
TSS (total suspended solids) A measure of the filterable matter in a water sample.
TVA Tennessee Valley Authority.
upland Any area that is not an aquatic, wetland or riparian habitat; an area that does not have the hydrological regime necessary to support hydrophytic vegetation.
UWWTD The European Union Urban Waste Water Treatment Directive of 1991.
vegetation The accumulation of living plants within an area.
vertebrate An animal characterized by the presence of a spinal cord protected by vertebrae.
VF Vertical-flow system. Intermittently dosed reed bed system in which the flow is predominantly downflow. The system will be under-drained and because of its aerobic nature will be better for nitrification.
volatile Capable of being evaporated at relatively low temperatures.
VSB Vegetated submerged bed.
watershed The entire surface drainage area that contributes runoff to a body of water.
water table The upper surface of the groundwater or saturated soil.
weir A device used to control and measure water or wastewater flow.
weir gate Water control device used to adjust water levels and measure flows simultaneously.
wetland An area that is inundated or saturated by surface or groundwater at a frequency, duration and depth sufficient to support a predominance of plant species adapted to growth in saturated soil conditions.
wetland function A physical, chemical or biological process occurring in a wetland. Examples of wetland functions include primary production, water quality enhancement, groundwater recharge, organic export, wildlife production and flood intensity decrease.
wetland mitigation bank A preserved, restored, constructed or enhanced wetland that has been purposely set aside to provide compensation credits for losses of wetland functions caused by future human development activities as approved by regulatory agencies.
wetland structure The physical, chemical and biological components of a wetland. Wetland structural components typically include wetland soils, macrophytes, surface water, detritus and microbes, and wetland animal populations.
wetland treatment system A wetland that has been engineered to receive water for the purpose of decreasing the concentrations of one or more pollutants.
wetland values Structural and functional attributes of wetlands that provide services to humans.
WRc Water Research Centre.
WWAR Watershed:wetland area ratio.
zonation The development of a visible progression of plant or animal communities in response to a gradient of water depth or some other environmental factor.
zooplankton Microscopic and small animals that live suspended in the water column.

1 Introduction to constructed wetlands

The term wetland describes a diverse spectrum of ecological systems. Scientific consensus of what constitutes a wetland has been subjectively influenced by definitions that attempt to encompass regulatory and environmental concerns. These concerns have been heightened by historic conversions of wetlands to dry lands and the resulting losses of a variety of natural functions originally provided by the former wetlands. Scientific definitions of wetland types have also been refined as the various structural and functional aspects of these ecosystems have been better described through accelerated research efforts.

A basic understanding of wetland landform will increase an engineer's ability to design constructed wetlands successfully as part of water pollution control systems. This chapter provides a general description of what wetlands are, where they occur, and how they can be constructed for water quality treatment.

1.1 Wetlands in general

The technical meaning of the term wetland includes a wide range of ecosystems. Areas that are not flooded can still be classified as wetlands because of saturated soil conditions, where water is at or below the ground surface during part of a typical growing season. Wetland areas that are deeply flooded grade imperceptibly into aquatic ecosystems as water depth exceeds the growth limits of emergent or submergent vegetation. Figure 1.1 shows how wetlands lie on a continuum between dry lands (uplands) and deeply flooded lands (aquatic systems). Because this is a true continuum, with temporal and biological variability, there is no absolute hydrological demarcation between these ecosystems, and all definitions are somewhat arbitrary.

Figure 1.2 shows structural components typical of wetland ecosystems. Starting with the unaltered sediments or bedrock below the wetlands, these typical components are:

- **Underlying strata**: unaltered organic, mineral or lithic strata, which are typically saturated with or impervious to water and are below the active rooting zone of the wetlands vegetation
- **Hydric soils**: the mineral to organic soil layer of the wetland, which is frequently saturated with water and contains roots, rhizomes, tubers, tunnels, burrows and other active connections to the surface environment
- **Detritus**: the accumulation of live and dead organic material in a wetland, which consists of dead emergent plant material, dead algae, living and dead animals (primarily invertebrates) and microbes (fungi and bacteria)
- **Water**: standing water, which provides a habitat for aquatic organisms including fish and other vertebrate animals, submerged and floating plant species that depend on water for buoyancy and support, living algae and populations of microbes
- **Emergent vegetation**: vascular, rooted, hydrophytic plant species, which contain structural components that emerge above the water surface, including both herbaceous and woody plant species.

Natural wetlands usually have all of these attributes. Constructed wetlands can have less mature components, especially soil organic matter, which forms over an extended period of time. The structural components of natural wetlands are highly variable and depend on

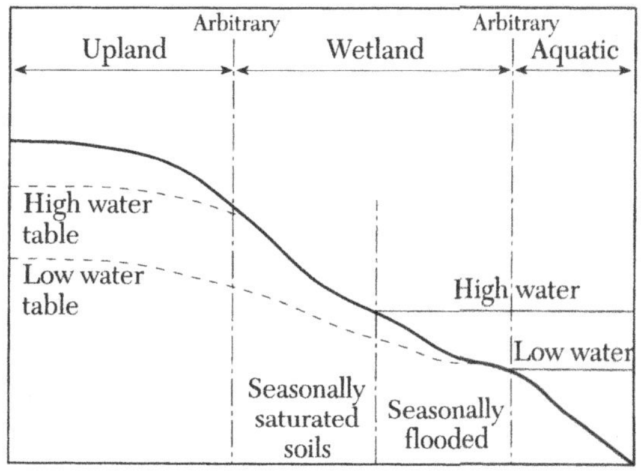

Figure 1.1. Wetlands are transitional areas between uplands, where excessive water is not a factor for plant growth, and aquatic ecosystems, where flooding excludes rooted, emergent vegetation (Kadlec & Knight 1996).

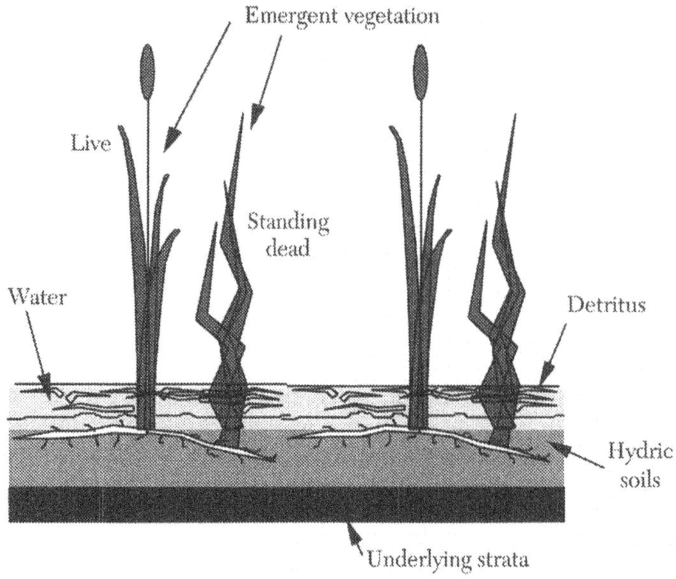

Figure 1.2. Structural components of a wetland.

hydrology, underlying sediment types, water quality, climate and successonal maturity.

1.1.1 Hydrology

The water status of a wetland defines its extent and is the determinant of species composition in natural wetlands (Mitsch & Gosselink 1993). Hydrologic conditions also influence the soils and nutrients, which in turn influence the character of the biota. The flows and storage volume determine the length of time that water spends in the wetland, and thus the opportunity for interactions between water-borne substances and the wetland ecosystem.

1.1.1.1 Water regime

The most consistent attribute of wetlands is the presence of water during some or all of an average annual period. Wetlands are areas in which the soil is saturated with water or in which shallow standing water results in the absence of plant species that depend on aerobic soil conditions. Wetlands are dominated by plant species that are adapted to growing in seasonally or continuously flooded soils with resulting anaerobic or low-oxygen conditions. At their upslope margin, wetlands can be distinguished from uplands by the latter's tendency to remain flooded or saturated for less than 7–30 days each year, a short enough period for oxygen and other soil conditions not to limit plant growth. At their downgradient edge, wetlands grade into aquatic systems that are flooded to a depth at which, or at a duration for which, emergent, rooted plants cannot survive. The average water depth that typically separates wetlands from adjacent aquatic ecosystems is on the order of 1 m.

The concepts of **hydroperiod** and **water regime** include two interdependent components: (1) hydroperiod, which is the duration of flooded or saturated soil conditions as a percentage of time, and (2) the depth of flooding (Figure 1.3). While hydroperiod refers to the duration of flooding, the term water regime refers to the combination of water depth and flooding duration (depth–duration curve). Although the regular and continual presence of water separates uplands from wetlands and aquatic ecosystems, the overall water regime is the most important contributor to wetland type or class (Gosselink & Turner 1978). The importance of this factor in wetland treatment system design and operation cannot be overstated because an incorrect understanding of the water regime requirements of wetland plant species is the most frequent cause of vegetation problems in natural and constructed wetlands. The duration and depth of flooding affect plant physiology because of soil oxygen concentration, soil pH, dissolved and chelated macronutrients and micronutrients, and toxic chemical concentrations. Predicting and controlling the water regime of a treatment wetland is relatively easy. Creating and maintaining a complex plant community during wetland treatment design and performance is more difficult.

1.1.1.2 Water budget

Water enters natural wetlands via streamflow, runoff, groundwater discharge and precipitation (Figure 1.4). These flows are extremely variable in most instances, and the variations are stochastic in character. Stormwater treatment wetlands generally possess this same suite of inflows. Treatment wetlands dealing with continuous sources of wastewater can have the same inputs, although streamflow and groundwater inputs are typically absent. The steady inflow associated with continuous source treatment wetlands represents an important distinguishing feature. A dominant steady

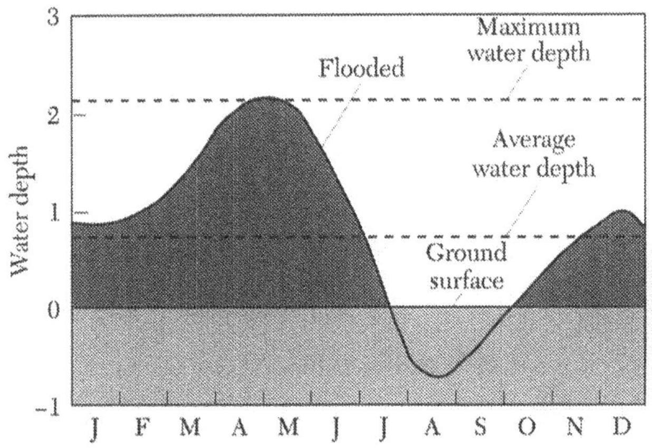

Figure 1.3. Components of wetland hydroperiod and water regime (Kadlec & Knight 1996). Hydroperiod = 9/12 = 75%; average depth = 0.8; maximum depth = 2.2.

inflow drives the ecosystem towards an ecological condition that is somewhat different from a stochastically driven system.

Wetlands lose water via streamflow, groundwater recharge and evapotranspiration (Figure 1.4). Stormwater treatment wetlands also possess this suite of outflows. Continuous source treatment wetlands would normally be isolated from groundwater, and most of the water would leave via streamflow in most cases. Evapotranspiration (ET) occurs with strong diurnal and seasonal cycles because it is driven by solar radiation, which undergoes such cycles. Thus ET can be an important water loss on a periodic basis.

Wetland water storage is determined by the inflows and outflows together with the characteristics of the wetland basin. Depth and storage in natural wetlands are likely to be modulated by landscape features, such as the depth of an adjoining water body or the conveyance capacity of the outlet stream. Large variations in storage are therefore possible, in response to the high variability in the inflows and outflows. Such periods of drying out have strong implications for the vegetative structure of the ecosystem. Constructed treatment wetlands, in contrast, typically have some form of outlet water level control structure. There is therefore little or no variation in water level, except in stormwater treatment wetlands. Drying out does not normally occur, and only those plants that can withstand continuous flooding will survive.

Temporal changes in depth, combined with an uneven topography of the wetland bottom, lead to vegetative pattern effects in natural wetlands. Constructed treatment wetlands usually have nearly uniform bottoms. Combined with controlled, steady water levels, this means uniform hydrological conditions and an absence of pattern effects. Pattern effects interact with water flows through the wetland, with preferential, sparsely vegetated channels carrying a disproportionately high fraction of the water. This in turn impairs the treatment potential because much of the wetland surface is not exposed to the water flow.

The important features of wetland hydrology from the standpoint of treatment efficiency are those that determine the duration of interactions between water and biota and the proximity of waterborne substances to the sites of biological and physical activity. There is a strong tendency in the wetland treatment literature to borrow the detention time concept from other aquatic systems, such as 'conventional' wastewater treatment processes. In purely aquatic environments, reactive organisms are distributed throughout the water, and there is often a clear understanding of the flow paths through the vessel or pond. However, wetland ecosystems are more complex and therefore require more descriptors.

1.1.2 Soils

1.1.2.1 Formation

Many wetland soils are characterized by a lack of oxygen, induced by flooding. Oxygen diffusion in flooded soils is nearly 10,000 times slower than in dry soils (Armstrong 1978). Well-aerated, upland soils rapidly experience a decline in soil oxygen and redox potential when they are flooded. Continuous or seasonal inundation combined with the production of large amounts of dead organic matter (litterfall) results in nearly perpetual soil anaerobiosis in many wetlands. The resulting lower dissolved oxygen level results in the accumulation of organic matter in wetland soils because of a decreased level of microbial activity and organic decomposition.

Under oxygen-deficient conditions created by extended and deep innundation and a high consumption rate of available electron acceptors, there is a net accretion of organic matter, over and above any sedimentation of incoming suspended matter. This is the process of peat accumulation if most of the material originates

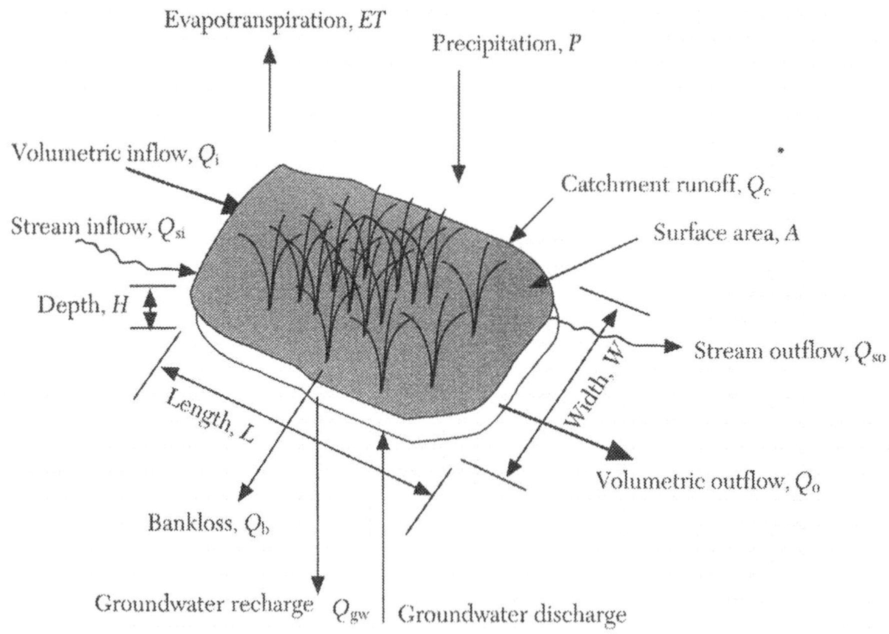

Figure 1.4. Components of the water budget and associated terminology.

from leaf and stem detritus of emergent macrophytes in marshes or from sphagnum mosses in bogs. This can range from 0.3 to 1.4 cm yr^{-1} in warm-climate freshwater marshes (DeLaune *et al.* 1978) and from 0.1 to 1.1 mm yr^{-1} for northern bogs in the UK (Durno 1961).

These processes also occur in treatment wetlands. However, the antecedent soils often undergo a transformation to a hydric status (Figure 1.5).

The build-up of mineral matter that settles from incoming stormwater, river water or wastewater is often slow. At the low end of the spectrum are the clean wastewaters from advanced treatment plants, which can have less than 10 mg l^{-1} of suspended matter. If this material is all inorganic and undecomposible, it can accrete in the treatment wetland. Under most circumstances this represents only a few millimetres per decade of solids buildup in the wetland. Typical loadings are less than 100 mg m^{-2} yr^{-1}.

1.1.2.2 Chemical environment

Wetland soils have a high trapping efficiency for a variety of chemical constituents; they are retained within the hydrated soil matrix by forces ranging from chemical bonding to physical dissolution within the water of hydration. The combined phenomena are referred to as **sorption**. A significant portion of the chemical binding is cation exchange, which is the replacement of one positively charged ion, attached to the soil or sediment, with another positively charged ion. The humic substances found in wetlands contain large numbers of hydroxyl and carboxylic functional groups, which are hydrophilic and serve as cation-binding sites.

Wetlands are ideal environments for chemical transformations because of the range of oxidation states that naturally occur in wetland soils. Free oxygen decreases rapidly with depth in most flooded soils because of the metabolism of microbes that consume organic matter in the soil and through the chemical oxidation of reduced substances. This decline in free oxygen is measured as an increasingly negative electric potential between a standard platinum electrode and a calomel electrode. The measure of electric potential is called reduction–oxidation or redox potential (E_h).

As long as free, dissolved oxygen is present in solution, the redox potential varies little (in the range of +400 to +700 mV). However, it is a sensitive measure of the degree of reduction of wetland soils after oxygen disappears, ranging from +400 mV down to −400 mV. The greater range of redox potentials for flooded soils than for aerobic soils is important. Wetland systems maintain a wider range of redox reactions than upland soils, and their most important function might be as chemical transformers. Wetlands are often the major reducing ecosystems on the landscape and therefore have great potential for processing nutrients and other materials.

Once a soil is flooded, the oxygen present is quickly consumed by microbial respiration and chemical oxidation. Subsequently, anaerobic microorganisms use a variety of substances to replace oxygen as the terminal electron acceptor during respiration. This electron transfer causes significant changes in the valence state of the chemical species used and the overall soil reduction. Reduction of a saturated soil is a sequential process governed by the laws of thermodynamics. Nitrate is the first soil component reduced ($NO_3^- \rightarrow O_2^-$, E_h 220 mV) after

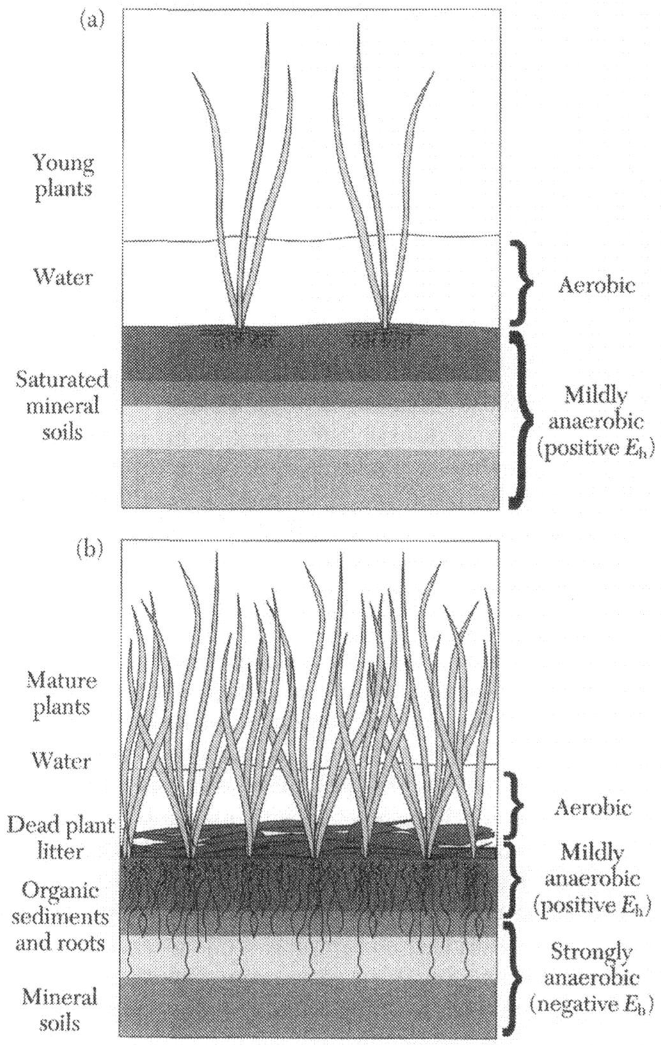

Figure 1.5. Stages in the maturation of constructed wetland soils: (a) newly planted, (b) mature (Kadlec & Knight 1996).

oxygen, although this process can proceed before oxygen is completely consumed. Manganese as manganic ions ($Mn \rightarrow Mn^{2+}$, E_h 200 mV) closely follows NO_3^- in the reduction sequence, even before NO_3^- has completely disappeared. Although the preceding reactions can and do overlap, the subsequent sequential reactions of ferric iron to ferrous iron ($Fe^{3+} \rightarrow Fe^{2+}$, E_h 120 mV), sulphate to sulphide ($SO_4^{2-} \rightarrow S^{2-}$, E_h −75 to −150 mV) and carbon dioxide to methane ($CO_2 \rightarrow CH_4$, E_h −250 to −350 mV) will not occur unless the preceding component has been completely reduced. Bacteria-mediated reduction and oxidation processes in waterlogged soils are summarized in an excellent review by Laanbroek (1990).

1.1.2.3 Microbial processes

Soil microbial populations have significant influence on the chemistry of most wetland soils. Important transformations of nitrogen, iron, sulphur and carbon result from microbial processes. These microbial processes are typically affected by the concentrations of reactants as well as the redox potential and pH of the soil.

Several nitrogen transformations occur in wetlands. Organic nitrogen is biologically transformed to ammonia nitrogen through the process of mineralization (= ammonification). Mineralization results as a consequence of the decomposition of organic matter, resulting from the actions of both aerobic and anaerobic microbes. Ammonia is in turn converted to nitrite and nitrate nitrogen through aerobic microbial processes called nitrification. Nitrate nitrogen can be further transformed to nitrous oxide or nitrogen gas in anoxic or anaerobic wetland soils by the action of another group of microbes (denitrifiers). Nitrogen gas can also be transformed to organic nitrogen by bacterial nitrogen fixation in some aerobic and some anaerobic wetland soils.

When the reduction of nitrate stops by depletion of this electron acceptor, the reduction of ferric oxide starts in waterlogged soils. Ferric oxides are assumed to be one of the most abundant electron acceptors in soils as well as in sediments. The direct enzymic oxidation of Fe^{2+} (and also Mn^{2+}) is confirmed to a restricted range of organisms; most bacteria cause precipitation of Fe and/or Mn by indirect means by altering E_h or pH, which in turn leads to chemical oxidation and precipitation (Grant & Long 1985).

Sulphate can be reduced to sulphide by obligate anaerobic bacteria in wetlands. The sulphate serves as an electron acceptor in the absence of free oxygen at low redox potentials. Sulphides can provide a source of energy for chemoautotrophic and photosynthetic bacteria in aerobic wetlands, resulting in the formation of elemental sulphur and sulphate. The sulphide is in turn capable of precipitating metal sulphides.

Organic soil carbon is degraded microbially to carbon dioxide by aerobic respiration when oxygen is available, and by fermentation under anaerobic conditions. In fermentation, organic matter serves as the terminal electron acceptor, forming acids and alcohols. Methane can be formed in wetland soils by the action of bacteria at very low redox potentials.

1.1.2.4 Plant-animal-soil interactions

Organic matter accumulation in some wetlands is a direct or indirect result of the primary fixation of carbon from the atmosphere by plants. Particulate macrophytic detrital material and dead algal cells contribute organic carbon, nitrogen and phosphorus to the wetland litter/soil layer in the form of cellulose, hemicellulose, lignin, proteins and phospholipids (Reddy & D'Angelo 1994). In some low-nutrient wetlands and in wetlands that are drained and exposed to the atmosphere, oxidation can result in no net accumulation of organic matter.

Wetland macrophytes further modify the texture, the hydraulic conductivity and the chemistry of the soil by the growth of plant roots and rhizomes. These plant structures initially serve as pathways for increased gaseous diffusion into and out of the wetland sediments. Gas-filled aerenchyma in wetland plants provides significantly less diffusional resistance, allowing some oxidation of soils in the immediate vicinity of the roots (rhizosphere) and the diffusion of carbon dioxide, hydrogen sulphide and even methane back to the atmosphere through the plants. Several of the important chemical transformations mentioned earlier occur on or within the aerated rhizosphere of wetland plants.

The top layer of soil contains the roots of the emergent, submergent and floating-leaved macrophytes. These most often occupy the top 20–30 cm of soil. A dense macrophyte stand will have a large amount of below-ground biomass in the form of roots and rhizomes, often in the range 500–5000 g of dry matter m^{-2}.

Fish such as carp are known to stir lake and wetland sediments in their continual search for prey organisms. Wading birds also will feed in aerobic sediments on macroinvertebrates and their resulting beak holes can number in the dozens per square metre in shallow wetland areas. Mammals can inhabit wetlands and either dig for crayfish, clams or other sediment-colonizing food organisms or build dens and burrows in or through the wetland sediments. When they occur, all of these faunal processes tend to increase the localized oxidation potential of wetland soils.

1.1.2.5 Treatment wetland soils

The sediments that form in surface flow treatment wetlands are often different from those that form in natural wetlands, for a number of reasons. First, the enhanced activity of various microbes, fungi, algae and soft bodied invertebrates leads to a greater proportion of fine detritus than leaf, root and stem fragments. There is a significant formation of low-density biosolids (sludge). Secondly, there can be a precipitation of metal hydroxides or sulphides, which add mineral flocs to the sediments. Finally, there is often a high ionic strength associated with effluents being treated, reflected in a high content of dissolved salt. The effect of high ionic strength is to alter the structure of the highly hydrated organic materials that comprise wetland sediments and soils. Some of the same types of material accrete in the pore spaces of subsurface flow (SSF) wetlands.

Some measure of performance control can be exerted by the use of specially tailored bed media for constructed treatment wetlands. If sands, soils or gravels are borrowed from natural sources, there will be a period of adaptation as hydric soil properties develop. However, a bed material can be chosen that is manufactured to have a very large phosphorus sorption capacity, such as an expanded clay (Jenssen *et al.* 1994). This design philosophy is now quite different from that for most treatment wetlands: the intent is to exhaust a short-term capacity, regenerate the wetland and repeat the cycle. This can be a feasible strategy in some cases, provided that the expense of regeneration coupled with its frequency are within acceptable economic bounds.

1.1.3 Vegetation

Macrophytic plants provide much of the visible structure of wetland treatment systems. There is no doubt that they are essential for the high levels of water quality improvement typical of most wetland treatment systems. The numerous studies measuring treatment with and without plants have concluded almost invariably that performance is higher when plants are

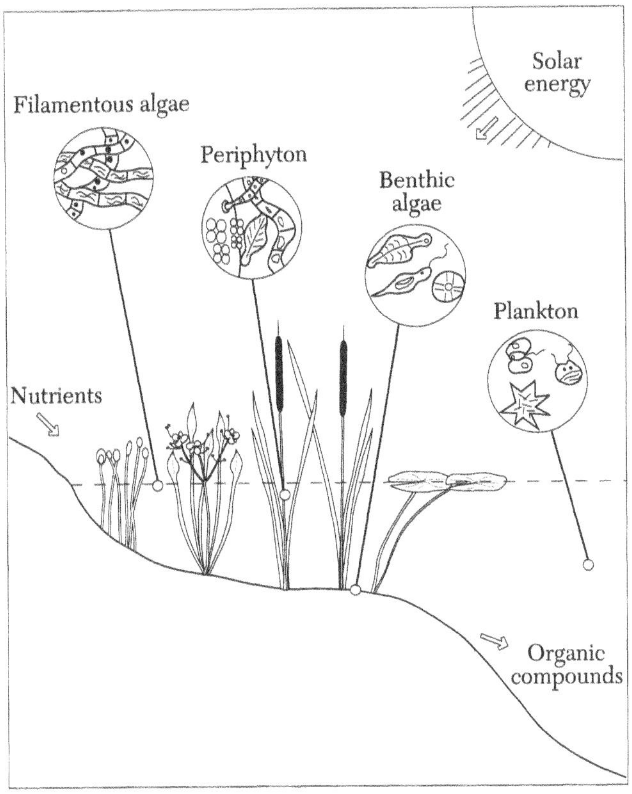

Figure 1.6. Algae and macrophytes in treatment wetlands (Kadlec & Knight 1996).

present. This finding led some researchers to conclude that wetland plants were the dominant source of treatment because of their direct uptake and sequestering of pollutants. It is now known that plant uptake is the principal removal mechanism only for some pollutants and some types of treatment wetland (for example with free-floating plants with a regular harvest) and only in lightly loaded systems. During an initial successional period of rapid plant growth, direct pollutant immobilization in wetland plants can be important. For many other pollutants, plant uptake is generally of minor importance compared with microbial and physical transformations that occur within most wetlands. Macrophytic plants are essential in wetland treatment systems because they provide structure and a source of reduced carbon for the microbes that mediate most of the pollutant transformations that occur in wetlands.

The diversity of wetland plant adaptations provides the wetland treatment system designer with numerous options and potential problems. Some plant species produce large amounts of carbon that are able to support heterotrophic microbes important in nutrient transformations. Other plant species provide shading of the water surface, in turn controlling algal growth and concentrations of suspended solids in the discharge from the wetland treatment system. An understanding of the ecological properties of these wetland plant species is essential for the successful design, construction and operation of wetland treatment systems.

1.1.3.1 Algae

Algae are unicellular or multicellular plants that do not have the variety of tissues and organs of higher plants. Algae are a highly diverse assemblage of species that can live in a wide range of aquatic and wetland habitats. Many species of algae are microscopic and are only discernable as the green or brown colour or 'slime' occurring on submerged substrates or in the water column of lakes and ponds.

Several functional algal groups are found in wetlands (Figure 1.6). Algae can be broadly classified as free-floating (phytoplankton) and attached (periphyton, benthos). Planktonic algae swim or are found in the water column. Planktonic algae are generally not important in wetland ecosystems through their direct action. However, in wetlands with open water their photosynthetic activity can result in high pH values (over 10 during the day).

Benthos and periphyton are composed of attached and bottom-dwelling organisms. There is some controversy between the use of terms benthos and periphyton. Benthos is sometimes confined only to organisms attached to the bottom, and periphyton was first used to refer only to organisms growing on objects placed in the water by people. Both terms, however, include the same division into groups according to the substrate to which the organisms are attached (Vymazal 1995):

- epilithon (attached to stones)
- epipelon (attached to mud or sand)
- epiphyton (attached to plants)
- epizoon (attached to animals)
- epipsammon (attached to sand particles).

Attached communities always include a variety of free-living algae (not attached to the surface, i.e. metaphyton), fungi, bacteria and protozoans. Attached algae can form a significant portion of the plant biomass in some wetland systems; dry biomass can amount to more than 1000 g m^{-2} (Vymazal 1995).

Algae can also form so-called floating mats that are formed mostly by filamentous algae. Quite often the floating mats are formed with epipelic filamentous species that get loose from the bottom or epiphytic species growing on small, submerged macrophytes.

1.1.3.2 Macrophytes

The term macrophyte includes vascular plants that have tissues that are easily visible (Figure 1.6). Wetland macrophytes are the dominant structural component of most wetland treatment systems. A basic understanding of the growth requirements and characteristics of these wetland plants is essential for the successful design and operation of treatment wetlands. Vascular plants differ from algae in their internal organization into tissues, resulting from the presence of specialized cells. A wide variety of macrophytic plants occur naturally in wetland environments. More than 6700 plant species of obligate and facultative wetland plant species are present in the USA. Obligate wetland plant species are defined as those that are found exclusively in wetland habitats; facultative species are those that can be found in upland or wetland areas. There are many guidebooks that illustrate wetland plants (for example, Hotchkiss 1972; Niering 1985).

Annual plant species survive for only one growing season and must be re-established annually from seed. Perennial plant species live for more than one year and typically propagate each year from perennial root systems or from perennial above-ground stems and branches. The terms emergent, floating and submerged refer to the predominant growth form of a plant species. In emergent plant species, most of the above-ground part of the plant emerges above the water line and into the air. Both floating and submerged vascular plant species can occur in wetland treatment systems. Floating species have leaves and stems buoyant enough to float on the water surface. Submerged species have buoyant stems and leaves that fill the niche between the sediment surface and the top of the water column. Floating and submerged species are more typical of deeper, aquatic habitats than of wetlands, but they can occur in wetlands when water depth exceeds the tolerance range for rooted, emergent species.

All vascular plant roots require gaseous exchange to supply oxygen for cell respiration and to exhaust gases such as carbon dioxide that might accumulate during metabolic processes. All plants also require water for numerous biochemical processes, including photosynthesis and transpiration; it assists with the intercellular transport of nutrients and metabolites. One adaptation to flooding is the development of aerenchymous plant tissues that transport gases to and from the roots through the vascular tissues of the plant above water and in contact with the atmosphere, providing an aerated root zone and thereby lowering the plant's reliance on external oxygen diffusion through water and soil (Armstrong 1978).

As with all plant species, wetland plants increase their numbers and density through asexual and sexual reproduction. Asexual reproduction refers to an increase in the number of individuals of a plant species through vegetative growth; it typically occurs through the growth of roots or rhizomes, with the subsequent emergence of new above-ground stems and leaves. Technically, a cattail bed that developed vegetatively from a single parent plant is a single plant. However, when these rhizomes are cut or decay, the individual daughter plants can remain viable and continue to spread vegetatively. In sexual reproduction, two individual plants, or male and female flowers from a single plant, contribute gametes to form seeds with new combinations of genetic material. Sexual reproduction is important in providing alternative strategies for plants to survive from year to year through seasonal extremes, to propagate the species over large distances, to colonize new habitats rapidly and to provide genetic variants that can adapt to changing environmental and competitive conditions.

The net primary productivity of freshwater marshes is estimated most frequently through the harvesting of annual peak standing stocks of live and dead plant biomass. When root biomass is measured, it is usually an important part of net annual plant production. Some researchers consider net primary productivity estimates that are made by peak standing stock to be underestimates because they do not account for biomass turnover during the growing season (Pickett *et al.* 1989). The range of net production rates in natural wetlands that are not subject to obvious anthropogenic nutrient enrichments vary from about 50 g of dry matter $m^{-2}\,yr^{-1}$ in arctic tundra to 3500 g of dry matter $m^{-2} yr^{-1}$ in marshes in semi-tropical climates. Most temperate freshwater marshes have net primary production rates of 600–3000 g of dry matter $m^{-2} yr^{-1}$.

Nutrients affect wetland plant growth. The maximum rate of plant growth is attained as nutrient levels are initially increased. However, at higher nutrient levels, plant growth levels off while luxury nutrient uptake continues; at higher nutrient concentrations, phytotoxic responses can be observed.

Over the life cycle of a vascular plant, all plant tissues are consumed, exported or eventually recycled back to the ground as plant litter. Litterfall and the resulting decomposition of organic plant material is an ecologically important function in wetlands. Wetland plant tissues fall at variable rates, depending on the survival strategy of the individual plant species. Herbaceous plant species typically recycle the entire above-ground portion of the plant annually in temperate environments. The growth season can vary from 10 months or more in subtropical regions to less than 3 months in colder climates. In addition, most herbaceous species lose a fraction of living leaf and stem material as litter throughout the growing season, so there is a continuous rain of dead plant tissues throughout the year with seasonal highs and lows of litterfall.

Litter decomposition rates vary widely between macrophyte species. Decomposition constants have been reported in the range 0.0005–0.16 d^{-1} for herbaceous wetland plants, with lower values for emergent species and higher values for submerged and free-floating species. Twigs, branches and roots of woody species have lower decay rates than herbaceous species. The half-lives (that is, the time for 50% decomposition) range between less than 20 d for free-floating and submerged species and more than 500 d for emergent species. The half-lives for parts of woody species are usually more than 1000 d (Vymazal 1995).

1.1.4 Fauna

Animals have a sometimes subtle but important role in wetlands used for water quality enhancement. From the tiniest microscopic protozoans to the largest mammals, animals consume energy-yielding biomass, convert part of this energy into new biomass, and recycle unused organic matter and nutrients. Nutrients spiral their way up the food chain and are continuously used and transformed so that they can be used again. Consumers keep nutrients in circulation and regulate the populations of lower trophic levels in a manner that maximizes system function (Odum 1983). Wetland ecosystems exposed to toxins or other factors that eliminate consumer populations have smaller nutrient cycling functions, which can in turn affect the performance of biological water quality treatment.

In most cases, the wetland designer does not need to be concerned with the nutritional and habitat requirements of the animal populations present in a wetland treatment system. The diversity of adjacent wetlands and aquatic systems is frequently adequate to provide faunal colonizers for constructed wetland treatment systems. When these natural colonizers are present, a diverse assemblage of organisms will establish in a newly constructed wetland in a few years or less and will create a balanced wetland ecosystem that has essential self-regulating functions. However, if a wetland treatment system is to be constructed where adjacent sources of adapted species are not present, the designer might need to promote colonization artificially through the importation of water, sediments and plants containing microscopic and minute wetland animals and microbes from more distant sources.

In cases in which the wetland designer wishes to achieve significant wildlife benefits in addition to water quality treatment benefits, greater consideration must be given to wildlife populations during the design, construction and operation of wetland treatment systems. The ancillary benefits potentially achieved when treatment wetlands are built to attract wildlife can be an added value with relatively little capital expenditure and operating cost.

Some animal activities can be detrimental to treatment functions or to the physical integrity of the constructed wetland. Beavers, nutria and muskrats are all capable of great damage by burrowing and herbivory. Grazing and foraging animals, such as deer, elk and wild pigs, can cause damage to vegetation. Several species of bottom-foraging fish can defeat solids settling.

1.2 Constructed wetlands

1.2.1 Technology description

A wetland is a complex assemblage of water, substrate, plants (vascular and algae), litter (primarily fallen plant material), invertebrates (mostly insect larvae and worms) and an array of microorganisms (most importantly bacteria). The mechanisms that are available to improve water quality are therefore numerous and often interrelated. These mechanisms include:

- settling of suspended particulate matter
- filtration and chemical precipitation through contact of the water with the substrate and litter
- chemical transformation
- adsorption and ion exchange on the surfaces of plants, substrate, sediment and litter

- breakdown, and transformation and uptake, of pollutants and nutrients by microorganisms and plants
- predation and natural die-off of pathogens.

The most effective treatment wetlands are those that foster these mechanisms. The specifics for the various types of wetlands and wastewater are discussed in Chapters 2 and 3.

Constructed wetlands are a cost-effective and technically feasible approach to the treatment of wastewater and runoff for several reasons:

- wetlands can be less expensive to build than other treatment options
- operation and maintenance expenses (energy and supplies) are low
- operation and maintenance require only periodic, rather than continuous, on-site labour
- wetlands are able to tolerate fluctuations in flow
- wetlands are able to treat wastewaters with low organic load (too low for activated sludge)
- they facilitate water reuse and recycling.

In addition:

- they provide habitat for many wetland organisms
- they can be built to fit harmoniously into the landscape
- they provide numerous benefits in addition to water quality improvement, such as wildlife habitat and the aesthetic enhancement of open spaces
- they are an environmentally sensitive approach that is viewed with favour by the general public.

Wetland treatment systems use water-tolerant plant species and shallow, flooded or saturated soil conditions to provide various types of wastewater treatment. The two basic types of wetland treatment systems include constructed free water surface (FWS) or surface flow (SF) wetlands, and constructed SSF wetlands.

Constructed wetlands mimic the optimal treatment conditions found in natural wetlands but provide the flexibility of being constructable at almost any location and can be used for treatment of primary and secondary wastewaters as well as waters from a variety of other sources including stormwaters, landfill leachate, industrial and agricultural wastewaters, and acid-mine drainage.

Surface flow wetlands are densely vegetated by a variety of plant species and typically have water depths less than 0.4 m. Open water areas can be incorporated into a design to provide for the optimization of hydraulics and for wildlife habitat enhancement. According to WPCF (1990), typical hydraulic loading rates are between 0.7 and 5.0 cm d^{-1} (between 2 and 14 ha per 1000 m^3 d^{-1}) in constructed surface flow treatment wetlands.

SSF wetlands use a bed of soil or gravel as a substrate for the growth of rooted emergent wetland plants. Pretreated wastewater flows by gravity, horizontally or vertically, through the bed substrate, where it contacts a mixture of facultative microbes living in association with the substrate and plant roots. The bed depth in SSF flow wetlands is typically between 0.6 and 1.0 m, and the bottom of the bed is sloped to minimize water flow overland.

Most frequently used species in SSF constructed wetlands are common reed (*Phragmites australis*), cattail (*Typha* spp.), bulrush (*Scirpus* spp.), reed canarygrass (*Phalaris arundinacea*) and sweet mannagrass (*Glyceria maxima*). Some oxygen enters the bed substrate by direct atmospheric diffusion and some through the plant, resulting in a mixture of aerobic and anaerobic zones. Most of the saturated bed is anoxic or anaerobic under most wastewater design loadings. According to WPCF (1990), typical hydraulic loading rates in SSF wetlands range from 2 to 20 cm d^{-1} (from 0.5 to 5 ha Dm^{-3} d^{-1}).

Wetlands have been found to be effective in treating biochemical oxygen demand, suspended solids, nitrogen and phosphorus, as well as for decreasing the concentrations of metals, organic chemicals and pathogens. Effective wetland performance depends on adequate pretreatment, conservative constituent and hydraulic loading rates, the collection of monitoring information to assess system performance, and a knowledge of successful operation strategies.

A common difficulty experienced by wetland treatment systems has been inadequate oxygen supply. When wetland systems are overloaded by oxygen-demanding constituents, or are operated with excessive water depth, highly reduced conditions occur in the sediments, resulting in plant stress and decreased removal efficiencies for biochemical oxygen demand and ammonia nitrogen. A common problem encountered in SSF constructed wetlands is an inadequate hydraulic gradient and resulting surface flows.

Constructed FWS wetlands require a capital expenditure typically between US\$10,000 and US\$100,000 ha^{-1}, primarily as a result of the earthwork costs. SSF wetlands are typically more expensive per unit area than FWS systems, with capital costs from US\$100,000 to US\$200,000 ha^{-1}. Operation and maintenance costs for constructed wetlands are primarily related to system monitoring and are generally very low (US\$0.03–0.09 m^{-3} (WPCF 1990)).

1.2.2 Historical development

1.2.2.1 Free water surface wetlands

Natural wetlands have been used as convenient wastewater discharge sites for as long as sewage has been collected (at least 100 years in some locations in the USA). Examples of old wetland sites include the Great Meadows natural wetland near the Concord River in Lexington, Massachusetts, which began receiving wastewater in 1912; the Brillion Marsh in Wisconsin that has received municipal wastewater discharges since 1923; and the Dundas sewage treatment plant, which began discharging to the Cootes Paradise natural wetland near Hamilton, Ontario, in 1919. When monitoring was initiated at some of these existing discharges, an awareness of the water quality purification potential of wetlands began to emerge. The FWS wetland 'technology' started in North America in the 1970s, with the ecological engineering of natural wetlands for wastewater treatment (Ewel & Odum 1984; Kadlec & Tilton 1979).

In 1973, the first intentionally engineered, constructed wetland treatment pilot systems in North America were constructed at Brookhaven National Laboratory near Brookhaven, New York. These pilot treatment systems combined a marsh wetland with a pond and a meadow in series and were designated as the meadow/marsh/pond (MMP) treatment system. Also in 1973, the Mt View Sanitary District in California constructed about 8.5 ha of wetland marshes for wildlife habitat and wastewater discharge. Industrial stormwaters and process waters were also applied to constructed pond/wetland systems as early as 1975 at Amoco Oil Company's Mandan Refinery in North Dakota (Litchfield 1989).

Currently, Florida has several of the largest constructed wetland treatment areas in the world, including the Lakeland and Orlando constructed wetlands, both of which were started in 1987. Each wetland has about 500 ha for the advanced treatment of municipal wastewater. Another large constructed wetland in Florida (1490 ha) has treated drainage from the Everglades Agricultural Area since 1994. The largest constructed treatment wetland is the 1800 ha Kis-Balaton project in Hungary, which has operated since 1985.

This historical perspective should help to illustrate the relatively recent development of wetland treatment technology and emphasize the youth of even the oldest operating, full-scale engineered wetland treatment systems (about 20 years in 1995). This relatively short period of experience in the design and operation of wetland treatment systems is cause for reflection and understanding and is not unlike many of the other wastewater treatment technologies used today.

1.2.2.2 Subsurface flow wetlands

The origins of SSF wetland technology are in the work of Seidel and co-workers at the Max Planck Institute in Germany, during 1960–80 (Seidel 1976; Kickuth 1977). The treatment process was called the root-zone method (RZM) (in German, *Wurzelraumentsorgung*). Since then, the technology has grown remarkably in many European countries and is finding worldwide application.

The British Water Research Centre (WRc) first became aware of reed bed treatment systems (RBTS) in mid-1985 and started to investigate the potential of the horizontal-flow RZM system, which had then just started to be applied in Denmark. The water authorities were interested in a system that would allow them to apply low-cost, low-maintenance systems to small village communities that either had inadequate treatment or no treatment at all. Their interest was typically for its use in villages with populations of 50–1000 person equivalents.

WRc staff became convinced that there was enough potential in the system to justify research and development work. It was, however, clear that there were several problems with the system that required solutions. To achieve rapid progress it was decided that all the authorities and WRc should work together and create the Water Services Association Reed Bed Treatment Systems Co-ordinating Group, which was formed in late 1985. The aim was to share information from the different pilot and demonstration systems that were built around the country. In addition, a number of contracts were placed with organizations outside the UK water industry to study specialist areas that lay outside the normal field of expertise in the water industry. The first UK systems went into operation in October 1985. Ten years later, there were more than 400 systems in the UK (Cooper & Green 1995). Severn Trent Water alone had more than 180 systems by 1998.

A five-year programme of development culminated in the International Conference on Constructed Wetlands, which took place at Cambridge in September 1990 (Cooper & Findlater 1990). At the same time, co-operation with European colleagues was developing, and in 1986 it was decided that an Expert Contact Group under the aegis of the European Community and the European Water Pollution Control Association should be formed. This allowed workers from nine European countries to exchange design and operational experience and resulted in the European Design and

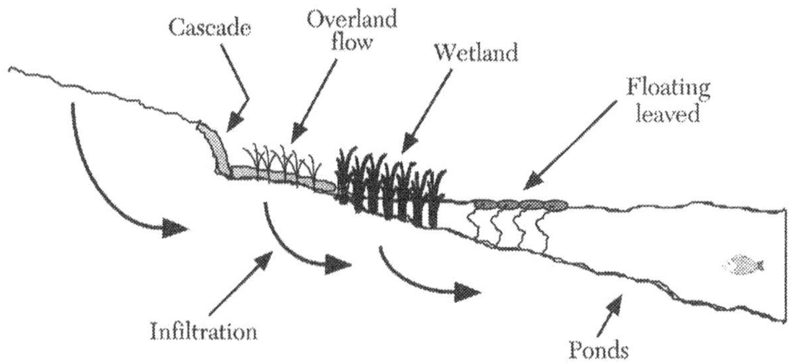

Figure 1.7. Natural systems: borrowing from the gradient.

Operations Guidelines for Reed Bed Treatment Systems, which was presented at the Constructed Wetlands Conference at Cambridge in September 1990 (Cooper & Findlater 1990). These guidelines are still widely used for the design of horizontal-flow systems, but they contain little on vertical-flow systems and nothing on tertiary treatment, stormwater treatment, or agricultural or industrial effluent treatment. Five years after the publication of those guidelines, WRc updated and broadened them in the light of experience gained (Cooper *et al.* 1996). They also produced a database and a bibliography containing approximately 800 references, together with abstracts and key words.

Besides Germany and the UK, SSF constructed wetlands were introduced in Austria, Denmark, France, Sweden, Switzerland, The Netherlands, North America, Australia and Africa in the 1980s. In the 1990s many SSF wetlands were also built in other countries in Europe (such as the Czech Republic, Poland, Norway and Slovenia) and Asia (for example China and India).

1.3 Companion natural technologies

Natural treatment systems for wastewater management are differentiated from conventional systems based on the source(s) of energy that predominate in the two treatment categories. In conventional wastewater treatment systems, non-renewable, fossil-fuel energies predominate in the treatment process. Whereas conventional treatment relies largely on transformations of naturally occurring, biological pollutants, these processes are typically enclosed in concrete, plastic or steel basins and are powered by the addition of forced aeration, mechanical mixing and/or a variety of chemicals. Because of the power intensity in conventional treatment systems, the physical space required for the biological transformations is decreased considerably compared with the area required for the same processes in the natural environment.

Natural treatment systems require the same amount of energy input as conventional biological treatment systems for every kilogram of pollutant that is degraded; however, the source of this energy is different in natural systems. Natural treatment systems rely (to a greater or lesser extent) on renewable, naturally occurring energies including solar radiation, the kinetic energy of wind, the chemical-free energy of rainwater, surface water and groundwater, and the storage of potential energy in biomass and soils. Natural treatment systems are land-intensive, whereas conventional treatment systems are energy-intensive.

Natural systems borrow their hydrologies from different portions of the upland aquatic gradient (Figure 1.7). Ponds are representative of the most aquatic end of the gradient. Land application and rapid infiltration are related to the upland end of that gradient. Overland flow represents the intermediate (upland runoff) position, whereas wetlands represent the nearly aquatic end of the hydrologic gradient.

1.3.1 Lagoons

Pond systems are one of the oldest and most widely used wastewater treatment technologies. Pond systems can be passive lagoons, dominated by renewable energies from the sun, wind and biota, or they can be highly sophisticated systems with liners and substantial forced aeration, in which case they are similar to conventional suspended-growth treatment systems.

Facultative ponds are designed to maintain a natural aerated surface layer over a deeper anaerobic layer. Natural aeration occurs because of the combined action of atmospheric oxygen diffusion and the release of oxygen during algal photosynthesis in the water column. The oxygen concentration can be highly variable over daily and seasonal periods within a facultative pond system. Excessive anaerobic conditions in a facultative pond are controlled by limiting the biochemical oxygen demand (BOD) loading rate. Typical design loading rates vary from about 14 to 50 kg BOD_5 ha^{-1} d^{-1} (where BOD_5 is the BOD at

5 d) with a detention time of between 80 and 180 d (WEF 1991).

Pond performance is typically a function of the effective hydraulic retention time, which in turn is related to flow dynamics and short-circuiting. Multiple cell ponds are typically more effective, and flow curtains or cell configuration can be used to increase the ratio between the actual and the theoretical residence times. A typical depth for facultative ponds is about 1.2–2.5 m. Typical hydraulic loading rates range from about 0.7 to 3.4 cm d^{-1} (WEF 1991).

Conservatively designed and carefully operated facultative ponds are effective in consistently achieving decreases in biochemical oxygen demand. However, because of their reliance on algal growth, ponds have a fundamental limitation on attaining low outflow concentrations of suspended solids. These elevated levels of suspended solids (up to and exceeding 100 mg l^{-1}) contain a fraction of decomposable organics and nutrients; facultative ponds therefore do not produce tertiary quality water. Facultative ponds also have some potential for total nitrogen removal (Reed 1985) but have little effect on total phosphorus concentrations.

1.3.2 Overland flow

Unlike other upland alternatives, overland flow treatment systems rely on low-permeability soils to restrict infiltration and consequently have a surface discharge (WPCF 1990; WEF 1991; Reed *et al.* 1995). Pretreated (primary or secondary) wastewater is applied intermittently to the top of sloped, vegetated terraces by gated pipes or by spray nozzles and allowed to flow by gravity down the slopes to a series of collection channels. As water flows through the dense vegetation on the slope, particulate pollutants settle and dissolved constituents are sorbed by plants and soils. Typically, wastewater application continues for 8–12 h out of every 24 h. During resting periods with no application, the organic fraction of the settled particulates is oxidized microbially and sorbed nutrients are incorporated into biomass (primarily inorganic nitrogen and phosphorus), transformed microbially (nitrification of ammonia nitrogen to nitrate nitrogen) or bound in the soil layer.

Typically, overland flow slopes from 1% to 6% are graded by laser technology and are between 30 and 60 m in length. The width of slopes varies to provide the necessary wetted area to accomplish treatment goals. Typical average hydraulic application rates to overland flow systems range from 1 to 10 cm d^{-1}

Overland flow systems are subject to operational problems in three areas: (1) maintenance of a viable cover crop, (2) maintenance of sheet flow and (3) violation of criteria for suspended solids. Ponding is likely to occur on overland flow terraces with low slopes, resulting in the depletion of soil oxygen and the eventual death of desired cover crops. Alternatively, on higher slope terraces, erosion is likely to occur and to result in high discharge concentrations of mineral sediments.

1.3.3 Rapid infiltration

High-rate land application systems use highly permeable soils for groundwater discharge (WPCF 1990; WEF 1991; Reed *et al.* 1995). High-rate land application systems are generally designed as relatively small or narrow, shallow basins or ponds with berm heights of less than 1.5 m. High-rate systems are typically loaded at hydraulic loading rates of between 1.6 and 25 cm d^{-1} over the bottom area of the basins. Because of groundwater mounding that occurs beneath high-rate land application basins, a sustainable infiltration rate is a function of the ratio between the length of the basin edges and the bottom surface area. Smaller basin areas and higher length-to-width ratios increase this infiltration rate. Multiple basins are typically used to allow drying down and resting. A careful rotational schedule can eliminate problems caused by overlapping groundwater mounds beneath basins. During resting periods, basin permeability can be renovated by rototilling or harrowing. Alternatively, a water-tolerant ground-cover crop can be planted in the basins to maintain soil texture and aeration.

At typical hydraulic loading rates, high-rate land application systems provide limited wastewater quality renovation. Whereas a significant fraction of the particulate organic matter and nutrients present in the pretreated wastewater is removed, soluble fractions are generally not diminished. One of the features of rapid infiltration systems is the oxidation of reduced nitrogen compounds in the aerobic soil zone, with the potential for elevated nitrate nitrogen concentrations in receiving groundwaters.

Because of the potentially low land-area requirements for high-rate land application systems and the relative ease of periodically applying wastewater to the basins, when it is technically feasible and permitted by regulations this technology is less costly (on a flow basis) than slow-rate land application and most other natural treatment alternatives.

1.3.4 Land application

Slow-rate land application of wastewaters uses irrigation of vegetated systems for wastewater polishing and ultimate disposal. Irrigation rates are generally low and intermittent, allowing the re-establishment of aerobic soil conditions at regular intervals. These aerobic conditions are essential for the growth of dry land vegetation, which is in turn essential for nutrient removal, filtering of wastewater solids, and maintenance of permeable soil texture. Slow-rate systems are used to treat and dispose of both municipal and industrial wastewaters. More than 800 slow-rate land application systems currently exist in the USA.

The slow-rate land application technology has a wide variety of process modifications and design criteria that depend on project goals (WPCF 1990; WEF 1991; Reed *et al.* 1995). In some cases, water disposal is the primary goal and the maximum wastewater volume compatible with site characteristics and groundwater criteria is applied to a given land area. These systems frequently use cover crops for partial nutrient removal through harvesting and by-product recovery. Commonly used cover crops include pasture grasses, corn, legumes and trees. The hydraulic loading rate to this type of land application system is limited by either long-term sustainable soil permeability or by the concentration of the most limiting wastewater constituent at the point of compliance with groundwater standards. The design hydraulic loading rate can be increased by adding soil underdrains; however, underdrains significantly increase system cost and convert this zero-discharge technology into an alternative with an intermittent or continuous surface discharge.

In other cases, slow-rate land application is used to irrigate golf courses and other human-contact landscaped areas after a high level of pretreatment. These systems use only enough water to satisfy the requirements of the cultivated plants and generally store or discharge excess wastewaters during periods of rainy weather. In areas with water shortages, treated wastewater becomes a valuable commodity to be conserved and is used sparingly for the irrigation of crops or landscaped areas.

Slow-rate land application systems are typically designed with hydraulic loading rates between about 0.15 and 1.6 cm d^{-1}. Wastewater is generally pumped to multiple irrigated areas and spread by using sprinklers, centre-pivot irrigators, or by ridge and furrow irrigation techniques. Individual irrigation areas can receive water from less than one to three times per week. Irrigation is generally ceased if surface runoff is observed from the application area.

The most common problems encountered with slow-rate land application systems are related to the overestimation of the long-term soil infiltration capacity during periods of sustained irrigation. Because of the high land-area requirements for slow-rate land application systems and because of the investment in piping and pumping necessary for wastewater distribution, these systems are generally the most costly of the natural system alternatives.

1.3.5 Natural treatment wetlands

As described above, the technology of constructed treatment wetlands is based largely on early research in natural wetlands receiving wastewater discharges. Although there are many types of natural wetland, they occur over a broad range of hydrologic regimes; only those wetland types with plant species adapted to continuous flooding are typically suitable for receiving continuous wastewater flows.

Natural treatment wetlands are being engineered and permitted in limited geographical areas such as the US southeastern coastal plain, the glaciated US upper midwest, and eastern Australia. Owing to their protected regulatory status, natural treatment wetlands can be difficult to permit and, when permitted, are generally used only for final polishing after substantial pretreatment.

According to WPCF (1990), typical hydraulic loading rates to natural treatment wetlands range from 0.4 to 4.0 cm d^{-1} (from 2.5 to 25 ha per 1000 m^3 d^{-1}). When available as a viable natural treatment alternative, natural treatment wetlands are typically the least expensive option, requiring very low capital expenditures other than land costs. Operation and maintenance costs for natural treatment wetlands are also quite low and are dominated by the costs for monitoring.

1.4 Integrated natural systems

The various natural treatment systems have often been used as 'stand-alone' process elements. However, the use of multiple natural system units in series and parallel provides a greater flexibility for tailoring the treatment to the specific problem goals. Therefore the broader view of natural systems treatment is that of integrated systems, which are composed of unit ecosystems.

Vertical and horizontal SSF wetlands can be used effectively in series and can accommodate recycling to increase efficiency. For example, vertical-flow beds are followed by horizontal-flow beds at St Bohaire, France (Lienard *et al.* 1990). Free water wetlands are effective polishing units for facultative lagoons and can be sized to provide nutrient removal that cannot be achieved in the lagoon (Kadlec 1996).

Many urban stormwater treatment systems contain sedimentation basin forebays (pond elements), followed by emergent FWS wetlands (Strecker *et al.* 1992; Schueler 1992). This series combination has the advantage of

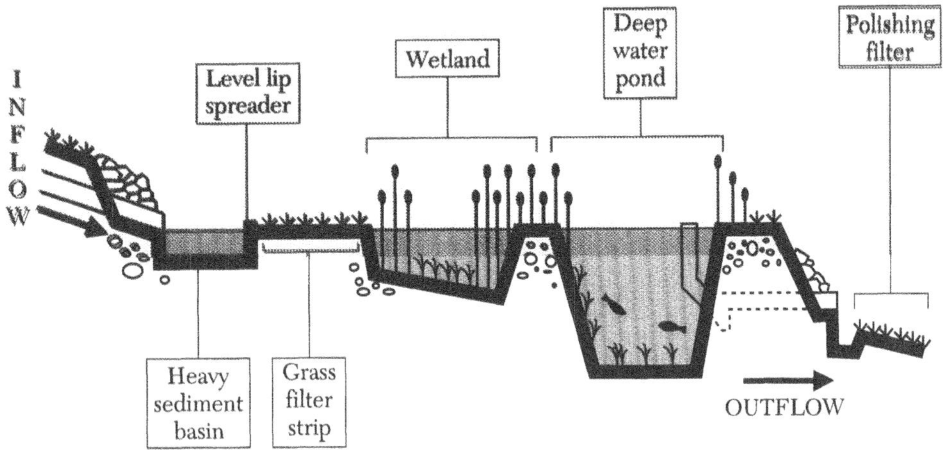

Figure 1.8. An NSCS for row crop runoff control. This natural system consists of five units arranged in series and has proved extremely effective in improving runoff quality from potato fields in Maine, USA.

providing the easy removal of solids accumulations, together with further processing in the marsh. In the agricultural landscape, more complex series arrangements have proved effective. The nutrient–sediment control system (NSCS) advocated by the US Natural Resources Conservation Service (NRCS) is composed of five elements in series: sedimentation basin, overland flow, FWS wetland, pond and overland flow (Figure 1.8).

A second example is an integrated natural system for treating potato-processing (french fries) wastewater (Kadlec *et al.* 1996). The first element is a FWS wetland that functions primarily as a solids trap, and to decrease chemical oxygen demand (COD). A second FWS wetland completes the ammonification of the organic nitrogen, and further decreases COD. Parallel downflow wetlands convert ammonium nitrogen to nitrate nitrogen. FWS wetlands then volatilize nitrate nitrogen via denitrification. Treated effluent is then stored in ponds in winter and finally discharged to land application on fodder crops.

There are, of course, many other innovative process flow arrangements that take advantage of the strengths of the various natural systems technology elements.

2 Types of constructed wetland

2.1 Free water surface treatment wetlands

The FWS wetland technology started with the ecological engineering of natural wetlands for wastewater treatment (Ewel & Odum 1984; Kadlec & Tilton 1979). Constructed FWS treatment wetlands mimic the hydrological regime of natural wetlands. In surface flow (SF) wetlands, water flows over the soil surface from an inlet point to an outlet point or, in a few cases, is totally lost to evapotranspiration and infiltration within the wetland.

FWS treatment wetlands have some properties in common with facultative lagoons and also have some important structural and functional differences. Water column processes in deeper zones within treatment wetlands are nearly identical to ponds with surface autotrophic zones dominated by planktonic or filamentous algae, or by floating or submerged aquatic macrophytes. Deeper zones tend to be dominated by anaerobic microbial processes in the absence of light. However, shallow emergent macrophyte zones in treatment wetlands and aerobic lagoons can be quite dissimilar. Emergent wetland plants tend to cool and shade the water. Net carbon production in vegetated wetlands tends to be higher than that in facultative ponds because of high gross primary production in the form of structural carbon, accompanied by resistance to degradation and low rates of decomposition of organic carbon in the oxygen-deficient water column. This high availability of carbon and the short diffusional gradients in shallow vegetated wetlands result in differences in biogeochemical cycling compared with ponds and lagoons.

During the process of elemental cycling within the wetland, chemical free energy is extracted by the heterotrophic biota, and fixed carbon and nitrogen are lost to the atmosphere. A smaller portion of the phosphorus and other non-volatile elements can be lost from the mineral cycle and buried in accreting sediments within the wetland. Wetlands are autotrophic ecosystems, and the additional fixed carbon and nitrogen from the atmosphere is processed simultaneously with the pollutants introduced from the wastewater source. The net effect of these complex processes is a general decrease in pollutant concentrations between the inlet and outlet of treatment wetlands. However, because of the internal autotrophic processes of the wetland, outflow pollutant concentrations are seldom zero, and in some cases for some parameters they can exceed inflow concentrations.

2.1.1 FWS treatment wetlands with emergent macrophytes

Many natural wetlands contain conspicuous plants (macrophytes) that have parts that extend above the wetland waters (emergent). Treatment wetlands make use of these same species (Figure 2.1).

A FWS wetland consists of a shallow basin constructed of soil or other medium to support the roots of vegetation, and a water control structure that maintains a shallow depth of water (Figure 2.1). A second commonly used name is SF wetland. The water surface is above the sediment, litter and soil, but live and standing dead plant parts are above water. FWS wetlands look and act much like natural marshes, and they can provide wildlife habitat and aesthetic benefits as well as water treatment. In FWS wetlands the near-surface layer is aerobic, whereas the deeper water and substrate are usually anaerobic. Typical water depths range from a few centimetres up to a metre.

FWS treatment wetlands function as land-intensive biological treatment systems. Inflow water containing particulate and dissolved pollutants slows and spreads through a large area of shallow water and emersed vegetation. Particulates, typically measured as total suspended solids (SS), tend to settle and are trapped due to lowered flow velocities and sheltering from wind. These particulates contain BOD components, fixed forms of total nitrogen (TN) and total phosphorus (TP), and trace levels of metals and organics. These insoluble pollutants enter into the biogeochemical element cycles within the water column and surface soils of the wetland. At the same time, a fraction of the dissolved BOD, TN, TP and trace elements are sorbed by soils

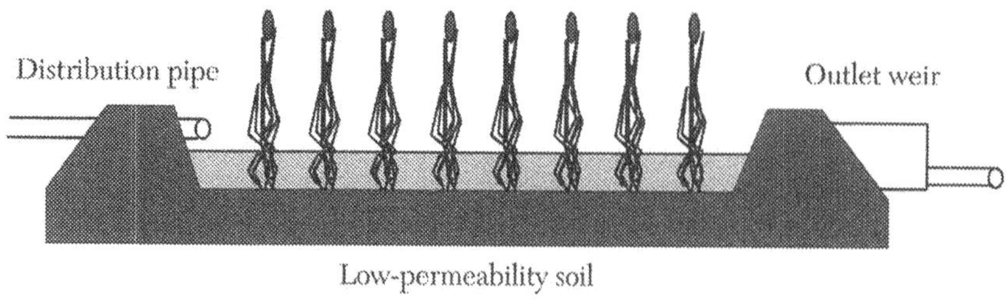

Figure 2.1. Diagram of FWS wetland containing emergent macrophytes.

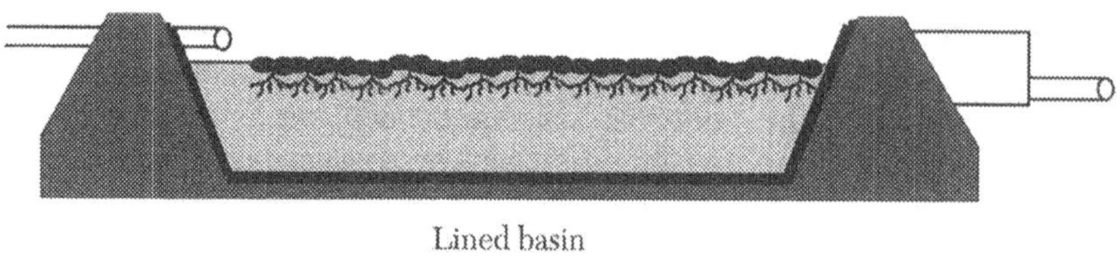

Figure 2.2. Diagram of FWS wetland containing floating plants.

and active microbial and plant populations throughout the wetland environment. These dissolved elements also enter the overall mineral cycles of the wetland ecosystem (Kadlec & Knight 1996).

Settleable organics are rapidly removed in FWS wetlands by quiescent conditions, deposition and filtration. Attached and suspended microbial growth is responsible for the removal of soluble BOD. The major oxygen source for these reactions is reaeration at the water surface. FWS systems effectively remove SS. In municipal systems, most of the solids are filtered and settled within the first few metres beyond the inlet (Watson *et al.* 1989).

Nitrogen is most effectively removed in FWS systems by nitrification and denitrification. Ammonia is oxidized by nitrifying bacteria in aerobic zones, and nitrate is converted to free nitrogen or nitrous oxide in the anoxic zones by denitrifying bacteria. FWS systems provide the sustainable removal of phosphorus but at relatively slow rates. Phosphorus removal in FWS systems occurs from adsorption, absorption, complexation and precipitation. However, the major process in phosphorus removal (precipitation with ions of Al, Fe and Ca) is limited by little contact between water column and the soil.

The plants that are most often used in FWS constructed wetlands are persistent emergent plants such as bulrushes (*Scripus* spp.), spikerush (*Eleocharis* spp.), sedges (*Cyperus* spp. and *Carex* spp.), rushes (*Juncus* spp.), common reed (*Phragmites australis*), reed canarygrass (*Phalaris arundinacea*), sweet mannagrass (*Glyceria maxima*) and cattails (*Typha* spp.). Not all wetland species are suitable for wastewater treatment because plants in treatment wetlands must be able to tolerate the combination of continuous flooding and exposure to wastewater or stormwater containing relatively high and often variable concentrations of pollutants.

For FWS wastewater treatment wetlands, the particular species selected are less important than establishing a vigorous stand of vegetation. Any species that will grow well can be chosen. For stormwater wetlands, species are often chosen to mimic the communities of plants of nearby natural wetlands. For both wastewater and stormwater wetlands, native, local species are preferred because they are adapted to the local climate, soils and surrounding plant and animal communites, and they are likely to do well.

FWS wetlands have been built to treat domestic wastewater, mine drainage, urban and agricultural runoff, leachate and a variety of industrial wastewaters. The advantages of FWS wetlands are that their capital and operating costs are lower and that their construction, operation and maintenance are straightforward. The main disadvantages of FWS systems are that they generally require a larger land area than other systems and that the water is exposed to potential human contact.

2.1.2 FWS treatment wetlands with free-floating macrophytes

Floating aquatic plant (FAP) treatment systems consist of one or more shallow ponds in which one or more species of water-tolerant, floating vascular plants are grown (Figure 2.2). The shallower depths and the presence of aquatic macrophytes in place of algae are the major differences between aquatic treatment systems and stabilization ponds. The presence of plants is of great practical significance because the effluent from aquatic systems is of higher

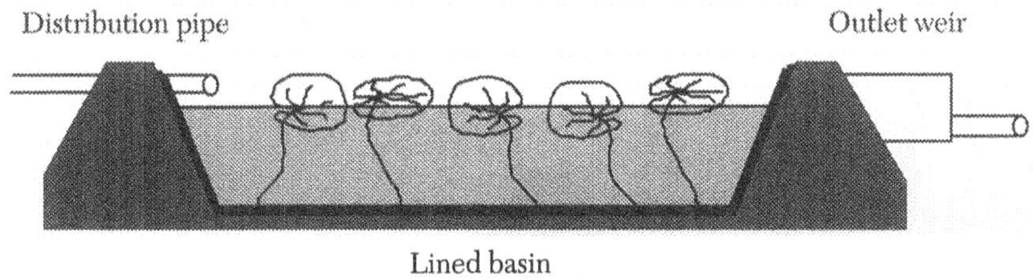

Figure 2.3. Diagram of FWS wetland containing rooted, floating leaf plants.

quality than the effluent from stabilization pond systems for equivalent or shorter detention times.

In FAP systems used for municipal wastewater, the carbonaceous BOD (cBOD) and SS are removed principally by bacterial metabolism and physical sedimentation. In systems used to treat cBOD and SS, the plants themselves bring about very little actual treatment of the wastewater. Their function is to provide components of the aquatic environment that improve the wastewater treatment capability and/or the reliability of that environment. In aquatic treatment systems designed to remove nutrients (N and P), plant uptake can contribute to the removals, especially where plants are harvested frequently.

The principal floating plant species used in FAP treatment systems are water hyacinth (*Eichhornia crassipes*) and duckweed (*Lemna* spp.) These and other floating species, such as water lettuce (*Pistia stratiotes*) and mosquito ferns (*Azolla* spp.), can occur in any FWS wetland. Water hyacinths have been used in a variety of experimental and full-scale systems for treating wastewater (see, for example, Reddy & Smith 1987). The use of water hyacinths has been limited in geographic location to warm-weather regions because of the sensitivity of water hyacinth to freezing conditions. Duckweed systems have been developed in colder climates because of the greater temperature tolerance of duckweed species. Both duckweed and water hyacinth systems have most often been used for nutrient removal after secondary treatment.

The photosynthetic parts of floating plants exist at or just above the water surface, and their roots extend down into the water column. In photosynthesis, floating aquatic plants use atmospheric carbon dioxide and produce oxygen. Nutrients are taken up from the water column through the roots. These roots provide an excellent support medium for the growth of bacteria and for the filtration/adsorption of SS. Root development is a function of nutrient availability in the water and the nutrient demand (that is, the growth rate) of the plant. Thus, in practice, the density and depth of treatment medium (that is, the plant roots) will be affected by wastewater quality pretreatment and factors affecting plant growth rate such as temperature and harvesting. With floating plants, the penetration of sunlight into the water column is decreased, and the transfer of gas between water and atmosphere is restricted. As a consequence, floating plants tend to keep the wastewater nearly free of algae and nearly anaerobic, depending on design parameters such as BOD loading rate, detention time, and the species and coverage density of floating plants. An observation of interest is that molecular oxygen produced by photosynthetic tissue is translocated to the roots and can keep root zone microorganisms metabolizing aerobically, even if that surrounding water is anaerobic/anoxic.

2.1.3 FWS treatment wetlands with floating-leaved, bottom-rooted macrophytes

Some macrophytes are rooted in the soils under the wetland waters, but their leaves float on the surface of the water (Figure 2.3). Water lilies (*Nymphaea* spp.), lotus (*Nelumbo* spp.) and cowlily (*Nuphar* spp.) are all capable of this growth mode. Some treatment systems have been operated in this fashion; one, for example, is the lotus cell in the wetland treatment system in Bainikeng, China. *Nymphaea* and *Nuphar* are common in more open water systems, such as the Des Plaines River demonstration wetlands near Wadsworth, Illinois, USA.

2.1.4 FWS treatment wetlands with floating mats

Some emergent wetland macrophytes are capable of forming floating mats, even though their individual plants are not capable of such existence (Figure 2.4). Cattails (*Typha* spp.), giant sweetgrass (*Glyceria maxima*), pennywort (*Hydrocotyle umbellata*) and common reed (*Phragmites australis*) are all capable of growing in mats. Treatment systems have been operated in this fashion (Kalin 1996; van Oostrom 1995; Worrall 1995; Hiley 1990).

Wastewater wetlands can accumulate large amounts of plant litter. This was so for SF wetlands planted with *Glyceria maxima*, which grows rapidly and has a high leaf turnover rate

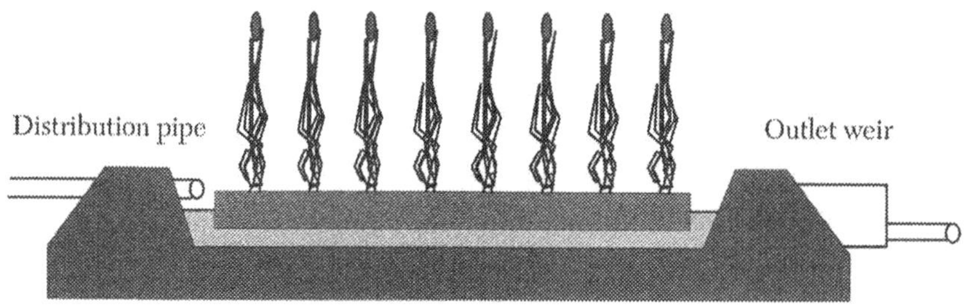

Figure 2.4. Diagram of FWS wetland with a floating emergent macrophyte mat.

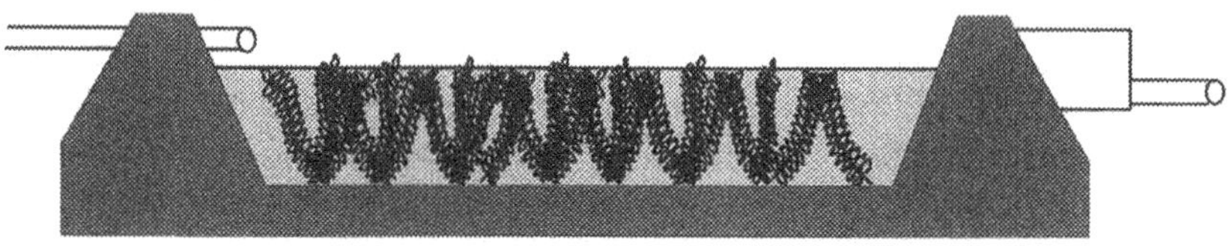

Figure 2.5. Diagram of FWS wetland containing submerged macrophytes.

(van Oostrom & Cooper 1990). Although this plant is normally established by planting it in soil in the bottom of a wetland, within two years most of the plants were rooted in a floating mat of decaying leaf litter on the wetland surface. This mat grew to a thickness of more than 200 mm.

Cattails (*Typha* spp.) can also form floating mats in treatment wetlands (Kadlec & Bevis 1990). Rhizomes and roots become woven together and accumulate plant detritus to form the mat. The mat is stable as long as it retains sufficient areal extent. If small portions of the 'raft' detach, the plants are top-heavy and tip over.

2.1.5 FWS treatment wetlands with submersed macrophytes

Submersed aquatic plants such as waterweed (*Elodea* spp.), water milfoil (*Myriophyllum* spp.) and naiads (*Najas* spp.) have sometimes been used to treat wastwater (Bavor *et al.* 1988). These submersed plants are buoyant and suspend in the water column, and might or might not be rooted in the bottom sediments (Figure 2.5). Typically, most of the photosynthetic plant tissue is suspended in the water column, but many submersed plants have aerial portions that extend above the surface of the water for flowering and increased light availability.

Various experiments have proved that minerals can be taken up by shoot tissues of submersed plants. However, there is also no question about the uptake capability of nutrients by the roots of these plants (Vymazal 1995). The potential for use of submersed aquatic vegetation for the treatment of primary or secondary effluent is limited by their tendency to be shaded out by emergent or floating plants and by their sensitivity to anaerobic conditions (WPCF 1990). In addition, the turbidity of the water must not be so high as to prevent light transmission to the plants to support their photosynthetic activity (Reed *et al.* 1988). The mechanisms by which submersed plants are able to remove ammonia from the water column is related to their high photosynthetic rates, which add oxygen to the water column, thus facilitating nitrification, and the fact that they utilize carbon dioxide from the water, thus raising the pH and driving ammonia to its volatile unionized form that can diffuse into the atmosphere. At night these plants respire (that is, use oxygen) in competition with the aquatic fauna. This category of constructed FWS wetland has not had widespread usage, but purposely planted submersed plant species are present in many natural treatment wetlands and are invaders in constructed wetlands that have deep-water zones.

2.2 Subsurface flow treatment wetlands

Many of the earliest treatment wetlands in Europe were SSF systems constructed to treat mechanically pretreated municipal wastewaters. Soil- and gravel-based SSF wetlands are still the most prevalent application of this technology in Europe (Cooper *et al.* 1996; Brix 1994; Vymazal *et al.* 1998). SSF wetlands that use gravel substrates have also been used extensively in the United States (Reed 1992). This technology is generally limited to systems with low flow rates and can be used with less than secondary pretreatment.

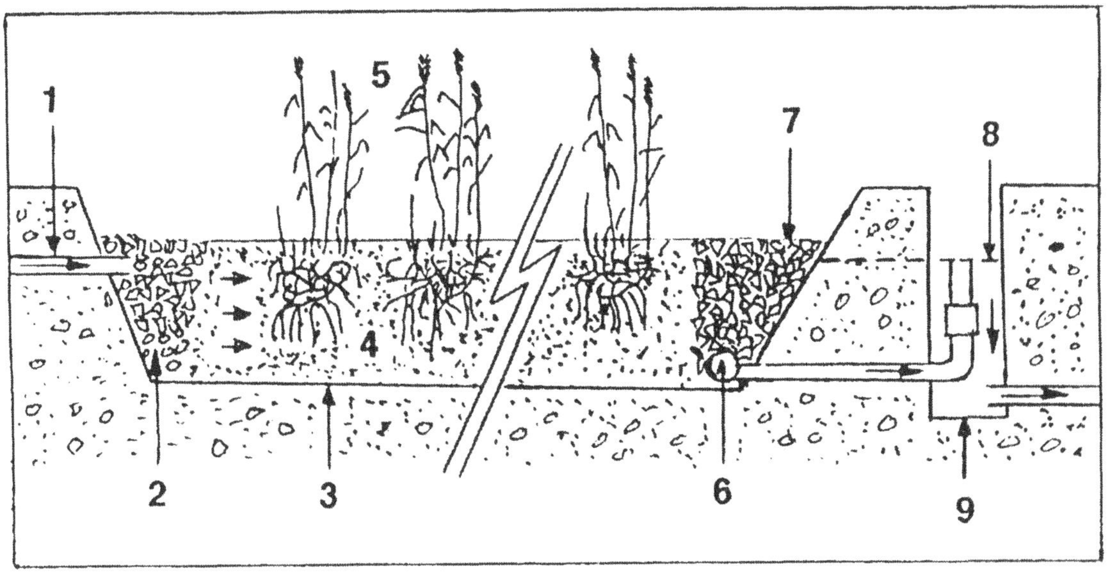

Figure 2.6. Longitudinal section of a constructed wetland with horizontal SSF. Key: 1, inflow of mechanically pretreated wastewater; 2, distribution zone filled with large stones; 3, impermeable liner; 4, medium (e.g. gravel, sand, crushed stones); 5, vegetation; 6. outlet collector; 7, collection zone filled with large stones; 8. water level in the bed maintained with outlet structure; 9, outflow (Vymazal 1997).

2.2.1 Horizontal-flow systems

Figure 2.6 shows a typical arrangement for the constructed wetland with a horizontal flow (HF). It is called 'horizontal flow' because the wastewater is fed in at the inlet and flows slowly through the porous medium under the surface of the bed in a more or less horizontal path until it reaches the outlet zone, where it is collected and discharged at the outlet (see Figure 2.6). During this passage, the wastewater will come into contact with a network of aerobic, anoxic and anaerobic zones. The aerobic zones occur around roots and rhizomes that leak oxygen into the substrate. During the passage of the wastewater through the rhizosphere, the wastewater is cleaned by microbiological degradation and by physical and chemical processes (Brix 1987; Cooper *et al.* 1996). In Europe, the most common term for HF constructed wetlands is the RBTS, because a frequently used plant is common reed (*Phragmites australis*). However, reed canarygrass (*Phalaris arundinacea*), sweet mannagrass (*Glyceria maxima*) and cattails (*Typha* spp.) are also used in Europe. In the USA, bulrushes (*Scripus* spp.) are also used. In North America the term vegetated submerged bed (VSB) is also used.

The concept of treating wastewater in constructed wetlands with horizontal subsurface flow was developed in Germany in the 1970s. The first operational constructed wetland was started in 1974 in Othfresen in Germany, and the treatment process was called the RZM (Kickuth 1977). The RZM system consists of a plastic-lined bed containing emergent macrophytes growing in soil. However, these soil-based systems, as a result of low hydraulic conductivity of the soil media, suffered from surface runoff, preventing the wastewater from coming into contact with the rhizosphere. The problem of surface runoff was overcome by the use of more porous media such as gravel (Cooper 1990).

Organic compounds are degraded aerobically as well as anaerobically by bacteria attached to plant underground organs (that is, roots and rhizomes) and media surfaces. The oxygen required for aerobic degradation is supplied directly from the atmosphere by diffusion or oxygen leakage from the macrophyte roots and rhizomes in the rhizosphere. Numerous investigations have shown that the oxygen transport capacity of the reeds is insufficient to ensure aerobic decomposition in the rhizosphere and that anoxic and anaerobic decomposition are important in HF constructed wetlands (see, for example, Brix 1990).

Settleable and SS that are not removed in pre-treatment systems are effectively removed by filtration and settlement. Settlement will take place in quiescent areas of any HF constructed wetland (Cooper *et al.* 1996).

Nitrogen is removed in HF constructed wetlands by nitrification and denitrification, volatilization, adsorption and plant uptake. The major removal mechanism of nitrogen in HF constructed wetlands is nitrification and denitrification. Ammonia is oxidized to nitrate by nitrifying bacteria in aerobic zones, and nitrates are converted to gaseous nitrogen by denitrifying bacteria in anoxic zones (Cooper *et al.* 1996). Field measurements have shown that the oxygenation of the rhizosphere of HF

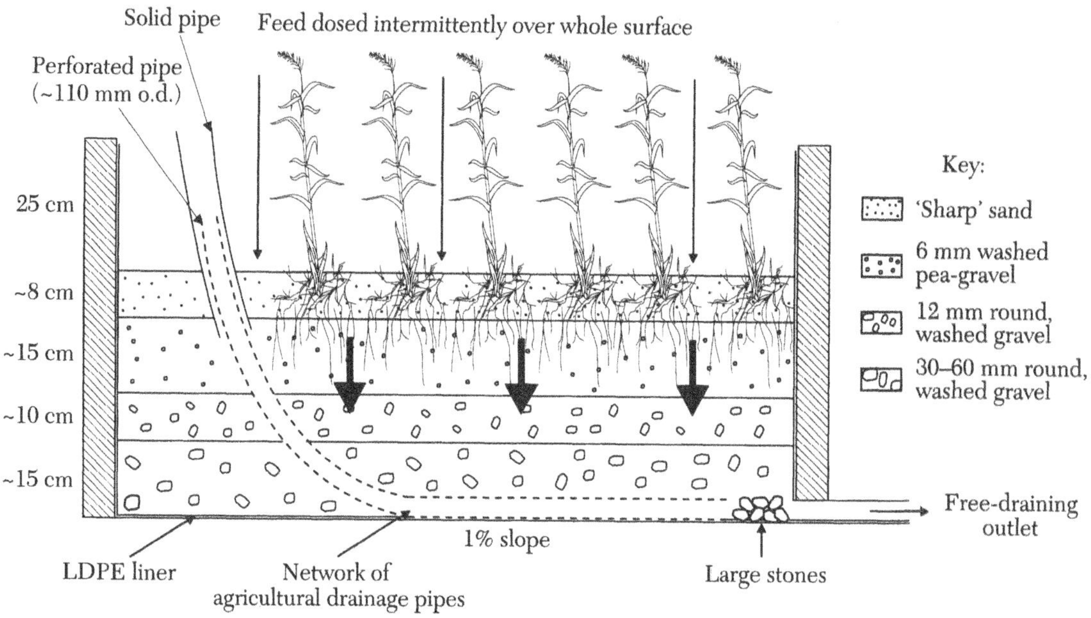

Figure 2.7. Typical arrangement of a VF reed bed system (Cooper 1996).

constructed wetlands is insufficient and that incomplete nitrification is therefore the major cause of limited nitrogen removal. Volatilization, plant uptake and adsorption are much less important in nitrogen removal.

Phosphorus is removed from wastewater in HF wetlands primarily by ligand exchange reactions, in which phosphate displaces water or hydroxyl ions from the surface of Fe and Al hydrous oxides. However, media used for HF wetlands (such as pea gravel or crushed stones) usually do not contain great quantities of Fe, Al or Ca, and therefore the removal of phosphorus is generally low.

2.2.2 Vertical-flow systems

Vertical-flow (VF) treatment wetlands are frequently planted with common reed. Other emergent wetland plants such as cattails or bulrush can also be used. VF reed beds typically look like the system shown in Figure 2.7. They are composed of a flat bed of gravel topped with sand, with reeds growing at the same sort of densities as HF systems. They are fed intermittently. The liquid is dosed on the bed in a large batch, flooding the surface. The liquid then gradually drains vertically down through the bed and is collected by a drainage network at the base. The bed drains completely free, allowing air to refill the bed. The next dose of liquid traps this air and this together with the aeration caused by the rapid dosing on the bed leads to good oxygen transfer and hence the ability to decompose BOD and to nitrify ammonia nitrogen (Cooper *et al.* 1996).

As with the HF systems, the reeds in VF systems will transfer some oxygen down into the rhizosphere, but it will be small in comparison with the oxygen transfer created by the dosing system.

VF treatment wetlands are very similar in principle to a rustic biological filter (Cooper *et al.* 1996). They are less good at the removal of SS and in most cases will be followed by a HF bed as part of a multistage treatment wetland system.

The earliest form of VF system is that of Seidel in Germany in the 1970s, sometimes called the Max Planck Institute Process (MPIP) or the Krefeld Process. Interest in the particular process seemed to wane, but it has been revived in the past six years because of the need to produce beds that nitrify. Operators and designers were disappointed in the ability of the early HF systems to oxidize ammonia to nitrate. In retrospect, this was clearly related to the fact that the ability of the reeds to transfer oxygen was greatly overestimated. Most HF systems have very low levels of dissolved oxygen in the effluent. Under these circumstances there will be no oxygen remaining to oxidize the ammonia nitrogen to nitrate. Because of this poor performance, designers and researchers started looking for alternative designs of reed bed that could oxidize the ammonia nitrogen.

3 Applications of the technology

There are an expanding number of application areas for constructed wetlands technology. During the early years (pre-1985) of the development of the technology, virtually all emphasis was on the treatment of domestic and municipal wastewater. In recent years there has been a branching to include a very broad spectrum of wastewaters, including industrial and stormwaters.

3.1 Domestic and municipal wastewaters

There are several roles for constructed wetlands in the treatment of domestic and municipal wastewaters. They can be positioned at any of several locations along the water quality improvement path. The commonly accepted terminology for describing that path is as follows (Metcalf & Eddy 1991):

- **Preliminary treatment** of wastewater is defined as the removal of wastewater constituents that might cause maintenance or operational problems with the treatment operations, processes and ancillary systems. Examples of preliminary operations are screening and comminution for the removal of debris and rags, grit removal for the elimination of coarse suspended matter that might cause wear or clogging of equipment, and flotation for the removal of large quantities of oil and grease.
- In **primary treatment**, a portion of the SS and organic matter is removed from the wastewater. This removal is usually accomplished with physical operations such as screening and sedimentation. More advanced methods of primary treatment include those that also provide a partial biodegradation of organic compounds. Frequently used units are primary clarifiers for larger flows and septic and Imhoff tanks for smaller applications. The effluent from primary treatment will ordinarily contain considerable organic matter and will have a relatively high BOD.
- **Secondary treatment** is directed principally towards the removal of biodegradable organics and SS. The most common secondary treatment technologies include activated sludge process, rotating biological contactors (so-called biodiscs), oxidation ditches and trickling filters (bacterial beds). Disinfection is frequently included in the definition of conventional secondary treatment in the USA but not in Europe, where it is less frequently applied.
- **Advanced treatment** of wastewater is defined as the level of tertiary treatment required beyond conventional secondary treatment to remove constituents of concern including nutrients, decreased levels of nitrogen (ammonia), toxic compounds and increased amounts of organic material and SS. Disinfection is typically regarded as tertiary treatment in Europe.

Constructed wetland technology is generally applied in two general themes for domestic and municipal wastewaters: for accomplishing secondary treatment and for accomplishing advanced treatment.

3.1.1 Wetlands for secondary treatment

3.1.1.1 Subsurface flow

Constructed SSF wetland treatment systems can provide secondary treatment of municipal or domestic wastewater after mechanical pretreatment consisting of a combination of screens, grit and grease chambers, sedimentation, septic and Imhoff tanks. The number of SSF constructed wetlands in operation in Europe is at present *ca.* 5000. In Germany alone, nearly 3500 systems are in operation (Börner *et al.* 1998). Many systems are also in operation in Denmark (200–400), the UK (400–600), Austria (*ca.* 160), Czech Republic (*ca.* 80), Poland (*ca.* 50), Slovenia (*ca.* 20) and Norway (*ca.* 10). In general, most European SSF treatment wetlands are designed to treat domestic or municipal wastewaters from sources of less than 500 population equivalent (PE). However, most systems are designed for small sources of pollution (less than 50 PE) and many systems are designed for single households. Only a small number of systems were designed for larger sources of pollution (more than 1000 PE) (Vymazal *et al.* 1998*a*).

A common local problem faced by home

Table 3.1. Performance data for a northern-climate CW treating septic tank effluent

Season	T(°C)	BOD ($mg\ l^{-1}$)		TSS ($mg\ l^{-1}$)		TP ($mg\ l^{-1}$)		TN ($mg\ l^{-1}$)		FC (no./100 ml)	
		In	Out	In	Out	In	Out	In	Out	In	Out
Winter	2	251	34	40	9	12	9	86	64	157,250	5798
Spring	5	252	33	39	8	12	9	80	66	268,750	5644
Summer	15	176	14	33	9	11	4	71	26	314,938	259
Autumn	11	288	17	41	6	12	5	81	35	209,438	2632
Annual	8	242	25	38	8	12	7	80	48	237,594	3583

Unpublished results from Northeastern Regional Corrections Center near Duluth, MN, USA. Abbreviations: CW, constructed wetland; FC, faecal coliforms.

owners and others in rural and non-sewered areas is poor site conditions that do not permit the installation and satisfactory performance of conventional on-site systems such as septic tank drain-fields. Practical solutions are needed, and there is great interest and desire in abating water pollution with effective, simple, reliable and affordable wastewater treatment processes. In recognition of this need, the Tennessee Valley Authority (TVA) began a demonstration of the constructed wetlands technology in 1986 as an alternative to conventional, mechanical processes, especially for small communities. Constructed wetlands can be scaled down from municipal systems to small systems, such as those for schools, camps and even individual homes. The systems are effective, simple, affordable, aesthetically pleasing, and educational. Guidelines have been developed by TVA to provide state-of-the-art and simple instructions for designing, constructing and operating constructed wetlands for small wastewater flows (TVA 1991).

An on-site SSF wetland can be a discharge system (i.e. discharges to a surface waters) or a non-discharge system (discharges to surface waters are eliminated by percolation, aided by evaporation and transpiration). A non-discharge system is used where conventional on-site methods are ineffective owing to poor site conditions (e.g. low soil percolation, shallow soils, high groundwater table or Karst topography). A non-discharge system is classified as 'on-site' if it is located within the property boundaries of the owners producing wastewater. The smallest systems are for single houses with limited wastewater. Different rules often apply to systems treating lower flows. A frequent use of on-site SSF wetlands is to replace failed adsorption fields or as an alternative to conventional systems where percolation rates are low. The technology might also be an alternative to low-pressure mound systems by constructing the on-site SSF system on top of bedrock, impermeable clay or high groundwater. Systems can be reliably designed to meet 'secondary' level permit limits.

Performance of on-site SSF wetlands can readily meet or exceed the goal of quality improvement for infiltration enhancement. This can be achieved even in extreme climatic conditions, such as the extreme cold of northern Minnesota, USA. Table 3.1 shows seasonal results for a (replicated) horizontal flow SSF wetland receiving septic tank effluent. Temperatures at this site attained minima of less than −40 °C during the winter.

In the early 1980s, when SSF constructed wetlands were introduced, the system usually consisted of only one bed, regardless of size. Hydraulic problems led to a changed approach. At present, for larger systems (larger than *ca.* 50 PE or 5 $m^3\ d^{-1}$), a multicell configuration is used. The most frequently used configurations are presented in Figure 3.1. The cells are usually rectangular with aspect ratios (length:width ratios) of between 0.3 and 3.

At present, most systems use coarse media (pea gravel, crushed stones) with a size fraction between 5 and 32 mm. Common reed (*Phragmites australis*) is the most frequently used plant in Europe, but reed canarygrass (*Phalaris arundinacea*), cattails (*Typha* spp.) and sweet mannagrass (*Glyceria maxima*) are also used either singly or in combination with common reed.

The treatment performance obtained in constructed wetlands with subsurface horizontal water flow is good in terms of removal of SS and BOD but lower in terms of nutrient removal (Tables 3.2 and 3.3). However, the treatment capacity of the SSF systems in terms of nutrient removal is low but comparable to the treatment efficiency of conventional treatment systems without a special regime for nutrient removal (i.e. nitrification and denitrification, phosphorus precipitation). Selection of system design should carefully consider the desired final effluent quality. Where only the removal of SS and BOD is required and where land is readily available and inexpensive, SF systems and one-unit SSF systems can be used. In sites with more stringent effluent quality demands, including demands for the removal

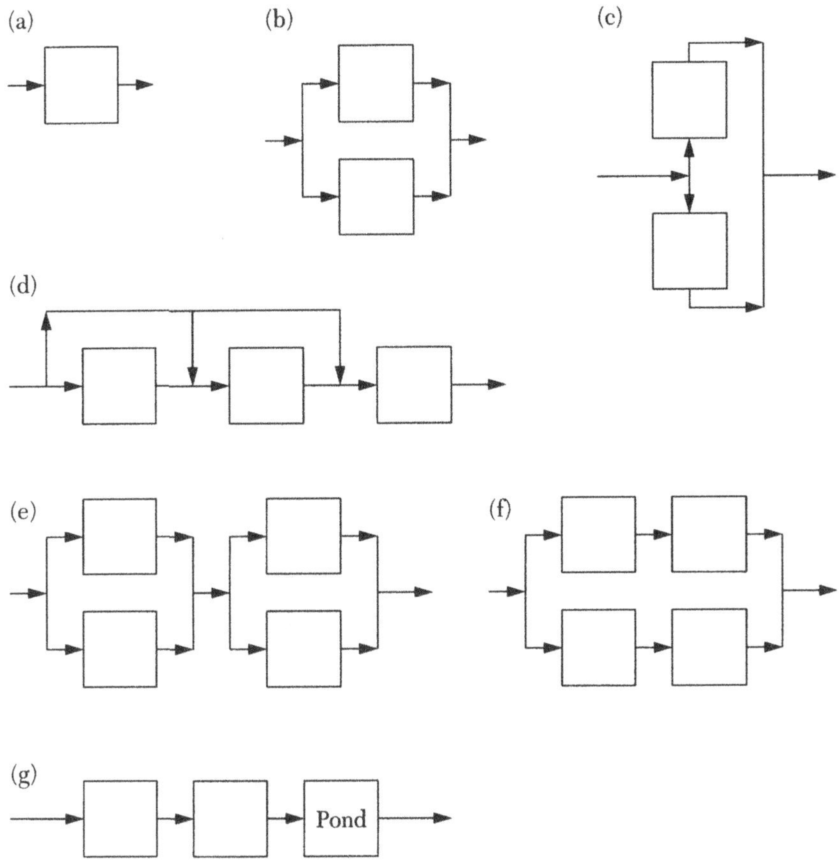

Figure 3.1. Alternative multi-cell SSF configurations: (a) single bed; (b, c) two parallel cells; (d) beds in series with a bypass; (e) two parallel cells in a series; (f) two series cells in parallel; (g) pond as a final step (Vymazal 1998a).

of nitrogen and phosphorus, combined systems consisting of VF beds with intermittent loading followed by horizontal SSF beds should be selected. The medium in the beds should be selected on the basis of hydraulic conductivity and phosphorus-binding capacity. Such multi-stage systems are more expensive in terms of construction, operation and maintenance than one-unit SF and SSF systems, but they might still be significantly cheaper than 'high-technology' alternatives.

3.1.1.2 Surface flow

Constructed FWS wetlands are typically not used at this time for complete secondary treatment of municipal wastewater. However, there are applications for secondary treatment in the USA and elsewhere that might serve as models by which to judge the success of this application. It is sometimes advantageous to supplement an undersized conventional secondary treatment plant with wetlands to bring the combination back to compliance with secondary standards. This was the design goal for Columbia, Missouri (Brunner *et al.* 1993), and for Wetwang, UK (Hiley 1990).

Facultative lagoons can provide secondary effluents, but they suffer from operational problems that might sometimes be best solved by adding a constructed wetland. For instance, warm summer temperatures can create algal populations that create TSS in excess of secondary standards. Wetlands can provide the TSS removal to bring the system into compliance. An example of this use of FWS wetlands is the Ouray, Colorado, USA, system (Andrews & Cockle 1996). The lagoon treatment at that site is over-taxed during the summer months and cannot meet a 30 mg l^{-1} standard on a seasonal basis, although it is achieved on an annual basis (Figure 3.2). The addition of a treatment wetland lowers the annual BOD and decreases the seasonal values below 30 mg l^{-1}.

3.1.2 Tertiary and higher

Data from North American treatment wetlands receiving secondary or better influents were summarized by the NADB (1993) and Kadlec & Knight (1996). Table 3.4 summarizes these data for SSF and SF treatment wetlands.

3.1.2.1 Subsurface flow

A vast quantity of data on tertiary treatment systems is now available via the database gained from the Severn Trent Water and UK water companies' experiences. Green & Upton (1995) described the effluent quality in BOD, TSS, ammonia and total organic nitrogen performance for 29 sites for the calendar year 1993. On the basis of these data, it is clear that

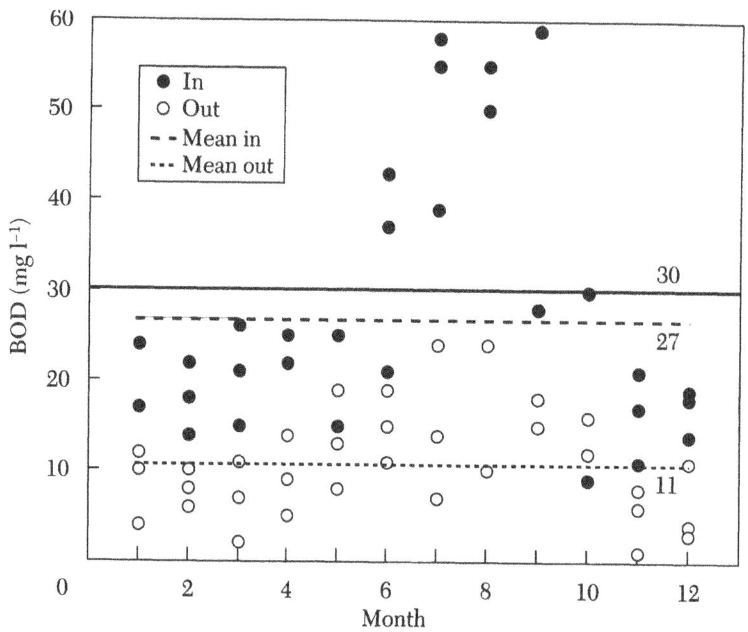

Figure 3.2. The use of FWS wetlands in Ouray, Colorado, USA, in 1993–95 to treat lagoon effluent (based on data from Andrews & Cockle (1996)).

Table 3.2. Average influent and effluent concentrations and mass loading rates of various pollutants in soil-based subsurface HF constructed reed beds in Europe

Parameter	*n*	Influent Mean	Influent SD	Effluent Mean	Effluent SD
Concentrations (mg l^{-1})					
SS	77	98.6	81.6	13.6	11.1
BOD_5	80	97.0	81.0	13.1	12.6
TN	73	28.5	14.7	18.0	10.7
TP	67	8.6	4.5	6.3	3.5
Mass loading rates (g m^{-2} d^{-1})					
SS	51	5.22	6.37	1.06	1.50
BOD_5	66	4.80	5.97	0.89	1.34
TN	57	1.15	0.79	0.78	0.77
TP	50	0.33	0.27	0.26	0.26

Summarized by Brix (1994*b*) with data from Coombes (1990) and Schierup*et al.* (1990*a*).

Table 3.3. Average influent and effluent concentrations and mass loading rates of various pollutants in gravel-based subsurface HF constructed reed beds in the Czech Republic (Vymazal 1998b)

Parameter	*n*	Influent Mean	Influent SD	Effluent Mean	Effluent SD
Concentrations (mg l^{-1})					
SS	37	71.9	47.2	10.8	7.1
BOD_5	39	87.4	65.7	11.9	11.4
TN	26	46.1	18.5	27.6	9.7
TP	27	6.4	3.8	3.1	2.1
Mass loading rates (g m^{-2} d^{-1})					
SS	31	3.34	3.11	0.44	0.42
BOD_5	35	3.36	2.86	0.53	0.67
TN	26	1.39	0.91	0.80	0.16
TP	24	0.30	0.18	0.18	0.16

Table 3.4. Summary of North American treatment wetland database operational performance

Parameter	Type	Average concentration (mg l^{-1})				Average mass (kg ha^{-1} d^{-1})°			
		In	Out	EFF (%)	Count (*n*)	Loading	Removal	EFF (%)	Count (*n*)
BOD_5	SF	30.3	8.0	74	182	7.2	5.1	71	133
	SSF	27.5	8.6	69	34	29.2	18.4	63	29
	All	29.8	8.1	73	216	10.9	7.5	68	162
TSS	SF	45.6	13.5	70	198	10.4	7.0	68	139
	SSF	48.2	10.3	79	34	48.1	35.3	74	29
	All	46.0	13.0	72	232	16.8	11.9	71	168
NH_4-N	SF	4.88	2.23	54	220	0.93	0.35	38	141
	SSF	5.98	4.51	25	19	7.02	0.62	9	15
	All	4.97	2.41	52	239	1.46	0.38	26	156
NO_2 + NO_3-N	SF	5.56	2.15	61	187	0.80	0.40	51	125
	SSF	4.40	1.35	69	13	3.10	1.89	61	13
	All	5.49	2.10	62	200	0.99	0.54	55	138
Org-N	SF	3.45	1.85	46	118	0.90	0.51	56	76
	SSF	10.11	4.03	60	11	7.28	4.05	56	11
	All	4.01	2.03	49	129	1.71	0.95	56	87
TKN	SF	7.60	4.31	43	144	2.20	1.03	47	94
	SSF	14.21	7.16	50	12	9.30	3.25	35	12
	All	8.11	4.53	44	156	2.99	1.29	43	106
TN	SF	9.03	4.27	53	175	1.94	1.06	55	114
	SSF	18.92	8.41	56	12	13.19	5.85	44	12
	All	9.67	4.53	53	187	2.98	1.52	51	126
P_i	SF	1.75	1.11	37	148	0.29	0.12	41	112
	SSF	n.d.	n.d.	n.d.	–	n.d.	n.d.	n.d.	–
	All	1.75	1.11	37	148	0.29	0.12	41	112
TP	SF	3.78	1.62	57	191	0.50	0.17	34	134
	SSF	4.41	2.97	32	8	5.14	1.14	22	8
	All	3.80	1.68	56	199	0.73	0.22	31	142

° kg ha^{-1} d^{-1} × 0.892 = lb $acre^{-1}$ d^{-1}. Abbreviations: EFF (%), efficiency of concentration reduction or mass removal; n.d., no data; NO_2 + NO_3-N, nitrite plus nitrate nitrogen; P_i, orthophosphate; Org-N, organic nitrogen; TKN, total Kjeldahl nitrogen.
Modified from Kadlec & Knight (1996).

designing a tertiary treatment SSF wetland at 1 m^2 per PE will achieve an effluent of less than 5 mg BOD_5 l^{-1} and 10 mg TSS l^{-1}, and in many cases will achieve very substantial nitrification. In Severn Trent it has become standard practice to use 0.7 m^2 per PE for tertiary treatment (Green & Upton 1995); smaller areas per head are used for some short-term or remedial applications.

In 1989, the decision was made to build the first of a new generation of tertiary treatment reed beds at Leek Wootton to provide reassurance for a decision that had been taken to make this the standard system for tertiary treatment for works serving populations of up to 1500 (in 1990 this was raised to 2000). The site was chosen because it came in the desired population range, was scheduled for asset renewal and had space on site for reeds.

The catchment area of the sewage works includes the villages of Leek Wootton and Hill Wootton, where there is a resident population of about 900 together with two village inns, a golf club and a training college. During the refurbishment of the treatment works, the inlet pumping station was replaced, the existing percolating filter was fitted with new distributors and clad with a wall of precast concrete sections, a new humus tank with automatic desludging was provided, and two tertiary treatment reed beds were built together with necessary chambers. Confirmation of the ability of reed bed systems to remove ammonia nitrogen and total oxidized nitrogen (TON) when used in the tertiary treatment mode was given by a survey performed at Leek Wootton (Table 3.5). Concentrations in the reed bed effluent varied little from an average of 19.4 mg l^{-1} during the whole period. The overall removals were 88% for ammonia N and 37% for TON.

3.1.2.2 Surface flow

SF wetlands in North America normally receive municipal water of approximately secondary quality or better. This is in contrast with the subsurface technology of northern Europe, which typically treats settled or primary influents. There are several hundred FWS treatment wetlands in the USA that are polishing

Table 3.5. Annual average performance data for Leek Wootton, UK, tertiary treatment HF RBTS (Cooper et al. *1996)*

	BOD_5 (mg l^{-1})		COD (mg l^{-1})		TSS (mg l^{-1})		NH_4-N (mg l^{-1})		TON (mg l^{-1})	
Year	In	Out	In	Out	In	Out	In	Out	In	Out
1990/91	11.6	4.8	75.7	32.1	27.6	6.1	7.6	5.8	32.8	23.4
1991/92	11.9	2.0	76.7	34.0	19.1	3.7	5.4	1.9	29.7	20.8
1992/93	15.4	2.7	109.0	55.5	24.2	5.3	7.0	2.8	20.4	8.7
1993/94	9.1	1.5	93.8	48.3	16.3	4.4	7.2	3.0	25.6	16.8
1994/95	9.1	1.0	82.1	46.6	18.4	4.5	6.6	1.9	25.7	18.4

Abbreviation: TON, total oxidized nitrogen (NO_2-N + NO_3-N).

secondary or tertiary wastewaters (NADB 1993).

In the USA, from a regulatory standpoint, there is fairly strong emphasis on creating FWS treatment wetlands of a moderately high quality. The unstated principle is one of minimizing the exposure of wildlife to poor-quality waters and habitat. As a result, most of the available FWS design data from the USA are in the lower ranges of concentration. It implies that incoming water quality is near enough to wetland background for those base-line numbers to influence design.

Reducing phosphorus levels is one of the least efficient processes in wetland treatment. Low TP concentrations can be decreased still further, but a large P load removal requires a large wetland area. Consequently, some pre-treatment for decreasing high TP concentrations is normally cost effective. The addition of iron and alum are the most frequent choices. For FWS wetlands, the point of addition needs to be upstream of the wetland because there is no effective way of providing chemical contacting in the wetland itself. The SSF wetland has an advantage in this regard because the media can be amended with the P-removing chemicals. The oxidation of ammonium nitrogen is more efficient when there is a small diffusional resistance to providing the oxygen to the dissolved or sorbed nitrogen. Neither horizontal SSF nor FWS wetlands are particularly good in this regard because the reaeration potential of the water sheet is relatively low. However, this tendency towards anaerobiosis is quite beneficial for the reduction of oxidized nitrogen to nitrous oxide and nitrogen gas. The greatest efficiency for N reduction is therefore achieved when the wetland is assisted by some form of nitrification pretreatment. Mechanical nitrification devices, planted or unplanted sand or gravel filters, or VF wetlands are candidates for the provision of supplemental nitrification.

3.2 Combined sewer overflows

Combined sewer overflows (CSOs) occur during rain events, when large infiltration and/or storm sewer flows are added to the normal domestic flows. The design capacity of the treatment facility is sometimes inadequate for the provision of treatment, and often it might not be hydraulically capable of passing the high flow. The principal contaminants, pathogens and solids, are passed directly to receiving waters. An add-on wetland system is intended to decrease SS and pathogenic bacteria before discharge to receiving waters. Removal of both of these pollutants is performed efficiently by both surface and subsurface wetlands.

Severn Trent Water in the UK has design and operating experience with CSO reed beds (Green & Martin 1994). The most common application of reed bed treatment for storm sewage overflows is at small (less than 2000 population) sewage treatment works. They have adopted reed bed treatment as a standard process at such works, where multiples of flow can exceed six times dry-weather flow. At such small works there is considerable variation in sewage flows in catchments with combined sewers, either with gravity flow or district or inlet pumping stations. It is unusual for CSOs to be provided within the system, although pumping stations can have overflows. Flows exceeding six times dry-weather flow spill over storm weirs at the inlet to the works and hence to screened outfalls via storm treatment reed beds.

The process flow sheet is determined by the nature of the discharge consent applied to the effluents from full treatment and from storm treatment. Figure 3.3 shows one typical flow sheet for the use of storm reed beds, where the UK requires that treatment at least equal to storm tank settlement is provided and where the final effluent has a consent more restrictive than 15 mg l^{-1} BOD and 25 mg l^{-1} TSS as the 95th centile.

There is a FWS system operating at Houten-Oost, The Netherlands. CSO waters cascade into a series and parallel set of wetlands, vegetated by submerged macrophytes and *Phragmites*. Early data showed excellent performance for pathogen decrease.

3.3 Urban stormwater

Uncontrolled urban stormwater has been identified as a major contributor to the non-point source (NPS) pollution of surface waters. Stormwater runoff originates from a wide range of sources: as runoff from parking lots, roadways, roofs and other impervious surfaces; as runoff across exposed soils such as construction sites and denuded landscapes; and as runoff from vegetated surfaces such as lawns and golf courses.

A great variety of pollutants – most importantly sediments, nutrients, trace metals and organic compounds – are carried to streams, lakes and estuaries by stormwater. Increasing urbanization has led to large increases in the pollutant loads delivered to natural receiving waters. In an average year, urban land has been estimated to contribute three times more nitrogen and 13 times more phosphorus for a given surface area than forest land and 1.2 times more nitrogen and 2.8 times more phosphorus than agricultural land (Linker 1989).

Small volumes of stormwater often carry large amounts of pollutants. For example, whereas local stormwater runoff is responsible for only a small percentage of the total flow to San Francisco Bay, this runoff contributes more than a third of all of the heavy-metal pollution that enters the bay (Silverman 1989).

Three approaches to controlling urban stormwater are dry detention ponds, wet detention ponds and stormwater wetlands. Dry detention ponds collect water during storms and release it within a day or so and are usually dry between storms. Wet ponds and stormwater wetlands typically contain water. Wet and dry ponds remove contaminants, primarily particulate matter, by sedimentation in deep basins, whereas wetlands provide a shallow pool of intense biological activity that removes and/or transforms a variety of pollutants through a complex array of biochemical pathways, in addition to the sedimentation that occurs in the wetland. Because wetlands are shallower than detention ponds, wetlands occupy more space. However, for the same reason, construction costs are often lower for wetlands than for ponds. Detention ponds and wetlands can be combined as needed to fit the requirements of a particular site and can also be combined with deep ponds.

Stormwater can carry a wide variety of urban NPS pollutants. Runoff from impervious surfaces, such as parking lots and roadways, can contain rubbish, suspended particulate matter, nutrients (especially nitrogen and phosphorus) from atmospheric deposition and from vehicle exhaust, trace metals from metal corrosion,

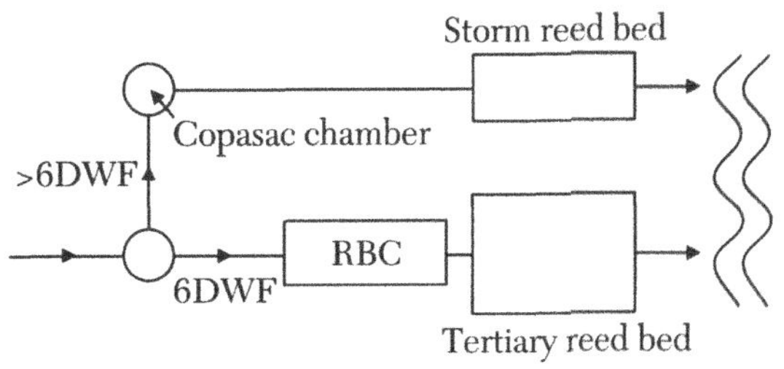

Figure 3.3. CSO reed bed treatment (from Green & Martin (1994)).

material from worn brake linings and tyres, de-icing salts and a wide array of complex hydrocarbons (such as motor additives, pesticides, rubber, and oil and grease). Runoff from exposed soils, such as construction sites, can carry large amounts of sediment. Runoff from vegetated areas can contain sediment, nutrients, pesticides, fertilizer and organic debris such as leaves. The types and amounts of pollutants in stormwater vary widely with the land uses in the contributing watershed, with higher pollutant concentrations associated with more intensive development and greater surface imperviousness (Livingston 1989).

The water quality of stormwater also varies widely with frequency and intensity of rainfall. In some instances, the quality of stormwater might not be correlated with the volume of flow (Silverman 1989). Rather, water quality effects might result from the 'first flush'. In the early stages of a storm, accumulated pollutants in the watershed, particularly on impervious surfaces such as streets and parking lots, are flushed clean by rainfall and runoff. The flushing action and inflow of the first inch of stormwater carries *ca.* 90% of the pollution load from a storm event to the receiving water (Livingston 1989), resulting in shock loading of the receiving system. Treatment of the first few centimetres of runoff therefore has a significant effect on the water quality consequences of the storm.

The use of wetlands for the treatment of stormwater has been examined in numerous studies over the last 15 years. However, the specific design concepts have perhaps not been as well examined as the use of treatment wetlands for more steady-state wastewater streams. The fundamental concepts of using wet settling basins for water treatment have, of course, long been accepted and practised. The design of such basins or tanks to accommodate the settling process is well understood and documented and is applied to steady-state wastewater streams every day. However, design guidance for stormwater treatment basins was generally not available until the early 1980s, when the US Environmental Protection Agency's Nationwide Urban Runoff Project was

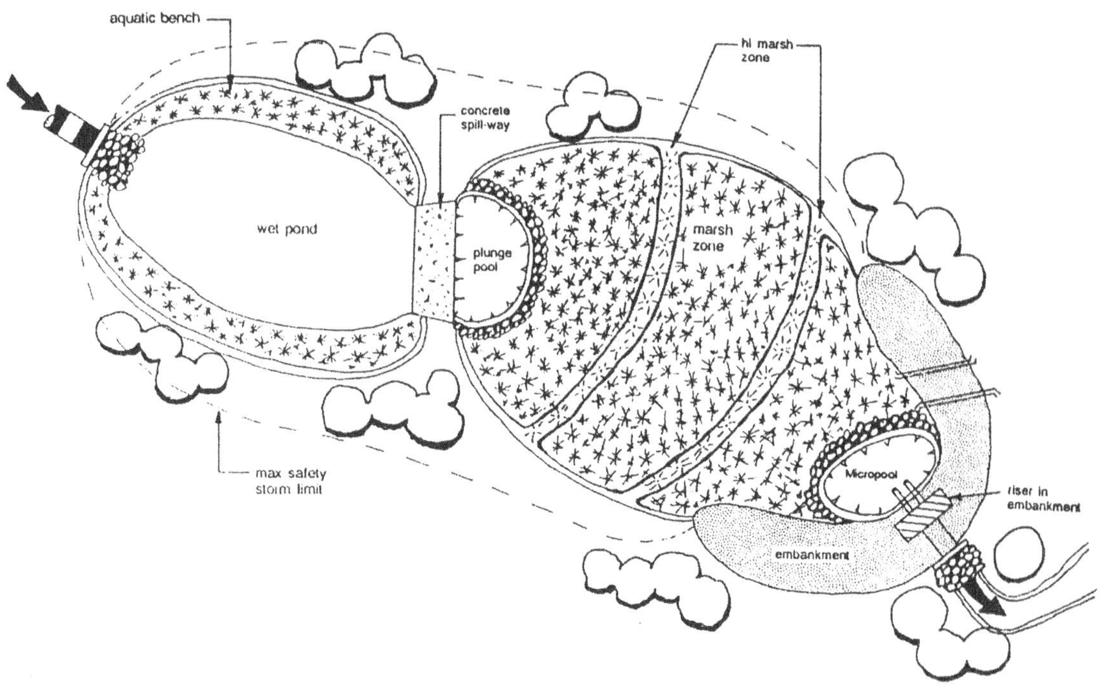

Figure 3.4. Layout of a stormwater treatment wetland (from Schueler (1992)).

completed. Results from this national research and demonstration effort provided the data needed to develop treatment basin sizing rules based on field data. The results of this effort have yielded many local and national guidance documents such as Driscoll (1983) and Schueler (1987). These documents provide rule-of-thumb approaches to preliminary sizing of wet detention basins to maximize the removal of pollutants.

Some guidance on the design of wetlands for stormwater treatment has been included in conference proceedings. As data have become more plentiful, more recent guidance has been published, including Strecker *et al.* (1992) and Schueler (1992). Stormwater wetland systems consist of shallow FWS wetlands (see Figure 3.4), in which the dense vegetation and almost level gradients slow the velocity of the stormwater and dampen peak flows. As the water finds its complex path through the vegetation, a number of physical, chemical and biological mechanisms remove contaminants in the stormwater or convert them to more innocuous compounds.

Stormwater is retained temporarily in the wetland by increasing the water level in the wetland and by spreading it out over the shoreline. The water is gradually released. Areas of deeper water are often included to increase retention times and to provide fish and wildlife habitats.

Pretreatment by sediment ponds is recommended, to slow the influent stormwater and decrease sediment loads before the stormwater enters the wetland. Floating rubbish can be removed in the pretreatment unit by rubbish racks. If oil and grease might be carried by the stormwater, they must be removed in the pretreatment unit because they can interfere with air/water interchanges in the wetland.

A deep pool at the outlet is included so that water will be discharged from below the water surface, thereby avoiding the organic-rich bottom sediments and the debris and plant wrack floating on the surface.

Schueler (1992) examined the performance of nearly 60 stormwater wetlands, including constructed and natural wetlands and pond-wetland systems, and estimated the long-term removal rates for stormwater wetlands in the mid-Atlantic region of the USA as:

total SS	75%
total nitrogen	25%
total phosphorus	45%
organic carbon	15%
lead	75%
zinc	50%
bacteria	2 log (10^{-2}) decrease

3.4 Agricultural

Agricultural and urban land uses often contribute to NPS pollution of lakes and streams, resulting in the impairment of fisheries and related recreational values. Reducing pollution by intensive but conventional treatment of the land by conservation practices can be difficult, costly and impracticable. These factors make the interception and removal of sediment, nutrients and other pollutants from runoff before it reaches waterways a necessary option. Seasonal variation of pollutant loads and hydrology seem to be suited to treatment by constructed wetlands. An effective treatment system that incorporates wetland values with pollutant-removal features is needed.

3.4.1 Animal wastewaters

Farmers and ranchers have responded to concerns about water quality by taking steps to decrease the amount of pollution that they generate and release into the environment. The greatest progress has been in the control of soil erosion, which has decreased the quantity of solids entering water bodies. For many confined animal feeding operations, the challenge is to prevent manure from being discharged with the water. The organic matter and nutrients in the manure are important resources that need to be recycled to the land to maintain high crop productivity. However, when released into natural water bodies, the organic matter and nutrients promote algal growth and deplete dissolved oxygen, leading to further problems.

To decrease pollution while maintaining or increasing productivity, confined animal operators need practical ways to either prevent wastewater from entering surface water and groundwater or treat the water before it leaves the farm. Operators want wastewater management systems that are affordable, reliable, and practical to build and operate. Today, various technologies are available to treat wastewater in ways that use the natural chemical, physical and biological processes of the environment and that rely on nature's energies. One of these technologies is a constructed wetland system. Constructed wetlands can be used in confined animal feeding operations before discharge or the application of wastewater to the land.

Results from existing constructed wetlands on farms suggest that these systems can help in several ways (DuBowy & Reaves 1994; DuBowy 1997). Constructed wetlands can remove solids and nutrients from wastewater so that more effluent can be applied to a given area of land or discharged to surface water. Constructed wetlands also help to minimize odour problems, decrease labour costs associated with hauling and applying effluent, and provide aesthetic and wildlife benefits. Constructed wetlands can be integrated into the farm in a way that benefits the operator and neighbours.

Constructed wetlands have been improving water quality at confined animal feeding operations for years (CH2M HILL & Payne Engineering 1997). Research has shown that these systems function even under ice and snow. A literature review identified information for 68 different sites using constructed wetlands to treat wastewater from confined animal feeding operations. Overall, the wetlands decreased the concentration of wastewater constituents such as BOD_5, total SS (TSS), ammonia nitrogen (NH_4-N), total nitrogen (TN) and total phosphorus (TP). Table 3.4 shows the average treatment performance.

Of the 68 sites identified, 46 were at dairy and cattle feeding operations. The herd sizes ranged from 25 to 330 animals, with an average of 85. Dairy wastewater often included water from milking barns and from feeding/loafing yards with varying characteristics. Cattle feeding wastewaters typically came from areas where animals were confined. Usually, dairy and cattle wastewaters were pretreated or diluted before being discharged to constructed wetlands.

Swine operations accounted for 19 of the wetland sites in the study. Swine wastes were collected by using flushwater from solid-floor barns and paved areas or directly from slatted floors in farrowing or nursery barns. In many cases, the wastewater was pretreated in lagoons and then discharged to a wetland system to decrease concentrations further to a level that could be applied to the land.

For poultry and aquaculture farms, the study found published information on constructed wetlands at one poultry site and two aquaculture sites.

Although decreases of 42–65% are impressive, the average outflow concentrations in Table 3.6 are not low enough to allow discharge to surface water; instead, the effluent is usually collected and applied to the land. However, by decreasing pollutant loadings to constructed wetlands by increasing the pretreatment or the wetland area, effluent pollutant concentrations can approach those typical of municipal treatment wetlands.

3.4.2 Crop runoff

Prototype systems have been designed and installed by the USDA Natural Resources Conservation Service in the State of Maine, USA (Higgins *et al.* 1993). These NSCSs are in watersheds draining cultivated potato croplands. Each NSCS has in series a sediment basin, a grass filter, a wetland, a pond and a wet meadow (see Figure 1.8). Selecting and managing the vegetation in each component is as important as site selection and system size. Wildlife also benefit from the constructed wetland/pond combination. Over 90% of total phosphorus and SS were removed by the system during all monitored storm events during the spring, summer and autumn. Successful implementation in the cold climate of Northern Maine demonstrates that these treatment systems provide relatively low-cost, but effective, NPS pollution control in agricultural areas of the USA.

Wetland buffers have proved to be effective

Table 3.6. Average treatment wetland performance for removal of BOD_5, TSS, NH_4-N and TN in the Livestock Wastewater Treatment Wetland Database

Parameter	Wastewater type	Count (*n*)	Average inflow concentration (mg l^{-1})	Average outflow concentration (mg l^{-1})	Average concentration reduction (%)
BOD_5	Cattle feeding	14	137	24	83
	Dairy	374	442	141	68
	Poultry	80	153	115	25
	Swine	183	104	44	58
TSS	Cattle feeding	12	291	55	81
	Dairy	361	1111	592	47
	Swine	180	128	62	52
NH_4-N	Cattle feeding	12	5.1	2.2	57
	Dairy	351	105	42	60
	Poultry	80	74	59	20
	Swine	183	366	221	40
TN	Dairy	32	103	51	51
	Poultry	80	89	70	22
	Swine	164	407	248	39

Source: CH2M HILL & Payne Engineering (1997).

in the control of phosphorus runoff from the Everglades Agricultural Area in southern Florida, USA (Moustafa *et al.* 1997). A 1500 ha constructed wetland has decreased phosphorus concentrations from 113 to 22 µg l^{-1} for a mean flow of 625,000 m^3 d^{-1} over a 29-month period.

In contrast with low-rate nutrient-control wetlands, it is possible to control sediments with relatively small wetland systems. In-line wetlands have been tested at hydraulic loadings of up to 5 m d^{-1}, which correspond to wetland:watershed area ratios on the order of 0.05. This aspect of the technology has undergone extensive testing in Norway (Braskerud 1997). To the extent that nutrients and other pollutants are bound to or part of these sediments, partial removal might occur.

Vertical/horizontal subsurface flow wetlands can also be used in field runoff control, especially for nitrogen control (Davidsson 1997). This variant of the technology is in use in Scandinavia, where it is called 'water meadows'. These are predominantly infiltrating subsurface flow systems, but overland flow can occur at times (Figure 3.5). During dry periods, cattle grazing and other agricultural uses might be allowed. The same technology in North America is called riparian buffer strips.

3.5 Industrial

3.5.1 Mining

3.5.1.1 Coal mine waters

A very large application area for constructed wetlands is the treatment of acid coalmine drainage. Hundreds of wetlands are now in operation serving this function (Wieder 1989). The contaminants of interest are typically pH, iron and manganese. Despite the large number of such wetlands, no clearly stated design methodology is yet available for acid mine drainage.

The mining of coal can result in drainage that is contaminated with high concentrations of dissolved iron, manganese, aluminium and sulphate. Passive treatment offers a low-cost alternative to conventional chemical treatment. The treatment of mine drainage by wetlands has evolved from simple SF wetlands to sequential treatment in a variety of wet environments. Early constructed wetlands were built to mimic the peat (*Sphagnum*) wetlands that first showed that the quality of mine water was improved as it passed through these wetlands. However, *Sphagnum* wetlands were difficult to establish and maintain, and the design was replaced by one in which emergent plants, most often cattails, are the dominant vegetation.

Recently, passive treatment options have been expanded to include anoxic limestone drains, which add alkalinity to the drainage before wetland treatment, and successive alkalinity-producing systems, which decrease the amount of surface area needed to generate alkalinity. Often, several treatment options are used sequentially. A number of natural processes decrease the impacts of mine drainage on receiving waters. Metals react with oxygen in aerated water and precipitate as oxides and hydroxides. Dissolved iron (Fe) precipitates as an orange oxyhydroxide, dissolved manganese (Mn) precipitates as a black oxide or oxyhydroxide, and dissolved aluminium (Al) as a white hydroxide. The low pH that is common to many mine drainages is raised either by mixing with alkaline or less acidic water or through contact with carbonate rocks.

The goal of constructed wetland treatment is to have these processes occur in the wetland

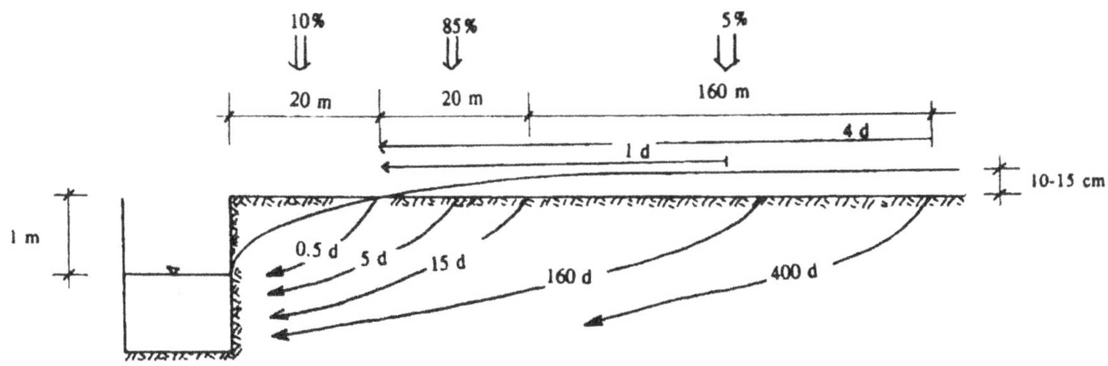

Figure 3.5. A water-meadow nitrogen control wetland (Davidsson 1997), showing a simulated distribution of infiltration, subsurface and overland flow travel times (days) at the Vomb water meadows.

rather than in the receiving water. Passive treatment systems function by retaining contaminated mine water long enough for chemical, physical and biological processes to decrease contaminant concentrations to acceptable levels. Efficient passive systems create conditions that promote the processes that most rapidly remove contaminants. Thus, the design of efficient passive systems must be based on an understanding of mine drainage chemistry and how different passive technologies affect this chemistry.

Analyses by the Office of Surface Mining and others (for example Wieder 1989; Wieder *et al.* 1990) question the feasibility of the constructed wetland concept. In contrast, many constructed wetland systems have worked quite well for several years (Brodie *et al.* 1993; Taylor *et al.* 1993; Hedin *et al.* 1994; Stark *et al.* 1994). Hundreds of constructed wetlands are now being used to decrease concentrations of contaminants from active, reclaimed and abandoned mines before the water is released, although not all of the systems can consistently treat water to effluent standards.

3.5.1.2 Metal mine waters

A small but growing application area for constructed wetlands is the treatment of various metal-mine drainage waters. Wetlands are now in operation treating waters from lead, zinc, silver, gold, copper, nickel and uranium mines (Noller 1994; Eger *et al.* 1993).

A number of metals are biologically essential at trace concentrations, but many metals become toxic to sensitive organisms at moderately low concentrations. For a few metals, biochemical transformations and chemical characteristics can lead to 'biomagnification', a phenomenon in which increasing concentrations occur in consumers along a food chain. Biomagnification can have devastating effects at top consumer levels, including humans. However, although most metals are more concentrated in biological tissues and soils than they are in surface water, biomagnification does not usually occur for most of the metals of interest. Metals in wastewater must be removed before final discharge to protect the environment from toxic effects, but the use of wetlands to accomplish this goal must be examined cautiously. The potential for economical treatment is nevertheless quite attractive (Dunbabin & Bowmer 1992) and the concept of metal removal in wetlands has undergone considerable investigation.

Chemical precipitation, ion exchange and plant uptake remove metals. Individual metals have been the target of specific research at mining sites. For example, quite a lot is known about wetland treatment of copper (Eger *et al.* 1993), aluminium (Reily & Wojnar 1992) and zinc (Haffner 1992). Noller *et al.* (1994) report on the removal efficiencies of a number of wetlands receiving a variety of mine effluents. Moderately high efficiencies, usually in the range 60–90%, are reported from full-scale and pilot-scale projects.

Metals are removed by cation exchange to wetland sediments, precipitation as sulphides and other insoluble salts, and plant uptake. The anaerobic sediments provide sulphate reduction to sulphide and facilitate chemical precipitation. As a result, good removals of metals are reported for operating wetland facilities. For example, for zinc removal the following have been reported:

Report	*Concentration reduction efficiency (%)*	*Detention*	*Type*
Eger *et al.* (1993)	90–96	22–34 h	SF
Sinicrope *et al.* (1992)	71–79	24–31 h	SSF
Haffner (1992)	60–96	8 d	SSF

Similar removals are obtained for other metals; for instance, chromium levels are decreased by 70% in *ca.* 70 h in SF wetlands (Srinivasan & Kadlec 1995).

Many treatment wetland field studies have investigated the removal efficiencies for multimetal wastewaters, mostly at low to moderate concentrations in domestic/industrial combinations (Crites *et al.* 1995; Sinicrope *et al.* 1992; Delgado *et al.* 1993) and urban stormwaters

(Strecker *et al.* 1992; Zhang *et al.* 1990). Moderate efficiencies, usually in the range 60–90%, are also reported in these studies.

Laboratory and mesocosm scale studies bear out these results under more controlled, but less realistic, conditions; for example, fast and large decreases in copper and chromium (VI) have been reported in mesocosms (Srinivasan & Kadlec 1995).

3.5.2 Food wastes

Food processing wastes are prime candidates for biodegradation. The attractive features of wetland systems are moderate capital cost, very low operating cost and environmental friendliness. The disadvantage is a large land requirement.

3.5.2.1 Vegetables

Most vegetables grown in the USA are processed, with the end product ranging from the packaged raw product to precooked specialty dishes. Processing involves steps such as washing, peeling, slicing and precooking. Large amounts of water are used; the result is often a high-strength wastewater. Processing plants are often located in sparsely populated regions, and natural systems are therefore suitable candidates for treatment and disposal. Current practice is to recover solids by clarification and filtration and application of the remaining water to the land. Historically, land application has been year-round, and rates have been high.

A combination of surface flow wetlands, intermittent vertical flow wetlands, and ponds and land application has been used for treatment of potato processing wastewater (Kadlec *et al.* 1996). A first pilot wetland was operated to determine operability, effectiveness and plant survival at high COD and nitrogen concentrations. A second pilot system of four wetlands in series was operated to obtain design and operating information. Two SF wetlands provided a decrease in TSS and COD and ammonified the organic nitrogen. Subsequently, nitrification occurred in the VF wetlands, followed by denitrification in a surface flow wetland. The design target was a balanced nitrogen and irrigation supply for application to crops. Winter storage was used to match the crop application period to the growing season. Both pilot projects met design objectives, and a full-scale system is now in operation.

3.5.2.2 Sugar production

Processing of either sugar beet or sugar cane results in a moderately high-strength wastewater, containing wash solids, cellulose and occasionally spilled sucrose. Both SF (Anderson 1996) and subsurface VF wetlands (Morris & Herbert 1997) have been employed to treat these wastes. Studies in Louisiana have shown that natural wetlands can attenuate the soluble organic carbon from this source (Gambrell *et al.* 1987).

3.5.2.3 Meat processing

Meat processing effluents contain high concentrations of nitrogen, typically 70–250 mg l^{-1}. TSS and COD can also be high, in the range 300–500 mg l^{-1}.

Performance of SSF, FWS and floating-mat constructed wetlands for TN removal from meat processing effluent has been extensively investigated in New Zealand (van Oostrom & Cooper 1990; van Oostrom 1995; Tanner & Sukias 1994). Three full-scale wetland facilities are now serving this function in New Zealand. Wetlands are also in use treating abattoir effluents in Australia (Finlayson *et al.* 1990) and the Czech Republic (Vymazal 1996*b*, 1998*a*). In Canada, the treated effluent from the Cargill meat processing plant is merged with treated effluent from the town of High River, Alberta, and discharged to a constructed wetland.

3.5.3 Petrochemicals

Constructed treatment wetlands hold considerable promise for managing some wastewaters generated by the petroleum industry. Several large-scale wetland projects currently exist at oil refineries, and numerous pilot studies (API 1998) of constructed treatment wetlands have been conducted at terminals, petrol and oil extraction and pumping stations, and refineries.

3.5.3.1 Refinery effluents

Most of the major petroleum companies in the USA have either pilot-scale or full-scale treatment wetland projects following traditional wastewater treatment systems (such as oil/water separators and aerated bio-oxidation lagoons). Constructed wetlands can be used to polish secondarily treated refinery wastewaters to attain more stringent water quality objectives and decrease or prevent discharges exceeding National Pollutant Discharge Elimination System (NPDES) values at the site discharge location (Litchfield & Schatz 1990; Litchfield 1993).

A full-scale constructed wetland has been used at Amoco's Mandan, North Dakota, USA, refinery for more than 20 years. The treatment wetland consists of an earthen canal that distributes water from the secondary treatment bio-oxidation lagoon into a series of cascading ponds and ditches before discharging to the Missouri River (Litchfield & Schatz 1990). The NPDES permit for the Mandan facility requires regular monitoring of the following parameters: BOD_5, COD, NH_4-N, sulphides, phenols, oil and grease, hexavalent and total chromium and TSS. During 1990, BOD_5 (an indicator of overall organic loading) in the secondarily treated effluent was decreased by more than 88% within the treatment wetlands. Similarly, phenols and oil plus grease were both decreased by 94% within the treatment wetlands (Litchfield 1993). The Mandan constructed wetlands have demonstrated the ability to effect significant decreases in all of the NPDFS permit parameters on a sustainable basis. In addition, the constructed wetlands constitute a valuable wildlife habitat resource for the refinery.

The Chevron refinery in Richmond, California, USA, also has a full-scale surface flow treatment wetland that is used to polish wastewaters before they enter San Francisco Bay (Duda 1992). In addition to significant decreases in other wastewater contaminants, the treatment wetlands have decreased BOD_5 by 51%. Toxicity tests of the wetland effluent with rainbow trout have shown 0% mortality.

Wetlands have been used to treat wastewaters at petroleum refineries outside the USA. In Hungary, a wetland system was established for polishing the wastewaters from a petrochemical plant in 1979. The system consists of a series of ponds (algal pond, fishpond and ponds with emergent vegetation) and gravel beds (Lakatos 1998). In China, the Jinling Petrochemical Company reported small decreases in several effluent quality parameters, including phenol and oil, by treatment with a floating-plant (water hyacinth) wetland (Tang & Lu 1993). Decreases in trace levels of organic compounds by a water hyacinth wetland were also shown in a pilot-scale municipal wastewater project (Conn & Langworthy 1984), in which low-level phenol was decreased by 81%.

Full-scale treatment wetlands (Yanshan wetlands) and research-scale wetlands (Fangshan wetlands) in Beijing, China, were shown to decrease a number of pollutants associated with refinery wastewater. BOD_5, phenols, and oil and grease were decreased in the full-scale Yanshan wetlands by 60%, 63% and 65%, respectively (Dong & Lin 1994*a*). Phenol decreases in the Fangshan research wetlands ranged from 27.8% in the winter to 36.7% in the summer. The research wetland studies also indicated that, of the variables tested, hydraulic loading rate had the most significant effect on decreasing contaminants (Dong & Lin 1994*b*).

3.5.3.2 Spills and washing

Tenneco, Inc., used a rock–reed wetland to treat wastewaters from a natural gas pipeline compressor station. This wetland treatment system was shown to decrease oil and grease in the effluent by about 90% (Honig 1988).

A subsurface flow wetland has been used to treat runoff from a 0.8 ha vehicle yard in Surprise, Arizona. Oil and grease have been decreased by between 54% and 92% by these treatment wetlands (Wass & Fox 1993).

At an unnamed oil terminal outside the USA, a 600 m^2 constructed rock-reed wetlands (primarily SF) was established in December 1992 to treat an oily water stream and a detergent-laden truck wash effluent. Preliminary results from 1993 to 1995 indicated an 80% decrease in BOD_5 and a 54% decrease in oil and grease in addition to decreases in other contaminants of interest. Phenols were also decreased, except in several cases that might have corresponded to high loading rates. Toxicity testing with Microtox™ organisms indicated a substantial decrease in effluent toxicity (98%) by the constructed wetlands (Farmer *et al.*, unpublished internal draft report, *ca.* 1996).

At an ongoing remediation project at a bulk petroleum storage terminal in Port Everglades, Florida, USA, a 720 m^2 SF constructed wetland was used to polish effluent from a conventional groundwater treatment system that consisted of an oil–water separator and an air stripper. The SF wetlands were shown to decrease trace amounts of aromatic hydrocarbons and polycyclic aromatic hydrocarbons (PAHs). Volatile organics, which were already at low levels in the air-stripper effluent, were decreased to trace levels in the treatment wetlands. Individual and total PAHs were all decreased by the treatment wetlands to levels below the limits of detection with analytical methods (Rogozinski *et al.* 1992).

3.5.3.3 Oil-sand processing water

A pilot-scale wetland was constructed in 1991 to treat wastewater from an oil-sand processing facility at Fort MacMurray, Alberta, Canada. Naphthenic acids (NAs), which are water-soluble hydrocarbons, were considered to be the primary toxicants of concern in the waste stream. Results indicated that NA and other contaminants were decreased by the treatment wetland, as was toxicity to *Daphnia magna* and Microtox™ (bacteria luminescence test). When total extractable hydrocarbons were used as a gross organic parameter, preliminary results showed removal efficiencies ranging from 35% to 70% under input loads of *ca.* 3 kg ha^{-1} d^{-1} (Bishay *et al.* 1995). The decrease in NA was shown to be greater in the summer than in the winter (Gulley & Nix 1993). Ammonia removal in the treatment wetlands was not limited by the presence of hydrocarbons in the treatment system (Bishay *et al.* 1995).

3.5.3.4 Produced water

The applicability of wetland treatment systems to produced waters from the processing of natural gas is being studied at the Argonne National Laboratory, Argonne, Illinois, USA (Hinchman *et al.* 1993).

Table 3.7. Summary of operational performance data for treatment wetlands receiving pulp and paper industry effluents

Parameter	Average HLR (cm d^{-1})	Average concentration (mg l^{-1})				Average mass (kg ha^{-1} d^{-1})			
		In	Out	EFF (%)	Count (*n*)	Loading	Removal	EFF (%)	Count (*n*)
BOD_5	19.9	26.1	13.6	48	30	28.6	8.3	29	19
TSS	19.9	42.5	12.5	71	30	41.6	28.5	68	19
NH_4-N	20.6	4.7	3.0	36	22	3.6	0.5	14	11
NO_2 + NO_3-N	5.2	1.4	0.14	90	6	0.49	0.42	86	6
TKN	5.2	7.8	3.5	55	6	3.6	1.1	31	6
TN	9.4	12.6	6.6	48	9	4.2	1.6	38	6
TP	21.1	2.3	1.7	26	20	1.0	0.3	30	9
Colour	10.7	1617	1581	2	23	2541	-115	-4	17

Abbreviations: EFF, efficiency of concentration reduction or mass removal; TKN, total Kjeldahl nitrogen.
Source: CH2M HILL (1994).

A pilot-scale treatment wetland project has been conducted by the Marathon Oil Company in conjunction with the Wyoming Department of Environmental Quality. The system uses bacterial ponds followed by a riffle channel flowing into a surface flow wetland to treat produced waters. The treatment system has been shown to decrease the concentrations of benzene and phenolics and can run in all seasons (Caswell *et al.* 1992).

3.5.4 Pulp and paper

Pilot projects for constructed treatment wetlands have been completed or are continuing at a number of pulp and paper mills in the USA (Pries 1994; Knight 1993). In general, mills that have investigated this technology have been those facing severe effluent discharge constraints because their receiving waters had limited dilution and mixing capacity. In some locations, wetlands have assimilated pulp and paper mill effluents for more than 20 years. In a number of cases, these natural basins are still integral to the mills' water pollution control programmes and are monitored regularly. These natural impoundments provide functions similar to natural or constructed treatment wetlands. During the past 10 years, as information accumulated about the success of wetland treatment systems for other types of wastewater, some pulp and paper mills experimented with constructing treatment wetlands. Because many of the pollutants in pulp and paper mill wastewaters are similar to pollutants in other wastewaters, researchers thought that the technology might be suitable for polishing paper mill effluents. Table 3.7 summarizes operational data from a number of these pilot-scale and full-scale pulp and paper facilities.

3.5.4.1 Australian paper manufacturers

Observations of pollutant assimilation in a partly vegetated facultative treatment pond at the Australian Paper Manufacturer's Maryville Mill in Victoria, Australia, led to the first published experiments using wetland plants to treat pulp and paper mill wastewaters (Allender 1984). In the experiment, five emergent wetland plant species were obtained from nearby ponds or wetlands. Researchers measured concentrations of lignosulphonate, colour, TSS, BOD_5 and foaming propensity in tubs with and without plants for three weeks. The concentrations of all these wastewater constituents were decreased in tubs with plants.

3.5.4.2 Weyerhaeuser

In 1985, Weyerhaeuser Company began pilot studies of SSF wetland systems to treat pulp and paper mill effluents. Two separate pilot studies were conducted between 1985 and 1990 (Thut 1989, 1990*a*,*b*, 1993). Treatment troughs were planted with cordgrass (*Spartina cynosuroides*), cattail (*Typha latifolia*) or common reed (*Phragmites australis*); control troughs were not planted. Secondarily treated mill effluent was applied to the troughs, and hydraulic residence time (HRT) was varied between 6 and 24 h.

This study concluded that SSF wetlands could effectively decrease concentrations of BOD_5, TSS and NH_4-N and could lower organic N (Org-N) and TP. Researchers did not observe colour removal or any significant removal of total organic chlorine. They found that higher treatment performance was related to increased HRT up to *ca.* 15 h, and that control troughs without plants were as effective as planted troughs for removing BOD_5, TSS and Org-N, apparently owing to the physical filtration of particulate pollutants.

Between 1988 and 1990, Weyerhaeuser conducted a larger-scale pilot study for an SSF wetland treatment system (Thut 1993). This system consisted of a single, in-ground, SSF cell of 0.375 ha (0.93 acres) with 40–50 cm of gravel planted initially with common reed and bulrush (*Scirpus californicus*). Average inflow to this system was 600 m^3 d^{-1}. This Weyerhaeuser large-scale pilot study demonstrated potential for removing BOD_5 (70–90% at an average influent concentration of 13 mg l^{-1}), TSS (50% at 34 mg l^{-1} in the influent) and NH_4-N (variable).

Weyerhaeuser subsequently installed a full-scale FWS treatment wetland at its mill in Columbus, Mississippi, USA.

3.5.4.3 Pope & Talbot

Pope & Talbot began experimenting with constructed wetland treatment systems in 1990 at its mill in Halsey, Oregon, USA (Moore & Skarda 1992; Hatano *et al.* 1992). The pilot facility consisted of ten parallel constructed SF cells, each with an area of 0.14 ha (0.35 acres). One of the cells was unplanted (gravel substrate); the remaining cells were planted with cattail or bulrush (*Scirpus acutus*). Five of the planted cells operated at an average HRT of 2 d, and five operated at 10 d. Significant pollutant removals were observed for BOD_5 and TSS. Lower removal efficiencies were related to shorter residence times and the absence of plants. Microbial studies indicated that populations of bacteria, actinomycetes and fungi were very high in wetland soils, in surface water and on plant stems. This study concluded that because of their high populations and diverse enzymic activities, actinomycetes might be the primary microorganisms responsible for degrading organic substances in the wetlands.

3.5.4.4 Stone Container

In 1989, Stone Container tested an innovative wetland system for the primary and secondary treatment of paper mill wastewater at Hodge, Louisiana, USA (Boyd *et al.* 1993). This system consisted of a 'spray header system' for effluent distribution and an SSF wetland planted with common reed. The authors did not give the area and configuration of the wetland cell. This system operated at 1900–19,000 m^3 d^{-1} for 72 d and provided between 67% and 84% BOD_5 removal efficiency at influent BOD_5 concentrations between 955 and 1620 mg l^{-1}.

3.5.4.5 Bowater

Between 1989 and 1990, six small SSF marsh pilot cells receiving bleached kraft mill effluent were tested for colour removal (Hammer *et al.* 1993). Each cell measured 1.9 m^2 and was filled with 30 cm of clay–loam topsoil and 15 cm of decomposed wood mulch. All cells were planted with cattails. Influent colour values between *ca.* 1200 and 2000 mg l^{-1} were decreased by 2–36% in all treatments with an overall average removal efficiency of 15%. HRT apparently did not affect colour removal in this study.

3.5.4.6 Georgia–Pacific

Beginning in 1989, Georgia–Pacific Corporation, in conjunction with the University of Southern Mississippi, began a pilot study of SF wetlands for final effluent polishing at its Leaf River Pulp Operations Mill near New Augusta, Mississippi, USA (Tettleton *et al.* 1993). This project used three pilot wetland cells of about 0.132 ha each, linked in parallel, to receive secondary wastewater from a bleach kraft mill. Tettleton *et al.* (1993) reported on the first two years of the pilot study, during which time these cells were planted with torpedo grass. Concentrations of BOD_5, TSS, NH_4-N, NO_3 + NO_2-N and TKN were decreased, but TP concentrations were not decreased significantly.

3.5.4.7 Champion International

Champion International Corporation operated a pilot wetland treatment system receiving secondary treated effluent for two years at its bleached kraft mill in Pensacola, Florida, USA (Knight *et al.* 1994). This pilot facility consisted of six parallel cells in three pairs with areas of 0.1, 0.2 and 0.4 ha. These FWS constructed wetland cells were planted with ten emergent wetland plant species arranged in zones perpendicular to the water flow direction. Each pair of cells had one cell with and one cell without two transverse deep-water zones intended to redistribute water and to increase HRT in the cells. In this study, water quality treatment performance was best at the lowest hydraulic loading rates (HLRs). Deep-water zones increased treatment performance when they occupied a relatively small portion of the entire cell area (less than 25%). Long-term average removal efficiencies at the lowest average HLR of 3.2 cm d^{-1} were between 67% and 91% for BOD_5, TSS, NH_4-N, TN and TP. Average removal efficiencies for colour and conductivity were much lower (less than 14%).

3.6 Landfill leachate

As landfills become larger, the enormous quantities of putrescible wastes that they contain have increased the potential to generate highly polluting leachates as they decompose anaerobically over many years. Landfill leachates contain various quantities of undesirable, and even toxic, organic and inorganic substances. Treatment of this highly polluting wastewater is becoming mandatory worldwide. Haulage or discharge to a sewage treatment plant is often difficult and/or expensive. On-site 'high-tech' leachate treatment systems are also avoided because of high costs of construction and operation.

Historically, aerated lagoons have been popular for the treatment of landfill leachate and have proved to be successful in the removal of

Table 3.8. Leachate pollutant reductions at Perdido Landfill, Florida, USA (DeBusk 1997)

Parameter	Loading (kg/d)	Removal (kg/d)	Removal (%)
BOD_5	17.8	13.8	77.4
TSS	67.5	63.1	93.5
TOC	33.9	13.4	39.7
TP	1.7	1.4	81.8
TN	10.2	6.2	61.3
Fe	3.9	3.5	89.7
Mn	1	0.04	52.0

Abbreviation: TOC, total organic carbon.

COD and NH_4-N (Mans & Harrison 1984). Wetland treatment of landfill leachates has been successfully tested at several locations. A facility at Ithaca, New York, USA, that has been operating since 1989 (Staubitz *et al.* 1989; Surface *et al.* 1993), has used SSF wetlands. SF wetlands have been operating successfully in Escambia County, FL, USA, since 1990 (Martin *et al.* 1993; Martin & Moshiri 1994, Martin & Johnson 1995). Cold-climate systems are functioning properly in Norway (Maehlum 1994), as well as at several locations in Canada: Sarnia, Ontario; Richmond, British Columbia; and Sackville; Nova Scotia (Birkbeck *et al.* 1990; Pries 1994). SSF reed beds are used to treat leachate in the UK (Robinson 1990), Slovenia (Urbanc-Bercic 1994; Bulc*et al.* 1997) and Poland (Agopsowicz 1991). The first International Conference on Wetland Treatment of Leachates was held in 1997 (Mullamoottil*et al.* 1998).

Characterization of the leachate is essential because it can contain high concentrations of BOD, ammonia, metals, high or low pH, and often priority pollutants of concern. In addition, supplemental nutrients such as potassium and phosphorus might be required because leachate composition depends greatly on the type and quantity of material dumped in the site, the duration, and the degree of infiltration water. Leachate quality can vary from relatively harmless to extremely hazardous waste. Landfill leachates are generally anoxic and usually contain high concentrations of organic carbon, nitrogen, chloride, iron and manganese. They can also contain high concentrations of heavy metals, pesticides, chlorinated and aromatic hydrocarbons and other toxic chemicals, depending on what materials were originally placed in the landfill. However, the leachate flows are small in comparison with municipal wastewater flows, with a typical range of 40–400 $m^3\ d^{-1}$.

The performance of leachate wetlands is not fundamentally different from those treating other wastewaters. Significant percentage decreases of many pollutants are achieved. For instance, Table 3.8 summarizes removals at the Perdido Landfill treatment wetlands in Florida, USA (DeBusk 1997).

Leachates often contain metals and organics in concentrations that constitute potential hazards to wildlife and other biota. It is therefore necessary to remain aware of the levels that can accrete in a treatment wetland and to take precautions to prevent deleterious contacts. Three methods for controlling contact are dilution, deep pond accretion and SSF wetlands. Dilution of incoming leachate with either another wastewater or recycled outflow from a downstream section of the system can decrease the concentration in the water but not the loading to the system. However, the greatest amounts of contaminants are often in the wetland sediments, which can create a hazard for sediment-grazing organisms.

The wetland sediments can be isolated from contact with higher organisms by ensuring that they accrete in isolation. In turn, that can be accomplished by using an SSF wetland or a floating vegetative mat wetland. The gravel bed hydroponic system (SSF), when operated properly, places the water and aquatic sediments below ground and out of reach of sediment foragers. The deep pond achieves the same effect by accreting residuals at a depth sufficient to be out of reach of those sediment foragers.

Constructed wetlands have the advantage of long-term, sustainable treatment with very low costs of operation and maintenance. This is especially important for leachate control, which often requires indefinite treatment lifetimes. It is also often important to build projects with guaranteed long-term stewardship. Passive constructed wetlands offer very long lifetimes, with little or no equipment replacement.

In contrast with chemical and physical processing alternatives, wetlands provide insurance against unanticipated new pollutants. For instance, conventional air stripping can be used effectively to decrease the concentrations of NH_4-N and other volatiles. However, that technology has no capacity to deal with metals. Wetlands have the capacity to deal with both. If in the lifetime of the leachate source it becomes a source of metals, the wetland will have some capacity to treat this new pollutant.

3.7 Sludge consolidation

The handling, dewatering and disposal of municipal wastewater sludges in an economical manner is an increasing problem. A system that in many cases uses existing equipment, is not labour intensive, is economical to install and operate, and meets regulatory agency requirements seems too much to ask for, but a VF

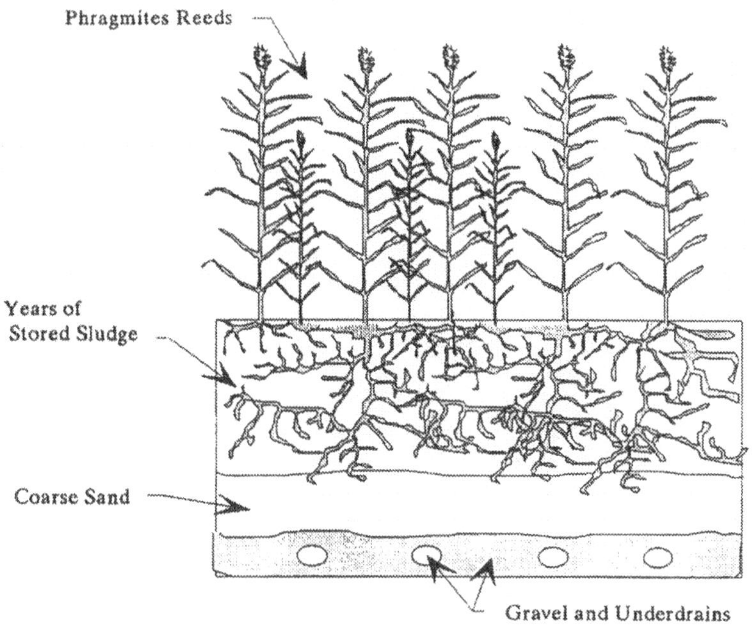

Figure 3.6. Cross section of a reed bed. Reeds in sludge consolidation beds are capable of adjusting their rooting horizon to match the accumulation rate (Sassaman & Kaufman 1992).

constructed wetland planted with common reed is all of these things. Reed systems are helping many municipalities to solve their sludge disposal problems in an economical and legal manner.

The vast majority of municipal wastewater treatment facilities in the USA are rated at less than 7570 $m^3 d^{-1}$ (2.0 million US gallons per day (MGD)). A reed bed process requires a significant area for installation and is consequently better suited to these smaller facilities. The consolidation of dilute suspensions by wetlands was first used in the USA by the US Army Corps of Engineers during the 1970s, for dewatering river dredgings. VF reed beds have since been successfully installed and operated in over 40 wastewater treatment facilities throughout the northeastern USA (Sassaman & Kaufman 1992). Many small industrial and municipal wastewater treatment plants built in the 1960s and 1970s already have sludge-drying beds. In these cases, often only minor modifications are required to convert these to sludge consolidation wetlands. In Denmark *ca.* 40 systems are in operation, the largest treating sludge from a 120,000 PE activated sludge treatment plant (Nielsen 1994).

Phragmites plants grow well in digested wastewater sludges and assist in the dewatering and stabilization processes. The root system grows throughout the upper sand layer and stored sludge (Figure 3.6). The plants bring oxygen to their root system, which harbours a rich bacterial microflora. These bacteria feed on the organic matter in the sludge. The plants use the nutrients in the sludge to promote vigorous plant growth. The reeds emerge from the root system as small shoots in early spring and grow to a height of 2–3 m in only three months. The root system also keeps channels open to the sand and gravel layers, which allows drainage of the beds by gravity. The reeds use large amounts of water for transpiration through their leaf system into the atmosphere.

Sludge application rates vary from 1630 to 2444 $l\ m^2 yr^{-1}$. The loadings depend on the type of digested sludge (aerobic or anaerobic), and the percentage concentration of solids. New installations usually consist of planting *Phragmites* root stock in the sand layer at *ca.* 30 cm spacing. Within the first growing season, sludge application rates are kept low while the roots spread rapidly and plants develop over the entire bed area. After the plants mature, the full sludge loading schedule can be started. Liquid sludge is applied to the beds at intervals all year round. Sludge application can be as often as weekly, with 100 $l\ m^{-2}$ being applied during the growing season. Sludge application continues throughout the winter at longer intervals.

Sludge accumulates in the beds until it reaches a depth of *ca.* 1 m. The accumulation of sludge to this depth can take 8–10 years. When a bed is full it is taken out of service and allowed to stand for 2–4 months to reach its maximum solids concentration. Solids concentrations can reach 40–50%. The sludge is then removed by mechanical methods such as a backhoe. The evacuated material is landfilled.

Typical decreases in biosolids are *ca.* 90% (Mellstrom & Jaeger 1994). Metals concentrations in the solids typically increase by 50–150%. Odours can be a problem, especially during the spring thaw in northern climates.

4 Framework for interpreting and predicting water quality improvement

Treatment wetlands are shallow vegetated basins, with or without a permeable substrate. One of their primary design purposes is to contact wastewater with reactive biological surfaces. Important aspects of the hydrological and thermal regimes in wetlands are summarized in this chapter. The water mass balance organizes information on inflows and outflows and is an essential component for determining mass pollutant removals in treatment wetlands. Water conveyance deals with water depth variation between the inlet and the outlet of a treatment wetland, and a knowledge of internal flow patterns and degree of mixing is essential for accurate modelling of pollutant decreases in these treatment systems.

The efficiency of pollutant removal is strongly related to the size of the wetland and is modified by the flow and thermal characteristics. Removal calculations can involve various degrees of complexity but must involve the concept of scaling to meet performance goals. Wetlands are composed of a large variety of living organisms and are subject to the vagaries of climate and meteorology. Pollutant removals therefore display probabilistic character, which must also be reckoned with in scaling.

4.1 Hydraulics

The water status of a wetland defines its extent and is the determinant of species composition in natural wetlands (Mitsch & Gosselink 1993). Hydrological conditions also influence the soils and nutrients, which in turn influence the character of the biota. The flows and storage volume determine the length of time that water spends in the wetland, and thus the opportunity for interactions between water-borne substances and the wetland ecosystem. The hydraulic profile of the treatment wetland is determined by flow rates and the characteristics of the wetland soils and vegetation, as is the degree of short-circuiting that can occur.

4.1.1 System hydrology

Water enters natural wetlands via streamflow, runoff, groundwater discharge and precipitation (see Figure 1.4). These flows are extremely variable in most instances, and the variations are stochastic in character. Stormwater treatment wetlands generally possess this same suite of inflows. Treatment wetlands dealing with continuous sources of wastewater can have these same inputs, although streamflow and groundwater inputs are typically absent. The steady inflow associated with continuous-source treatment wetlands represents an important distinguishing feature. A dominant steady inflow drives the ecosystem towards an ecological condition that is somewhat different from a stochastically driven system. Wetlands lose water via streamflow, groundwater recharge (infiltration) and evapotranspiration. Stormwater treatment wetlands also possess this suite of outflows. Continuous-source treatment wetlands would normally be isolated from groundwater, and most of the water would leave via streamflow in most cases. Evapotranspiration (ET) occurs with strong diurnal and seasonal cycles because it is driven by solar radiation, which undergoes such cycles. Thus, ET can be an important determinant of water loss on a periodic basis.

Wetland water storage is determined by the inflows and outflows, together with the characteristics of the wetland basin. Depth and storage in natural wetlands are likely to be modulated by landscape features such as the depth of an adjoining water body or the conveyance capacity of the outlet stream. Large variations in storage are therefore possible in response to the high variability in the inflows and outflows. Constructed treatment wetlands, in contrast, typically have some form of outlet water level control. There is therefore little or no variation in water level, except for stormwater treatment wetlands. Dryout does not normally occur, and only those plants that can withstand continuous flooding will survive.

4.1.1.1 Terminology

There are a few basic terms that serve as descriptors of treatment wetland hydrology (Figure 4.1). These are as follows.

Hydraulic loading rate (HLR). HLR is equal to the flow under consideration divided by the wetland surface area. It does not imply the physical distribution of water uniformly over the wetland surface. The wetted area is usually

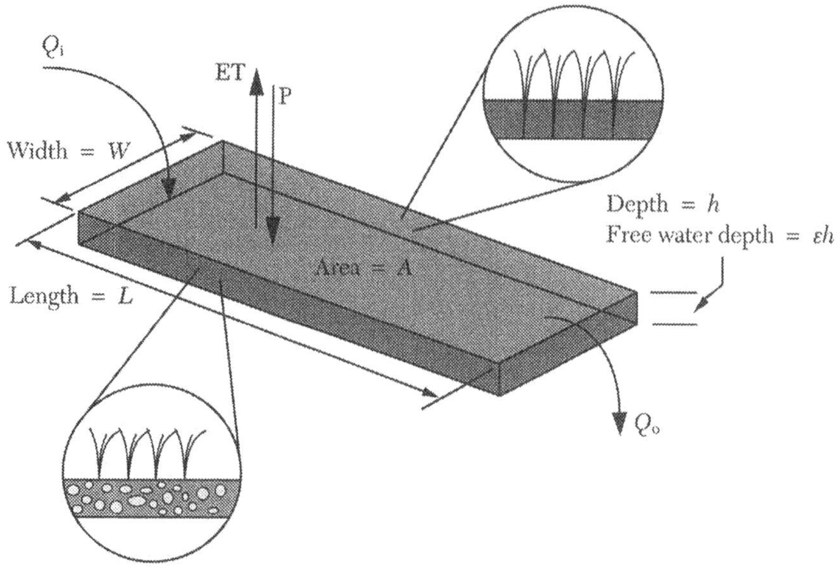

Figure 4.1. Water budget and content for controlled wetlands. For surface flow, solids in water = litter + biofilms + sludge + stems and leaves; volume fraction solids = 1 – ε. For subsurface flow, solids in water = gravel + biofilms + sludge + roots; volume fraction solids = 1 – ε.

known with good accuracy because of berms or other confining features. The definition is most often applied to the wastewater addition flow at the wetland inlet:

$$q_i = Q_i/A, \tag{4.1}$$

where
A = wetland top surface area (m^2)
q_i = HLR ($m\ d^{-1}$; often expressed as $cm\ d^{-1}$)
Q_i = wastewater inflow rate ($m^3\ d^{-1}$).

Depth. Wetland water depth (h) is easily measured for a small, level-bottomed wetland. For a constructed wetland without a contoured bottom, the mean water depth calculation can require a detailed survey of the wetland bottom topography, combined with a survey of the water surface elevation. The accuracy and precision must be better than normal, because of the small depths usually found in treatment wetlands. These difficulties have prevented accurate mean depth determinations in many treatment wetlands.

The amount of free water per unit wetland surface area is the free water depth, which is the porosity times the water depth:

$$h_f = h\varepsilon, \tag{4.2}$$

where
h = water depth (m)
h_f = free water depth (m)
ε = porosity ($m^3\ m^{-3}$).

Void fraction. The void fraction, ε, or porosity, of the wetland is also difficult to determine. It represents the fraction of the wetted volume that is occupied by free (drainable) water. In an SSF wetland, part of the wetted volume is occupied by gravel, roots, sludge and biofilms. Free water volume fractions are typically *ca.* 0.3–0.4 (30–40%). For an FWS wetland, porosity can vary spatially in the x- and y-directions, owing to pattern effects. It also varies strongly in the vertical direction, with smaller values near the bottom in the litter layer. The mean value is greater than 0.95 for cattails, but it can be lower (0.7–0.9) for dense stands of bulrushes.

Nominal detention time. Nominal detention time (= hydraulic retention time, HRT) is the volume of free water in the wetland divided by the volumetric flow rate of water:

$$\tau = \varepsilon LWh/Q, \tag{4.3}$$

where
L = length (m)
Q = flow rate ($m^3\ d^{-1}$)
W = width (m)
τ = nominal detention time (d).

There is obviously a possible ambiguity that results from the choice of the flow rate that is used in this equation. The inlet flow rate is most often used when there is no measurement or estimate of the outlet flow rate. There is a large degree of inaccuracy in the calculation of nominal detention time because it combines the uncertainties in depth and porosity. Nominal detention time is not necessarily indicative of the actual detention time (t_m) because it is based on the presumption that the entire volume of water in the wetland is involved in the flow. This can be seriously in error, with the usual result that actual, measured detention times are smaller than the nominal value.

Nominal detention time can also be expressed as free water depth divided by HLR:

$$\tau = \varepsilon h/q_i. \tag{4.4}$$

This relation shows that it is possible to increase detention time either by increasing depth (h) or by decreasing the HLR (q_i).

4.1.1.2 Water budget

Transfers of water to and from the wetland follow the same pattern for SF and SSF wetlands. In treatment wetlands, wastewater additions are normally the dominant flow, but under some circumstances other transfers of water are also important. The dynamic overall water budget for a wetland is

$$Q_i - Q_o + Q_c - Q_b + Q_{sm} + (P - ET - I)A = \frac{dV}{dt}, \qquad (4.5)$$

where

ET = evapotranspiration rate (m d^{-1})
I = infiltration to groundwater (m d^{-1})
P = precipitation rate (m d^{-1})
Q_b = bank loss rate (m^3 d^{-1})
Q_c = catchment runoff rate (m^3 d^{-1})
Q_o = output wastewater flow rate (m^3 d^{-1})
Q_{sm} = snowmelt rate (m^3 d^{-1})
t = time (d)
V = water storage in wetland (m^3).

Each term in these water budgets can be important for a given treatment wetland, but rarely do all terms contribute significantly. In most cases the storage within the wetland (V) will be determined by a weir setting and will remain constant. The most significant remaining terms in the water budget are then wastewater flows and possibly atmospheric additions and losses:

$$Q_o = Q_i + (P - ET)A. \qquad (4.6)$$

Evapotranspiration. Wetland treatment systems frequently operate with small HLRs. For 100 SF wetlands in North America, 1.00 cm d^{-1} is the 40th centile (Knight *et al.* 1993). For large wetlands, losses to ET approach a daily average of 0.50 cm d^{-1} in summer in the southern USA; consequently, more than half the daily added water can be lost to ET under those circumstances. However, ET follows a diurnal cycle, with a maximum during early afternoon and a minimum in the late night-time hours. Outflow can cease during the day for this extreme example.

The importance of evaporation and transpiration requires that methods be available for estimation. Several methods are explained in Kadlec & Knight (1996). The simplest estimator is that wetland ET is roughly equal to lake evaporation, which in turn is roughly equal to 80% of pan evaporation. Small wetlands of less than 0.5 ha can lose water up to twice as fast, and SSF wetlands might lose only half.

Water mass balance impacts on pollutant decreases. For purposes of contaminant mass balancing, an overall water balance is required. The time period over which averaging is done will generally be dictated by the frequency of water quality sampling. For instance, monthly water quality results would normally be combined with monthly average flows to determine mass removal rates. Seasonally variable wastewater flows can combine with seasonally variable rain and ET to produce large differences in hydrological functions.

Rain dilutes concentrations but decreases detention time. The combination can provide either poorer or better performance, depending on the wastewater loading rate. In very lightly loaded systems, concentration decrease (EFF) is likely to be poorer with rain additions; in heavily loaded systems, concentration decrease can be higher. In both cases, load decrease (percentage mass removal efficiency, RED) is poorer for high rainfall.

ET concentrates pollutants but increases detention time. The combination improves concentration decrease in very lightly loaded systems, but diminishes it in heavily loaded systems. In both cases, load decrease is better for high ET.

Infiltration. Mass balance equations can be developed for infiltrating flows (Kadlec & Knight 1996). The effect of infiltration is to slow the remaining water and increase concentration decrease. The load decrease is further enhanced by the loss of pollutant to infiltration. In this case it is a reasonable approximation to use a flow average in calculations.

Variability. In general, literature values of rate constants, or other measures of performance, have not been corrected for water losses and gains. In some instances, water budget information was not collected; in other cases, atmospheric losses and gains were not significant. Water mass balance effects are therefore the cause of some fraction of the variability in available data. The stochastic character of rainfall, and the periodicity and seasonal fluctuation in ET, are also responsible for a portion of the variability in the concentrations in wetland effluents.

4.1.2 Head loss

Head loss describes the establishment of water surface gradients necessary to drive water through the treatment wetland.

4.1.2.1 Free water surface

Water moves through SF treatment wetlands in response to a surface elevation gradient from inlet to outlet, impeded by drag created by submerged plants and litter (Figure 4.2). Depth is typically controlled via an outflow structure. The hydraulic profile is dictated by these factors, combined with the bottom slope and length:width ratio of the wetland. In many

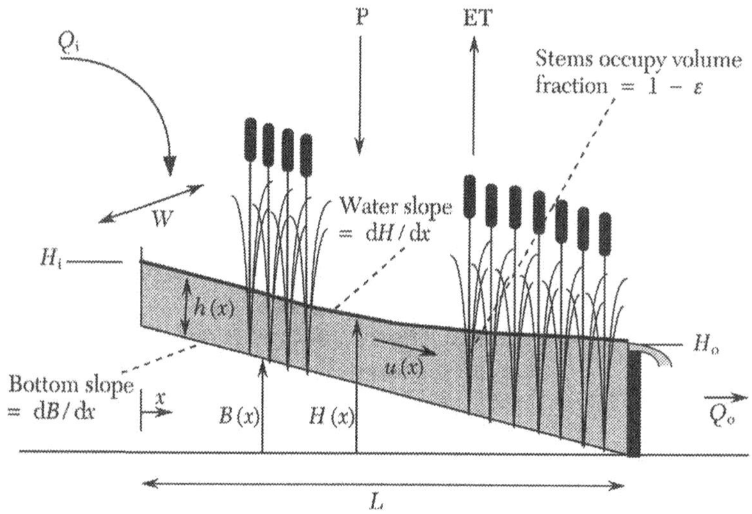

Figure 4.2. Hydraulic variables in overland flow in SF wetlands. This one-dimensional representation indicates the notation and the fact that the principal variables vary with distance along the flow path. (Kadlec & Knight (1996).)

cases, SF wetlands are not hydraulically constrained and can carry design flows and event flows within a small freeboard. In a few instances, SF wetland design has failed to account properly for head loss, with inlet overflooding as the result.

Two equations combine to describe water depth and flow rate as a function of position along the flow direction: (1) the water mass balance and (2) the friction equation, the latter being generically identified with Manning's equation. The mass balance relates flow, depth and water gains and losses. The equation for one-dimensional flow is

$$uhW = Q = Q_i + (P - ET)Wx, \qquad (4.7)$$

where

u = velocity (m d^{-1})
x = distance along the flow path (m).

Velocity (u) is the average superficial velocity, which is the volumetric flow divided by the full cross-sectional area. The friction equation relates flow rate and local head loss (dH/dx):

$$Q = aWh^b\left(-\frac{dH}{dx}\right)^c, \qquad (4.8)$$

where

a, b, c = constants
H = water surface elevation (m).

Equation 4.8 should not be used by itself for head loss calculations, because this procedure might be inaccurate owing to variations in h and dH/dx along the flow direction. At any location, the water surface (H) is at depth h above the wetland bottom (B):

$$H = h + B, \qquad (4.9)$$

where

B = bottom elevation (m).

The exponent b is in the range $1 \le b \le 3$, with the value $b = 3$ being applicable to wetlands with dense vegetation with a fully immersed litter layer, or for laminar flow in open channels. The exponent c ranges from 0.5 for turbulent flow in unvegetated channels to 1.0 for drag surfaces uniformly distributed in the water column or open channel laminar flow; the latter most closely corresponds to typical conditions in SF wetlands (Hammer & Kadlec 1986; Kadlec 1990). The conveyance coefficient (a) varies with vegetation type and density, as well as with site micro-topography. Research has shown that frictional resistance is proportional to stem density (Hall & Freeman 1994). The range for several wetlands is $10^7 \le a \le 5 \times 10^7\ m^{-1}\ d^{-1}$ (Kadlec & Knight 1996).

The values $a = 1/n$, $b = 1.67$ and $c = 0.5$ correspond to Manning's equation for turbulent flow in unvegetated channels, which is used for describing turbulent, open channel flows. Manning's equation can be used for describing friction in laminar, densely vegetated SF wetlands, provided that Manning's coefficient (n) is considered to be a function of depth and velocity. For $b = 3$ and $c = 1$, the appropriate relation is

$$1/n = \sqrt{a}\,h^{\frac{1}{3}}u^{\frac{1}{2}}. \qquad (4.10)$$

The depth dependence and velocity dependence have been confirmed by several studies (Hall & Freeman 1994; Shih *et al.* 1979).

Equations 4.7 to 4.9, together with aspect ratio, bottom elevation and exit water surface elevation, combine to describe the hydraulic profile. When applied to a typical range of flow rates, it is found that the absolute head loss, measured in centimetres of water, is approximately proportional to the length:width ratio, meaning that doubling the length:width ratio will double the head loss. For small wetlands of

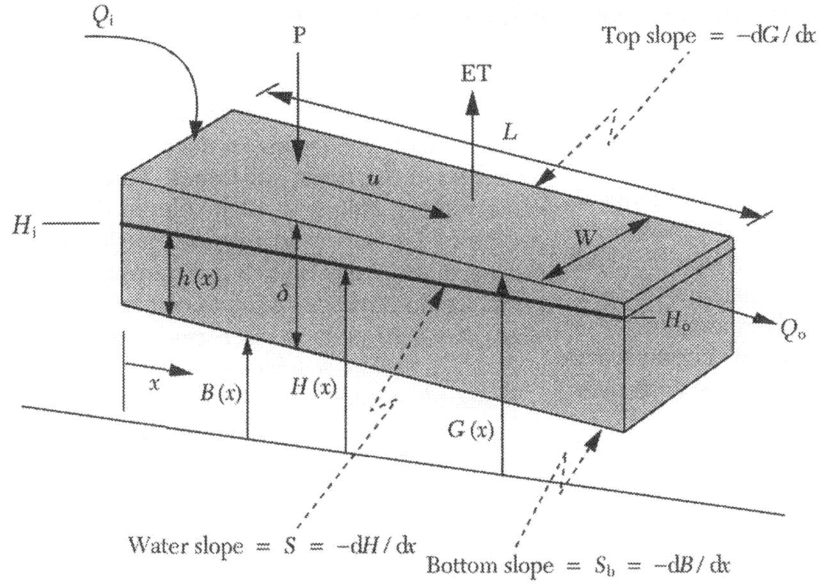

Figure 4.3. Notation for SSF bed hydraulics. The actual velocity of water is $v = u/\varepsilon$. (Kadlec & Knight (1996).)

linear geometry, simple solutions are available (Kadlec & Knight 1996). If the wetland is essentially linear but with a variable cross-section, a one-dimensional code such as HEC2 can be used (US Army Corps of Engineers Hydraulic Engineering Center code). If the wetland boundary is very irregular in plan view or if the wetland bottom topography is irregular, then a two-dimensional code is warranted, such as SMS (Surface Water Modeling System; Engineering Computer Graphics Laboratory, Brigham Young University, Provo, Utah, USA).

The hydraulic profile along the direction of flow is typically not a linear decrease, and depths are not constant (Figure 4.2). Consequently, approximate calculations, based on average depths and average gradients, might be inaccurate. Flows of uniform depth (defined as the normal depth) occur only when the bottom slope precisely matches the friction slope for the wetland and flow rate in question. Depth control, with an outlet structure set above the normal depth, leads to distance-thickening flows. If the outlet depth is set below the normal depth, the flow will be distance-thinning, with most of the head loss occurring near the wetland outlet (Kadlec & Knight 1996). In the extreme case of deep water, sparse vegetation and low flow rate, the wetland can operate with essentially a level pool.

4.1.2.2 Horizontal subsurface

SSF wetlands have less flexibility in design and operation because it is necessary to keep the water surface below ground at all locations. However, the water must not be too far below the top of the medium, or plant roots will not reach the water. Two new parameters arise: the dry zone thickness and the thickness of the medium (Figure 4.3).

The mass balances and geometrical definitions are the same as for FWS wetlands and have been presented in Equations 4.7 and 4.9. The porosity is lower in SSF wetlands, usually in the range 0.3–0.4 $m^3\,m^{-3}$ for sands and gravels, and there is the added geometry of a bed surface to consider. The elevation of the top surface of the medium is

$$G = B + \delta, \qquad (4.11)$$

where
G = elevation of the bed top above datum (m)
δ = thickness of the bed medium (m).

The headspace is defined as the distance from the top surface of the medium down to water:

$$f = \delta - h, \qquad (4.12)$$

where
f = headspace (m).

In general, the variables h, H, G, d, f and B are all dependent on distance from the bed inlet.

The simplest friction relation states that superficial velocity is proportional to the slope of the water surface:

$$u = -k_f \frac{dH}{dx}, \qquad (4.13)$$

where
H = elevation of the water surface (m)
k_f = hydraulic conductivity ($m\,d^{-1}$).

This is the one-dimensional version of Darcy's law. It is restricted to the laminar flow regime, but it can be extended if the hydraulic conductivity is adjusted to account for turbulence (Kadlec & Knight 1996). Values of clean-medium hydraulic conductivity should be measured; however, this can be estimated from

$$k_f = 12{,}600 D_p^{1.90}, \quad (4.14)$$

where

D_p = particle diameter (cm).

The bed will not maintain the conductivity of clean medium because of the deposition of solids and the blocking of pore space by plant roots. If one-third of the pore space is blocked, the hydraulic conductivity decreases to 10%. This is the typical amount of conductivity decrease observed in horizontal flow beds near the inlet.

Gravel bed SSF wetlands in the United States are frequently observed to be overflooded. The two probable causes are clogging of the media with particulates and improper hydraulic design. The same appears to be true for other countries as well (Brix 1994), especially the SSF wetlands that use a soil for the medium. The underlying cause of such hydraulic failure is the ad hoc procedure of designing to guessed values of hydraulic parameters.

4.1.2.3 Vertical subsurface

One method for enhancing oxygen availability involves vertical flow through a gravel bed wetland. Water is distributed uniformly over the bed surface and allowed to percolate downwards to a collection zone. Under-drains then convey the water from the bed. If the bed is operated periodically, or if the medium is highly permeable under steady operating conditions, there is atmospheric oxygen present in the pore space. Further, only thin films of water coat the grains of the medium, which has a very large surface area. Oxygenation of the water is thus promoted to a large degree.

Flow can be either unsaturated, trickle flow or saturated flow with completely filled voids. At the low average HLRs that are usually employed in treatment, a sand or gravel medium will not become saturated, and the flow will be percolation through voids partly filled with air. If the water is delivered in a short period, the instantaneous loading rate can exceed the drainage rate, and the medium will then fill with water. After the medium is full, if high loading continues, ponding will occur on the surface of the bed. After loading ceases, the pond depth will have reached its maximum value; drainage occurs thereafter.

Many vertical flow systems use a dosing period of several hours (Gray & Biddlestone 1995) rather than a series of many short dosing events each lasting only a few minutes (Seidel 1976). High-volume dosing events, also known as tidal inundation, ensure an even utilization of the available bed area. They are also thought to trap air in the substrate (Boller *et al.* 1993; Schwager & Boller 1997), resulting in higher available oxygen levels within the bed. Furthermore, Brix & Schierup (1990) showed that substrate air content declined rapidly once dosing began, continuing to decrease for up to four days in the finest substrates. Thus a series of very short, frequent, high-volume doses are preferred for improved oxygen transfer. If the dosing regime and bed design permit a more permanent saturation of the lowest layers of the sand, breakthrough is decreased and a more even effluent flow is achieved.

Hydraulic calculations require a complex computer code to simulate variably saturated water flow and solute transport, including hysteresis effects (Schwager & Boller 1997).

4.1.3 Dynamic effects

Most wetland systems are fed with a constant flow of wastewater. There is therefore a strong tendency to visualize a relatively constant set of system operating parameters – depths and outflows in particular. This is not necessarily true in practice. Effects of rain and evaporative losses are magnified by the porosity of an SSF bed to yield larger depth changes, and flows are very sensitive to depth and gradients. Thus, there can be significant outflow variability due to precipitation and ET.

As a case in point, cell no. 3 at Benton, Kentucky, USA, was operated in September 1990 at an HLR of 1.7 cm d^{-1}, corresponding to a nominal detention time of *ca.* 13 d. Evapotranspiration at this location and at this time of year was estimated to be *ca.* 0.5 cm d^{-1}. Consequently, ET forms a significant fraction of the hydraulic loading. Because ET is driven by solar radiation, it occurs on a diurnal cycle. The expected and observed effect was a diurnal variation in the outflow from the bed, with amplitude mimicking the amplitude of the combined (feed plus ET) loading cycle.

In such an instance, because the night outflow peak is nearly double the daytime minimum outflow, it is important to use diurnal timed samples of the outflow and to flow-weight them appropriately for the determination of water quality.

A sudden rain event, such as a summer thunderstorm, will raise water levels in the wetland. The amount of the change in level is magnified by porosity and catchment effects: a threefold magnification will be caused by a bed porosity of 33% and some further increase due to bank runoff. Thus, a 3 cm rain can raise bed water levels by more than 10 cm – if there is that much gravel head space. Overflooding of the bed might occur if there is insufficient head space. In any case, outflows from the system increase greatly as the rainwater flushes from the system (Figure 4.4).

The implications for water quality are not inconsequential. In the example of Figure 4.4, samples taken during the ensuing day represent flows much greater than average. Water has been pushed through the bed and exits about a day early, and it has been somewhat diluted. Velocity increases are great enough to move particulates that would otherwise remain anchored. Internal mixing patterns will blur the effects of the rain on water quality.

Sampling intervals are not normally small enough to define these rapid fluctuations. For instance, weekly sampling of Benton cell no. 3 would have missed all of the details of the effects of rain and ET in the illustrations above. It is therefore important to realize that compliance samples can give the appearance of having been drawn from a population of large variance, despite the fact that the variability is in large part due to deterministic responses to atmospheric phenomena.

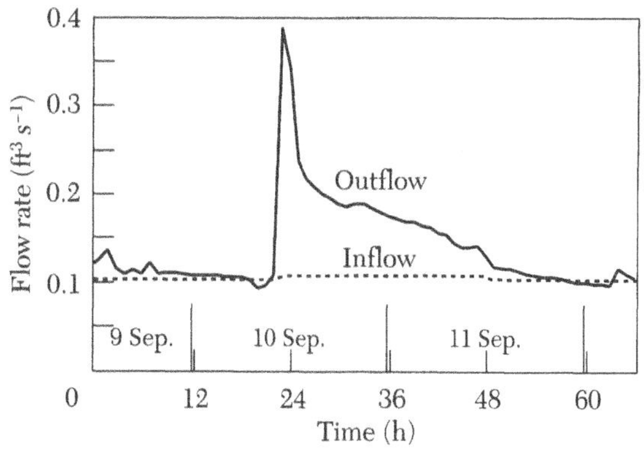

Figure 4.4. Flows into and out of an SSF wetland on 9, 10 and 11 September 1990. A 1.9 cm rain event caused the spike in the outflow. ($1\ ft^3\ s^{-1} = 28.32\ l\ s^{-1}$.)

4.2 Internal flow patterns

Water does not move through a wetland in lock-step from inlet to outlet. Several phenomena combine to produce a distribution of transit times for water parcels. Many treatment wetlands have been tested with tracers, and all exhibit a significant departure from plug flow (Kadlec 1994; Stairs 1993; King *et al.* 1997). This is very important for performance predictions because pollutant decreases are typically exponential in detention time.

The important features of wetland hydrology from the standpoint of treatment efficiency are those that determine the duration of water–biota interactions and the proximity of water-borne substances to the sites of biological and physical activity. There is a strong tendency in the wetland treatment literature to borrow the detention time concept from other aquatic systems, such as 'conventional' wastewater treatment processes. In purely aquatic environments, reactive organisms are distributed throughout the water, and there is often a clear understanding of the flow paths through the vessel or pond. However, wetland ecosystems are more complex, with areal distributions of plants and other biota, and therefore require more descriptors.

4.2.1 Horizontal SSF wetlands

In SSF wetlands, there are preferential paths through the medium. Roots occupy the upper portion of the bed, which slows the movement of water in that region (Fisher 1990). Flows then preferentially follow paths near the bottom. Further, there are significant non-uniformities within the wetland (Netter 1994; King *et al.* 1997).

The result of these non-idealities is a distribution of detention times within the wetland (termed the residence time distribution, or RTD). Some elements of incoming water travel much faster than average, whereas others travel much more slowly and compensate the mass balance (Figure 4.5). The fastest transit times in SSF beds typically range from 10% to 50% of the nominal detention time; the slowest are three to four times the nominal detention time.

4.2.2 FWS wetlands

There are also preferential paths through FWS wetlands. Water near the surface is less subject to bottom drag and moves faster than the mean at any location. Water must detour around plant bases, which are themselves stagnant pockets that exchange water with the adjacent channel by diffusion. Open water zones are subject to wind-driven surface flows, which contribute to mixing. The bottom topography can form deeper pathways, through which preferential flow causes short-circuiting.

The combined effect of these processes can be seen in the passage of an inert tracer through the wetland. An impulse of tracer, added across the flow width, moves with water through the wetland as an accelerated, spreading cloud. As with SSF wetlands, some water elements reach the wetland outlet much more quickly than indicated by the mean velocity; others are delayed and spend much longer times before exiting. The result is a distribution of detention times, similar to that shown in Figure 4.5. However, in FWS systems, the fastest transit times are shorter than for SSF wetlands. A typical first arrival is at 5–10% of the nominal detention time.

4.2.3 Impact on performance and design

Many of the available performance data are from wetlands without a quantified residence time distribution. For first-order rate equations, plug flow is the most efficient use of a

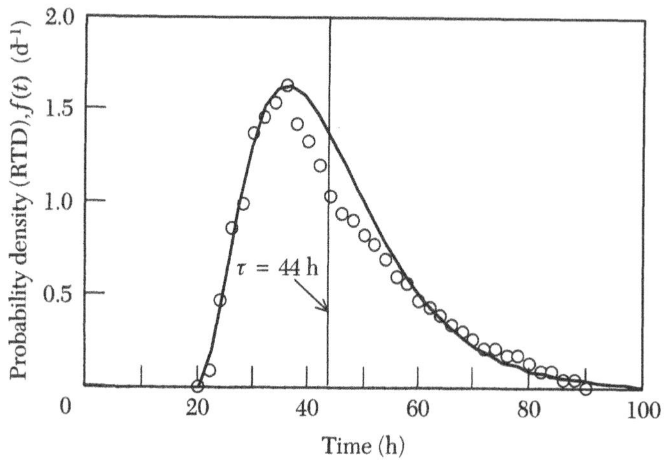

Figure 4.5. Tracer response for the SSF wetland in Carville, Louisiana, USA. The fraction of water leaving, after spending between t and t + Δt hours in the wetland, is f(t)Δt. A network model fits the data, with a plug flow reactor of 45% (20/44 of total volume) and three continuous-flow stirred tank reactors of 18% of the volume of each (data are from US Environmental Protection Agency (1993)). RTD, residence time distribution.

wetland because it produces the highest removal. It is therefore advantageous to configure the wetland to approximate plug flow as closely as possible. For high removal percentages, the plug flow assumption can result in design errors of as much as 100% (Kadlec & Knight 1996).

One intuitive method is increasing the wetland length:width (aspect) ratio, predicated on the hypothesis that long, slender systems will be closer to plug flow. Tracer tests show that this hypothesis is not true because tracer impulses show large amounts of dispersion even for high aspect ratios (Kadlec & Knight 1996). Aspect ratio is of secondary importance as a determinant of flow pattern. Velocity profile effects are more important and include large-scale and small-scale phenomena in both the vertical and horizontal directions.

The tools for accommodating non-ideal flow into process description include the tanks-in-series and plug-flow-with-dispersion models. Both are calibrated by means of a valid-impulse tracer test, which should involve 100% recovery of a non-reactive substance. Details of parameter estimation have been published (Kadlec 1994).

One of the important results of a tracer test is the determination of the tracer detention time, defined as the centroid of the response curve. It is equal to the actual interactive water volume divided by the volumetric flow rate, and thus represents a direct measure of actual detention time, t_m. It ends speculation about bathymetric accuracy and potential volume blockage.

The plug flow assumption is conservative for data analysis because it leads to the lowest possible estimate of a rate constant from any given data set. The plug flow assumption is conservative in design if the degree of non-ideality in the designed system is less than that in the data-generating wetlands. The treatment wetland literature typically provides only plug flow k values.

Temporal changes in depth, combined with an uneven topography of the wetland bottom, lead to pattern effects on vegetation in natural wetlands. Constructed treatment wetlands usually have nearly uniform bottoms. Combined with controlled, steady water levels, this means uniform hydrological conditions and an absence of pattern effects. Pattern effects interact with water flows through the wetland, with preferential, sparsely vegetated channels carrying a disproportionately high fraction of the water. This in turn impairs the treatment potential, because much of the wetland surface is not exposed to the water flow.

4.3 Thermal effects

Wetland water temperatures can influence some pollutant removal and conversion processes. In northern climates, ice formation can influence system hydraulics, but many treatment processes proceed in under-ice flow. The vehicle for understanding, correlating and predicting wetland thermal variables is the energy balance at the water surface.

4.3.1 Energy balance and water temperature

In summer, energy gains are primarily from solar radiation and to a smaller extent from warm air (they can be losses for cool air). Losses are primarily to evapotranspiration, and to a minor extent to the cool deep soils. In winter, gains are from soil storage, and loss is to the cold ambient air. At the snow surface, radiation, convection and sublimation create a

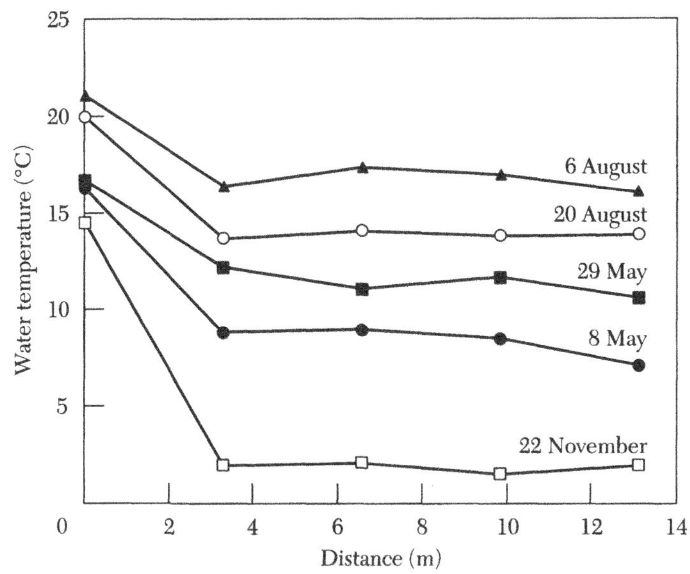

Figure 4.6. Water temperatures along the flow direction for an SSF wetland operating in a very cold climate (unpublished data from NERCC, Duluth, Minnesota, USA).

balance that dictates the snow surface temperature.

For a constant-width wetland, the energy balance is

$$cuh\frac{dT}{dx} = G - L, \qquad (4.15)$$

where

- c = thermal capacity (MJ m^{-3} °C^{-1})
- T = water temperature (C)
- G = energy gains (MJ m^{-2} d^{-1})
- L = energy losses (MJ m^{-2} d^{-1})
- x = distance along flow path (m).

In summer, large amounts of energy are supplied by solar radiation. A small portion of this recharges the soil energy storage, but most is lost by way of evaporation and transpiration. The thermal capacity, c, is typically small for the HLRs normally employed, and therefore the water temperature adapts to a balance condition where $G = L$ (Figure 4.6). The energy balance shows that, theoretically, that the balance point temperature is not too different from the mean daily air temperature (T_a) during the unfrozen seasons. This is borne out in practice, with linear regression yielding (Kadlec & Knight 1996):

$$\left.\begin{array}{l} T = T_a; \\ \text{mean } R^2 = 0.82; \\ N = 10 \text{ SF wetlands.} \end{array}\right\} \qquad (4.16)$$

During periods of ice cover, treatment wetland water temperatures are typically just above freezing, 0–2 °C. Figure 4.7 shows patterns of temperature in FWS wetlands.

Temperature swings at the wetland outlet can be as large as 10 °C from day to day because of the strong influence of meteorological factors. There is also a large diurnal cycle in water temperature, with daytime temperatures ranging from 5 to 15 °C higher than night-time temperatures.

4.3.2 Snow and ice

During the frozen season, the presence of insulating layers of snow and ice change the application of the energy balance considerably. Energy gains are now solely from deep soil storage, and losses are by heat conduction through the snow and ice to the cold air above and to ice formation. Incoming sensible heat is typically dissipated because losses are generally greater than gains. There is no longer a large radiation input to the water; the low winter insulation is reflected by the snow and absorbed by sublimation at the snow surface. Evaporation from the water layer is prevented by the ice cap. As a consequence, gains and losses do not totally dominate the energy balance as in summer, and temperature decline typically proceeds throughout the flow path. If heat losses are severe enough, ice formation occurs. Water is shallow and unstratified, so the effluent temperature for freezing conditions is close to 0 °C. Heat generation due to the oxidation of BOD and plant detritus is a very small but positive contribution to the energy balance.

The amount of ice formation is determined by climate conditions that vary greatly from one winter to another. The principal deterrent is the insulation provided by snow, which serves to prevent heat losses. Wetland vegetation is effective in trapping snow to greater extents than unvegetated areas. Ice thicknesses in wetlands are therefore much less than in adjacent lakes or frost depths in nearby uplands. The wetlands at Listowel, Ontario, Canada, experienced ice thickness on the

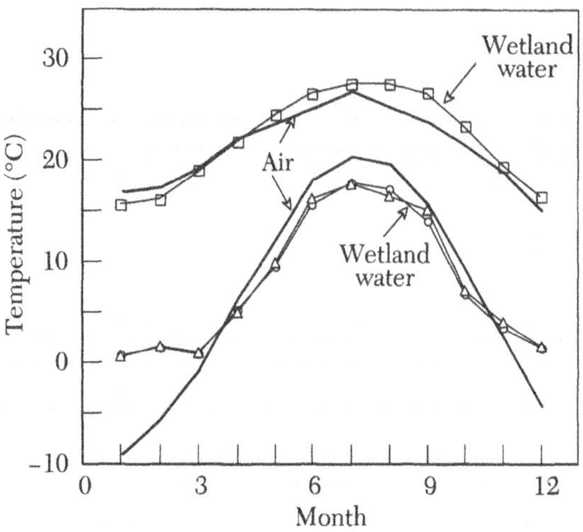

Figure 4.7. Both northern (lower curves) and southern (upper curves) systems show water temperatures that are strongly correlated with mean daily air temperature during warm months from nearby weather stations. However, during frozen months, water temperatures are just above the freezing point. The southern system was Orlando Easterly Wetlands, Florida, USA; $T = 0.96T_{air}$; $R^2 = 0.90$. The northern systems were Listowel Wetlands no. 3 (○) and 4(△); unfrozen $T = 0.86\ T_{air}$; $R^2 = 0.97$.

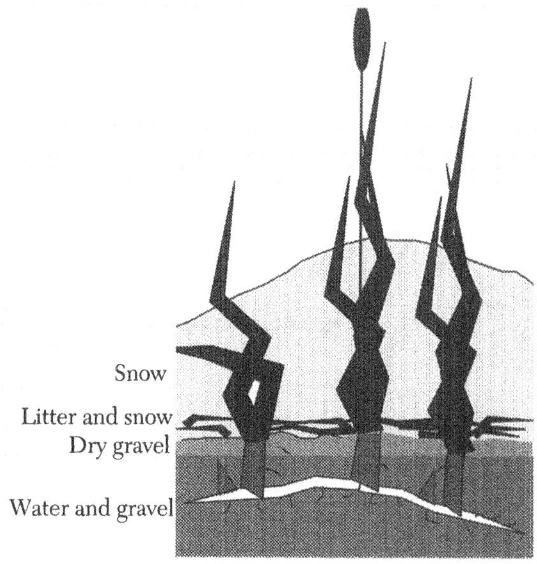

Figure 4.8. Cross section of an SSF gravel bed in winter.

order of 10–15 cm during flow conditions for a climate typified by a mean January air temperature of -9 °C. Ice or frost depths in the wetland at Houghton Lake, Michigan, USA, range from 0 cm (for copious early snow) to 20 cm for unvegetated pond zones with little snow. The mean January temperature is -8 °C, and there is no winter water flow.

Ice that forms in the upper horizons of an SSF bed is physically supported by the medium. Consequently, water levels can be dropped to create an air zone in the matrix above the water and below the ice (Figure 4.8). These various forms of insulation combine to allow winter operation of SSF wetlands in extremely cold climates. For instance, the Grand Lake horizontal SSF wetlands near Duluth, Minnesota, USA, have operated without any ice formation, even during prolonged periods in which temperatures dropped below -40 °C.

4.4 Performance considerations

Like other water quality treatment processes, wetlands perform within definable limits. These limits must be defined and summarized to allow the designer to scale a treatment wetland to decrease pollutant concentrations consistently from some inflow value to some desired outflow concentration. Regression equations and relatively simple first-order models are most commonly used to summarize wetland performance. On the basis of a general knowledge of performance expectations, the designer has the ability to determine the actual treatment efficiency to some extent by internal design features such as wetland area, water depth, cell configuration and the selection of media and plants.

The designer should also consider certain constraints associated with treatment wetlands because they are living ecosystems. The natural processes that occur in SF wetlands result in non-zero background concentrations of some chemicals that can, at higher concentrations, be the same constituents requiring treatment. Knowledge of these background concentrations is important to avoid unduly optimistic expectations for treatment wetlands. In addition, some statistical variability is inherent in wetland outflow constituent concentrations, some of which is due to environmental factors (such as seasonal temperature changes) outside the control of the wetland designer and operator. The inevitability of this noise must be incorporated into design to avoid violations of permits.

4.4.1 Wetland background concentrations

Wetland ecosystems typically include diverse autotrophic (primary producers such as plants) and heterotrophic (consumers such as microbes and animals) components. Most wetlands are more autotrophic than heterotrophic, resulting in a net surplus of fixed carbonaceous material that is buried as peat or is exported downstream to the next system (Mitsch & Gosselink 1993). This net production results in an internal release of particulate and dissolved biomass to the wetland water column, which is measured as non-zero levels of BOD, TSS, TN and TP. These wetland background concentrations are denoted by C^*. Enriched wetland ecosystems

Table 4.1. Long-term average annual outflow concentrations (mg l^{-1}) for lightly loaded surface flow wetlands in the NADB (NADB 1993)

System	BOD_5	TSS	NH_4 N	TN	TP
Eastern Service Area, Florida, USA	1.2	3.0	0.07	1.45	0.09
Iron Bridge, Florida, USA	2.0	2.8	0.18	0.95	0.08
Bear Bay, South Carolina, USA	1.9	2.7	0.27	2.35	0.40
DesPlaines, Illinois, USA	--	5.2	0.03	1.34	0.02
Hidden Lake, Florida, USA	3.0	13.0	0.05	0.66	0.16

are likely to produce higher background concentrations than oligotrophic wetlands because of the larger biogeochemical cycles that result from the addition of nutrients and organic carbon. Even the low levels of nutrients in precipitation result in some primary productivity in wetlands, and surface water concentrations in closed wetland basins with inflows dominated by precipitation represent the lowest wetland effluent concentrations observed.

Background concentration ranges in treatment wetlands can be estimated from systems that are loaded at a sufficiently low rate to result in asymptotic concentrations along a gradient of increasing distance from the inflow. Several examples exist in the North American Treatment Wetland Database (NADB) and in the tertiary systems of the Severn Trent Water Authority, UK. Tables 4.1 and 4.2 summarize long-term average annual outflow constituent concentrations for this selected group of treatment wetlands. Wetland systems typically have background concentrations within the following ranges:

BOD_5	1–10 mg l^{-1}
TSS	1–6 mg l^{-1}
Org-N + total nitrogen	1–3 mg l^{-1}
Faecal coliforms (FC)	50–500 FC/100 ml
NH_4-N	less than 0.5 mg l^{-1}
NO_3-N	less than 0.1 mg l^{-1}
TP	less than 0.1 mg l^{-1}

4.4.2 Performance equations

A vast quantity of operational performance data have been collected from treatment wetlands. These data were collected over wide ranges of inlet concentrations, mass loadings, flow rates and HLRs, HRTs, water depths, vegetation types and water temperatures. The advancement of treatment wetland technology and the ability of designers to harness wetland processes in predictable treatment systems hinges on the ability to summarize these diverse data sets into a small number of defining relationships. Types of descriptor that have been successfully applied to treatment wetland data include removal efficiencies, regression equations and first-order mass-decrease equations. Each of these methods of summarizing performance is described briefly below. Chapter 5 provides individual pollutant summaries based on these approaches.

Table 4.2. Long-term average annual outflow concentrations (mg l^{-1}) for lightly loaded subsurface flow wetlands in the Severn Trent area (Cooper et al. 1995)

System	BOD_5	TSS	NH_4 N
Knowbury	1.6	3.8	0.24
Hungarton	1.4	3.3	0.56
Ilmington	1.7	2.3	0.33
Napton	1.6	3.8	0.24
Four Crosses	1.7	3.0	0.27

4.4.2.1 Basic equations

The fundamental descriptors of wetland performance are inlet (C_i) and outlet (C_o) concentrations, volumetric flow rate (Q), area (A) and depth (h). Wetland water volume (V) is defined as area times depth times porosity (ε):

$$V = Ah\varepsilon. \tag{4.17}$$

One measure of relative flow rate is the nominal wetland detention time, τ,

$$\tau = \frac{Ah\varepsilon}{Q} = \frac{h\varepsilon}{q}, \tag{4.18}$$

where $q = Q/A$ is the HLR. It is noted here that detention time is also equal to the free water depth, εh, divided by the flow rate per unit surface area, q (also called the HLR). Thus q is the rainfall equivalent of the inlet flow of wastewater. Nominal detention to an interior point in the wetland is the overall travel time (τ) multiplied by the fractional distance through the wetland (y).

The pollutant loading rate, LR_i, at the wetland inlet, is defined as

$$LR_i = qC_i. \tag{4.19}$$

The percentage concentration decrease efficiency is

$$\text{EFF} = 100\frac{C_i - C_o}{C_i}. \tag{4.20}$$

The percentage mass removal efficiency is

$$\mathrm{RED} = 100\frac{LR_\mathrm{i} - LR_\mathrm{o}}{LR_\mathrm{i}}. \qquad (4.21)$$

The parameter RED embodies the overall water mass balance and explains the fate of the mass of entering pollutant. Both EFF and RED can be misleading for inlet concentrations and loadings that are close to zero, in which case very large positive (or negative) percentages can result. Both are sensitive to wetland background concentrations as well as the speed of reduction processes. These variables can be used in either regression equations or mass balance equations to describe a performance data set.

4.4.2.2 Regression equations

Regression equations are the most convenient choice for representing intersystem data sets, in the form of linear or log-linear equations. They must be accompanied by the ranges of the variables because regressions are unreliable outside the range of data that produced them. The correlation coefficient is another useful adjunct because it reveals the fraction of variability that is described by the regression equation. Regression equations do not directly account for factors in the water mass balance or the pollutant mass balance.

The most obvious regression variables are the concentrations in and out, HLR, EFF and RED.

4.4.2.3 First-order equations

Many pollutants decline exponentially to a background concentration (C^*) on passage through a wetland. At the same time, some substances are returned to the wetland water column through the complex chemical processes occurring in wetland soils, decomposing litter, and living plants and wildlife. Most removal and return processes involve solid surfaces, such as roots, litter and algal mats. The simplest removal equation that embodies a steep curved decline is first-order; the simplest return rate equation that embodies a non-zero background concentration is zero-order.

For first-order pollutant uptake (J_U),

$$J_\mathrm{U} = kC. \qquad (4.22)$$

For zero-order pollutant return (J_R) from the ecosystem to the water column,

$$J_\mathrm{R} = \text{constant} = kC^*. \qquad (4.23)$$

The net pollutant decrease rate (J) is the difference between the two:

$$J = k(C - C^*). \qquad (4.24)$$

When $C = C^*$, there is no net removal of the pollutant, although both destruction and production processes continue. The net pollutant decrease rate (J) is the mass removal per unit wetland surface area (g m^{-2} yr^{-1}). The global rate constant (k) is therefore proportional to the amount of active area (such as biofilms, plants and algae) per unit wetland area.

In many treatment wetland cases, infiltration is prevented, there is not significant atmospheric deposition, and $P \approx \mathrm{ET}$. Under this special set of conditions, Q = constant along the length of the wetland (y), and

$$-Q\frac{\mathrm{d}C}{\mathrm{d}y} = JA = kA(C - C^*). \qquad (4.25)$$

For a specified inlet concentration (C_i) this integrates to

$$\frac{C_\mathrm{o} - C^*}{C_\mathrm{i} - C^*} = \exp\left(-\frac{kA}{Q}\right) = \exp\left(-\frac{k}{q}\right). \qquad (4.26)$$

Equation 4.26 is the historically well known first-order plug flow concentration profile, for a non-zero background concentration. It relates concentrations within the wetland, including C_o, the concentration at the outlet point, to loading rate (q).

For those pollutants that have C^* values very close to zero, namely nitrate, phosphorus and ammonia nitrogen, Equation 4.26 reduces to

$$\frac{C_\mathrm{o}}{C_\mathrm{i}} = \exp\left(-\frac{kA}{Q}\right) = \exp\left(-\frac{k}{q}\right). \qquad (4.27)$$

Equation 4.27 can be rearranged to provide an estimate of the wetland surface area necessary to decrease an inflow pollutant concentration C_i to a target outflow concentration C_o:

$$A = -\frac{Q}{k}\ln\left(\frac{C_\mathrm{o} - C^*}{C_\mathrm{i} - C^*}\right). \qquad (4.28)$$

The two calibration parameters are k and C^*; this description is therefore termed the k–C^* model. A typical fit to data is shown in Figure 4.9.

Some pollutants, notably nitrogen (see Chapter 5), are linked by a sequential reaction pathway. In that case, the k–C^* concept is applied to each step, and production rates are included in the mass balances (Kadlec & Knight 1996). For those cases in which seepage and atmospheric losses or gains are significant, a more complex equation is required.

The exponent in Equation 4.27 is often regrouped to define a volumetric rate constant:

$$k_\mathrm{v} = k/\varepsilon h. \qquad (4.29)$$

For the volumetric case, Equation 4.10 can be modified to

$$\frac{C_\mathrm{o} - C^*}{C_\mathrm{i} - C^*} = \exp(-k_\mathrm{v}\tau). \qquad (4.30)$$

For those substances for which $C^* \approx 0$, Equation 4.14 reduces to

$$C_o/C_i = \exp(-k_v\tau). \qquad (4.31)$$

The following equation, first proposed by Kickuth (1980), has been widely used for the scaling of HSF systems for domestic sewage treatment in Europe:

$$A = Q(\ln C_i - \ln C_o)/k_{BOD_5}, \qquad (4.32)$$

where

A = surface area of bed (m^2)
Q = avarage flow ($m^3\ d^{-1}$)
C_i = influent BOD_5 (mg l^{-1})
C_o = effluent BOD_5 (mg l^{-1})
k_{BOD_5} = BOD_5 area-based rate constant (m d^{-1}).

On the basis of limited information from operational systems, it seems that the rate constant K_{v20} for a particular system might be related to the porosity of the medium used to construct the bed (Wood 1994). Wood (1994) reported the following values of k_{20} (d^{-1}) for different porosities: 1.84 (ε = 0.42), 1.35 (ε = 0.39) and 0.86 (ε = 0.35).

Kickuth proposed a value of 0.19 m d^{-1} for the constant k_{BOD_5} (Boon 1985). This resulted in a specific area of *ca.* 2.2 m^2/PE for sewage of 200 mg l^{-1} BOD_5, an effluent of 20 mg BOD_5 l^{-1} and a daily flow of 180 l/PE. However, the measurements in operational HSF wetlands have shown that k_{BOD_5} is lower. Schierup *et al.* (1990) reported a value of 0.083 m d^{-1} from 49 systems in Denmark. Cooper (1990) reported kBOD values in the range 0.067–0.1 m d^{-1} on the basis of measurements in systems in the UK. However, it seems that k_{BOD_5} increases with the age of the system (Brix 1998). Cooper *et al.* (1996) pointed out that k_{BOD_5} has generally been set at 0.10 for domestic sewage. This has generally meant that the bed surface area (A) has been approx. 5 m^2/PE.

4.4.2.4 Temperature equations

Temperature effects on k or k_V can be summarized by use of the modified Arrhenius equation:

$$k_T = k_{20}\theta^{(T-20)}, \qquad (4.33)$$

where k_T is the rate constant at temperature T °C and k_{20} is the rate constant at 20 °C. Values of the temperature correction factor (θ) have been estimated for data sets with adequate operational temperature data. The dimensions of k_V are reciprocal time, typically d^{-1}; those of k are velocity, m d^{-1}. Because of numerical magnitudes, the units of k are typically converted to m yr^{-1}.

As described earlier, the data from many treatment wetlands indicate non-zero values of

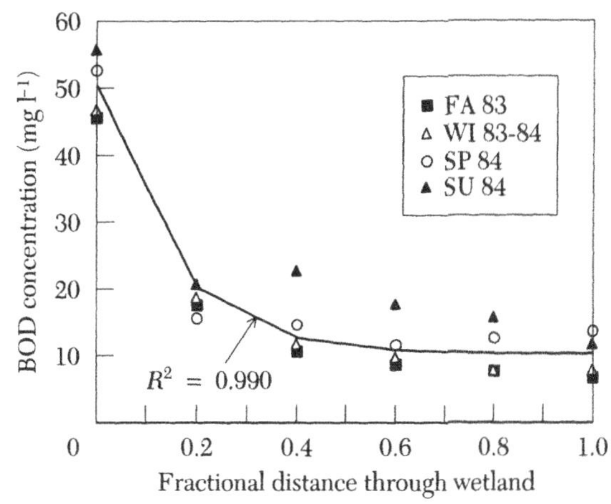

Figure 4.9. The progression of BOD concentrations with distance through a wetland (Listowel no. 4) operated in continuous-flow mode (data from Hershkowitz (1986)). The model line was determined independently from input–output data (k = 60; C° = 10.4).

C^* for some common pollutants (BOD, TSS, Org-N, TN, TP and faecal coliforms). However, most of the existing wetland literature makes the assumption that $C^* = 0$ and reports rate constants for the resulting one-parameter model (k_1 or k_{V1}). Rate constants determined on the basis of that assumption are always lower than the actual value by as much as a factor of 2 or 3 for light hydraulic loadings.

Either k or k_V can be used to represent a data set or be used in design. However, the use of k_V requires the accompanying information on water depth (h) because of the depth dependence indicated in Equation 4.29. This depth dependence also means that a larger detention time created by deeper water can be counteracted by a decrease in the volumetric rate constant. Data analysis and design with the use of volumetric coefficients therefore require a knowledge of the water depth. The use of areal coefficients does not require depth. For many SF wetlands, especially large ones, depth is not known to a reasonable degree of accuracy. For these reasons, the parameters k, C^* and θ are used to summarize operational performance data for treatment wetlands throughout this report.

In some instances there is a required supply of a limiting reactant: oxygen, carbon and alkalinity are three of the most common in treatment wetlands. In this assessment, any limitations of these essential reactants are implicit in the variability of system performances.

In general, literature values of rate constants have not been corrected for water losses and gains. In some instances, water budget information was not collected; in other cases atmospheric losses and gains were not significant. Therefore water mass balance effects are the

Table 4.3. Ratios of maximum monthly outflow values to annual values for NADB surface flow treatment wetlands (Kadlec & Knight 1996)

Parameter	Number of wetlands	Maximum monthly/annual
TP	43	1.8
Dissolved P	21	1.9
TN	30	1.6
TKN	36	1.5
NH_4-N	48	2.5
NO_x-N	46	2.5
Org-N	22	1.8
BOD_5	47	1.7
TSS	49	1.9
Faecal coliforms	23	3.0
Dissolved oxygen	32	1.9

Abbreviation: TKN, total Kjeldahl nitrogen.

cause of some fraction of the variability in rate-constant data.

It has been suggested that these effects can be effectively included in calculations by using an average flow (US Environmental Protetion Agency 1993; Reed *et al.* 1995). Unfortunately, this intuitively appealing method compensates only for the detention time effect and not the mass balance effects of dilution or concentration. In general, the percentage error resulting from use of inlet flow rate in calculations is no worse than the percentage error in the water mass balance due to P and ET.

4.5 Stochastic variability

Treatment wetlands demonstrate the same type of variability in water quality typical of other complex biological treatment processes. Although inlet concentration pulses are frequently dampened through the long hydraulic and solids residence times of the treatment wetland, there is still always significant spatial and temporal variability in surface water pollutant concentrations in wetlands.

The stochastic character of rainfall and the periodicity and seasonal fluctuation in ET are also responsible for a portion of the variability in the concentrations in wetland effluents.

One index of this variability is the ratio between average conditions and maximum conditions observed over shorter time periods. Table 4.3 presents a summary of ratios of maximum month values annual values for data sets in the NADB. Data were available from 22 sites with a total of 53 different wetland cells for this analysis. Average ratios range from *ca.* 1.6 for TKN and TN to 3.5 for faecal coliforms. Table 4.4 presents a similar data summary from HF wetlands in the Czech Republic. Each year of data is recorded separately. These results are generally in good agreement with the data for North American SF treatment wetlands. Some higher ratios in the Czech Republic might be influenced by the fact that these systems are small: the fluctuation in inlet concentration can be greater for small systems.

Table 4.4. Ratios of maximum monthly outflow values to annual average values for HF SSF constructed wetlands in the Czech Republic

Parameter	Number of wetlands	Maximum monthly/annual
TP	16	1.9
TN	13	1.5
TKN	7	1.8
NH_4-N	18	1.9
NO_x-N	12	2.7
Org-N	9	2.0
BOD_5	26	2.2
TSS	26	2.3
Faecal coliforms	5	3.4

Abbreviation: TKN, total Kjeldahl nitrogen.

5 Mechanisms and results for water quality improvement

5.1 Suspended solids

5.1.1 Processes

5.1.1.1 Surface-flow wetlands

Suspended solids are one manifestation of natural wetland processes, as well as being common contaminants in feed waters. Incoming particulate matter usually has ample time to settle and become trapped in litter or dead zones. The combination of removal processes is called filtration, although stem and litter densities are not typically high enough to be considered a filter mat. A number of wetland processes produce particulate matter: the death of invertebrates, the fragmentation of detritus from plants and algae, and the formation of chemical precipitates such as iron flocs. Bacteria and fungi can colonize these materials and add to their mass.

Wetland sediments and microdetritus are typically near neutral buoyancy, flocculent, and easily disturbed. Bioturbation by fish, mammals and birds can resuspend these materials and lead to high measured TSS in the wetland effluent. The oxygen generated by algal photosynthesis or methane formed in anaerobic processes can cause the flotation of floc assemblages. Resuspension due to fluid shear forces on bed solids is not usually a major process, except in the vicinity of a point discharge into the treatment wetland because of the low velocities normally used for treatment purposes.

Wetland particulate cycling is large and almost always overshadows TSS additions, with high levels of gross sedimentation and resuspension (Figure 5.1). TSS background concentrations are rarely irreducible leftovers from feed water; they are often the result of the wetland processes enumerated above. If TSS in added water is lower than this background, an increase in TSS concentration is seen. Most SF wetlands are large enough to approach background levels of suspendable materials. Typical long-term accretion rates for lightly loaded FWS wetlands are in the range 2–10 mm yr^{-1} (Craft & Richardson 1993).

High incoming TSS or high nutrient loadings that stimulate high production can eventually lead to measurable increases in bottom elevation (van Oostrom & Cooper 1990). However, no FWS treatment wetland has yet required maintenance because of solids accumulation, including some that have been in operation for 20 years or more. In situations of high incoming solids, a settling basin can be designed to intercept a large portion of the solids, provide for easier cleanout and extend the life of the inlet region of the wetland.

Animals can be strong determinants of wetland TSS by virtue of their physical activity. Some known examples of negative effects by stirring include the following:

- foraging carp
- spawning shad
- muskrats, nutria and beavers
- wild pigs, deer and elk
- foraging waterfowl.

5.1.1.2 Subsurface wetlands

The existence of a subsurface air/water interface causes sediment processing in the SSF wetland to differ considerably from that in SF wetlands. Macrophyte leaf and seed litter are mostly contained on the surface of the bed and do not interact with the water flowing in the interstices below. Most vertebrates and invertebrates do not interact with the water. Resuspension is not caused by wind or vertebrate activities.

However, many particulate processes do operate in the water-filled voids. Particles settle into stagnant micropockets or are strained by flow constrictions. They can also impinge on substrate granules and stick as a result of several possible interparticle adhesion forces. These physical processes are termed **granular medium filtration** (Metcalf & Eddy 1991). Higher velocities can dislodge adhering or deposited material, which forms the basis for the back-washing method of filter regeneration. Generation of particulate material can occur via all the mechanisms shown for FWS wetlands. Below-ground macrophyte parts – roots and rhizomes – die, decay and produce fine detrital fragments. Many other organisms are present in the bed that can contribute to TSS via the same route: algae, fungi and bacteria all die and contribute particulate matter to the water

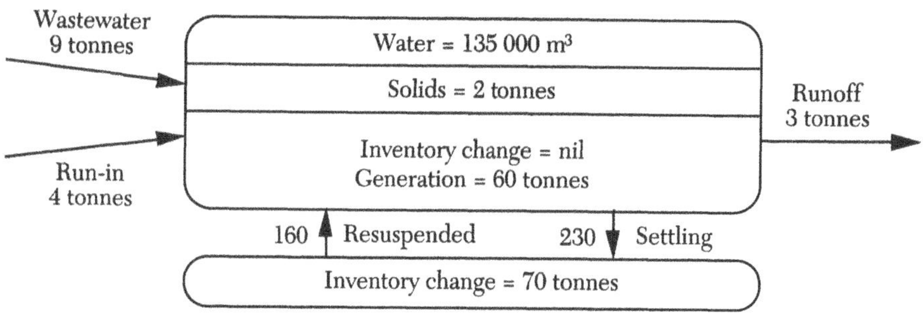

Figure 5.1. Budget of transportable solids for the discharge zone of the wetland at Houghton Lake, Michigan, USA.

flowing in the pore space. These microorganisms are unevenly distributed spatially within the gravel bed, with more organisms located near the inlet and near the bottom (Bavor *et al.* 1988).

5.1.2 Performance

Treatment wetlands are typically efficient in bringing about a net decrease of TSS, with removal efficiencies often in the 80-90% range. As a result of the combined processes discussed above, TSS declines along the flow path from inlet to outlet, down to the background level (Figure 5.2). The k-C* model provides a highly simplified description of the complex wetland solids interactions, and typically represents the decreasing profile quite well, accounting for over 90% of the intrasystem variability ($R^2 \approx 0.9$) (Kadlec & Knight 1996).

5.1.2.1 Surface flow

The value of k for TSS is theoretically the same as the settling velocity of the incoming particles, which can vary widely with the type of wastewater and its pretreatment. Some incoming solids, such as emulsions and planktonic debris, are very slow to settle. For instance, the planktonic solids in Muskego Lake (*ca.* 1 m deep and vegetated with submerged and emergent macrophytes) remain suspended for long periods. Values of k range from 0.1 m d^{-1} for plankton to 10 m d^{-1} for lagoon or river solids.

The wetland background TSS concentration is typically in the range 3-15 mg l^{-1} but depends on the strength of the wetland carbon cycle. High nutrient levels stimulate growth and hence accentuate the return flux and increase the resultant background concentration. C^* is therefore elevated for strong influents. The incoming TSS concentration can be used as a surrogate for incoming nutrient load in many cases and also indicates possible residuals.

Intersystem performance is not strongly sensitive to HLRs because many wetlands are over-designed with regard to solids removal. Therefore the TSS in the outlet stream is characteristic of wetland background. Data from several sites show a trend of increasing outlet concentrations with increasing inlet concentrations. A simple regression model explains the general trend, but the intersystem scatter in input-output data is large, leading to a low R^2 (Kadlec & Knight 1996):

$$C_o = 1.125 C_i^{0.58}, \quad (5.1)$$
$$R^2 = 0.38,$$
$$N = 460 \text{ quarterly averages},$$
$$1 < C_i < 800 \text{ mg l}^{-1},$$
$$0.5 < C_o < 200 \text{ mg l}^{-1}.$$

Specialized subsets of data, relating to one specific wastewater source, have tighter regressions. For example, regression of the decrease in animal wastewater TSS produces a much higher R^2 (CH2M HILL and Payne Engineering 1997):

$$C_o = 1.047 C_i^{0.818}, \quad (5.2)$$
$$R^2 = 0.78,$$
$$N = 28 \text{ wetlands},$$
$$\text{SE in } \ln[C_o] = 0.285,$$
$$84 < C_i < 545 \text{ mg l}^{-1},$$
$$23 < C_o < 191 \text{ mg l}^{-1}.$$

5.1.2.2 Subsurface flow

Data from Richmond, NSW, Australia (Sapkota & Bavor 1994), show that TSS profiles measured in SSF treatment wetlands display an exponential decrease to a background value (Figure 5.3). Data from laboratory columns determined the rate constant k as 23.1 m d^{-1}. Data from a large-scale pilot wetland determined the rate constant k as 31.6 m d^{-1}. There is a residual turbidity of 24%, corresponding to 3-6 mg l^{-1} of TSS. The rate constant of 32 m d^{-1} is very high, and signals high removals in a small wetland. Consequently, most wetland outlet data are representative of background concentrations and cannot be used to model the inlet zone of high removal. TSS removal rate constants in SSF wetlands in the Czech Republic are much lower (0.119 m d^{-1}, $N = 33$).

There is a trend to higher outlet concentrations as the inlet concentration increases. The regression of Severn Trent and NADB information for SSF wetlands produces the following correlation:

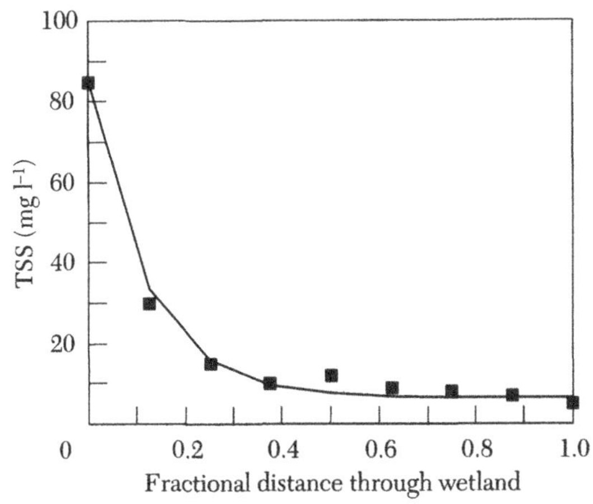

Figure 5.2. TSS profile through a compartmentalized wetland at Arcata pilot marsh, California, USA (Gearheart 1992). Each data point represents the average of samples collected twice a week over 9 months (N = 78). The line is a plot of the k–C model with values k = 343 m yr^{-1}, C* = 6.7 mg l^{-1} and R^2 = 0.96.*

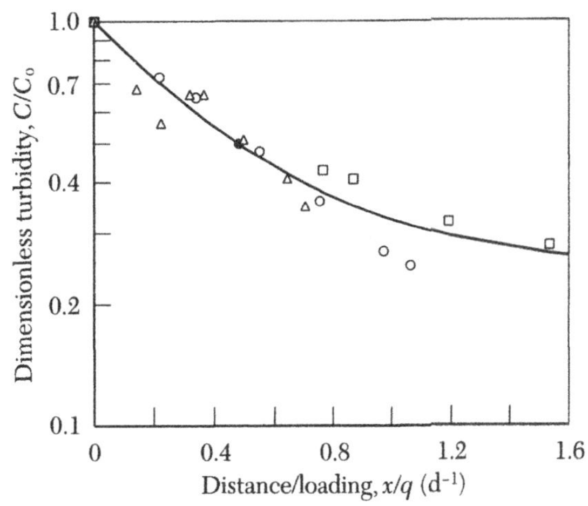

Figure 5.3. TSS (turbidity) profile through a horizontal SSF wetland at Richmond, New South Wales, Australia (data from Sapkota & Bavor (1994)). The bed was run at different hydraulic loadings (□, 8.4 m d^{-1}; ○, 13.3 m d^{-1}; △, 20 m d^{-1}). The line is a plot of $\ln[(C - C^*)/(C_i - C^*)] = -31.6x/qL)$ with values k = w = 31.6 m d^{-1} = 343 m yr^{-1}, C = 0.24 mg l^{-1} and R^2 = 0.95.*

$$C_o = 0.76C_i^{0.706}, \quad (5.3)$$
$R^2 = 0.55$, $N = 78$ wetlands,
SE in $\ln[C_o] = 0.6$,
$8 < C_i < 595$ mg l^{-1},
$2 < C_o < 58$ mg l^{-1}.

Similarly, a better regression equation was found for 77 Danish soil-based wetlands (Brix 1994):

$$C_o = 0.09C_i + 4.7, \quad (5.4)$$
$R^2 = 0.67$, $N = 77$ wetlands,
SE in $C_o = 15$,
$0 < C_i < 330$ mg l^{-1},
$0 < C_o < 60$ mg l^{-1}.

A regression found in the Czech Republic (Vymazal 1998*b*) for vegetated beds was

$$C_o = 0.021C_i + 9.17, \quad (5.5)$$
$R^2 = 0.018$, $N = 37$,
$13 < C_i < 179$ mg l^{-1},
$1.7 < C_o < 30$ mg l^{-1},
$0.6 < q < 14.2$ cm d^{-1};

for the whole system including pretreatment was

$$C_o = 0.0068C_i + 10.74, \quad (5.6)$$
$R^2 = 0.17$, $N = 37$,
$25 < C_i < 743$ mg l^{-1},
$1.3 < C_o < 32$ mg l^{-1},
$1.2 < q < 28.4$ cm d^{-1};

and for loadings of the vegetated beds was

$$L_o = 0.083\,L_i + 1.18, \quad (5.7)$$
$R^2 = 0.64$, $N = 30$,
$3.7 < L_i < 123$ kg ha^{-1} d^{-1},
$0.45 < L_o < 15.4$ kg ha^{-1} d^{-1},
$0.6 < q < 14.2$ cm d^{-1}.

Clogging

In an SSF bed, particulate matter accumulates in voids, blocking them. This clogging process is counteracted by the decomposition of organic particulates. At a minimum, the mineral content of the trapped solids contributes to pore blockage.

Root growth decreases the available pore space in SSF wetlands. Studies on beds with bulrushes have shown that roots and rhizomes are typically located in the upper 30 cm of the bed (US Environmental Protection Agency 1993). *Phragmites* roots and rhizomes have been reported to penetrate further in some instances (Gersberg *et al.* 1986), but other investigations show only 20–40 cm penetration (Schierup 1990; Saurer 1992). The below-ground biomass of *Phragmites* is on the order of 2000 g dry matter m^{-2}, which approximates a quarter of the void volume in a 30 cm root zone.

The end result of subsurface biological and vegetative activity is the build-up of solids within the pore spaces of the medium. That build-up is larger near the inlet and larger near the top of the bed (Tanner & Sukias 1994; Kadlec & Watson 1993). A significant portion of the pore volume can be blocked by accumulated organic matter, leading to increased hydraulic gradients and decreased retention times (Tanner & Sukias 1994). The deposits consist of low-density biosolids together with fine mineral particulates, which can have a very low bulk density.

Tanner & Sukias (1994) measured organic depositions in the inlets of *Schoenoplectus*

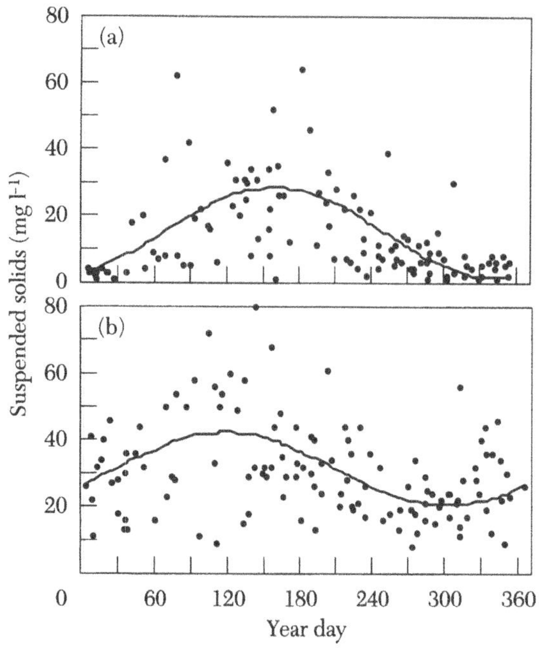

Figure 5.4. Annual patterns of effluent TSS from FWS (Airton) and horizontal SSF (Westow) wetlands providing secondary treatment in the UK (data from Cooper et al. (1996)). There are data for 4 yr for Airton, and for 7 yr for Westow.

gravel beds on the order of 5 kg m^{-2} over a two-year period, exclusive of live and dead roots and rhizomes. The tracer detention time at the end of the period was about half the nominal detention time, suggesting that *ca.* 50% of the voids were blocked. Estimates can be made of the potential accumulation rate of solids in SSF systems. Conley *et al.* (1991) estimate a service life of 100 years. However, these calculations were based on a solids density of 2.65 g cm^{-3} (Tanner & Sukias 1994). If live roots and rhizomes block a third of the pores, and the density is 0.2 g cm^{-3}, this estimate drops to less than 5 years.

5.1.2.3 Annual patterns

There are typically more TSS leaving treatment wetlands in spring and summer than in autumn and winter (Figure 5.4). The seasonal trend is discernible only if several years' data are available, because of the strong stochastic character of the data. The underlying cause of seasonality is presumably the larger generation of particulate matter owing to high productivity of macrophytes and algae during the warmer times of the year. It is unlikely that there is a lower rate of settling or trapping during warm months because warmer water favours faster settling.

5.1.2.4 Variability

The large data scatter in Figure 5.4 is characteristic of all treatment wetlands. The bandwidth of the scatter is approximately double the mean value of the trend line, which is in accordance with the performance ratio reported in Table 4.2. Both systems in Figure 5.4 produce secondary or better water on an annual average basis, but both have frequent individual measurements above 30 mg l^{-1}. In addition, the SSF system has poorer than secondary treatment in the spring of the year, on average over seven years.

5.2 Biochemical oxygen demand

5.2.1 Processes

Settleable organics are rapidly removed in wetland systems under quiescent conditions by deposition and filtration. Attached and suspended microbial growth is responsible for the removal of soluble organic compounds, which are degraded aerobically as well as anaerobically. The oxygen required for aerobic degradation is supplied directly from the atmosphere by diffusion or oxygen leakage from the macrophyte roots into the rhizosphere. The uptake of organic matter by the macrophytes is negligible compared with biological degradation (Watson *et al.* 1989; Cooper *et al.* 1996).

Basic to the understanding of any biological treatment mechanism is an understanding of the microorganisms undertaking the treatment. To continue to reproduce and function properly, an organism must have a source of energy, carbon for the synthesis of new cellular material, and inorganic elements (nutrients) such as nitrogen, phosphorus, sulphur, potassium, calcium and magnesium. Some organic nutrient can also be required. Often industrial effluents require the addition of nutrients such as phosphorus or nitrogen for effective biological treatment.

The two main sources of cell carbon are organic chemicals and carbon dioxide. Organisms that use organic carbon for the formation of cell tissue are called **heterotrophs**. Organisms that derive cell carbon from carbon dioxide are called **autotrophs**. Both groups use light or a chemical oxidation–reduction reaction as an energy source for cell synthesis.

If the major objective of treatment is a decrease in organic content (carbonaceous BOD), the heterotrophic organisms are of primary importance because of their requirement for organic material as a carbon source and their higher metabolic rate.

5.2.1.2 Aerobic degradation

The aerobic degradation of soluble organic matter is governed by aerobic heterotrophic bacteria in accordance with the following reaction:

$$(CH_2O) + O_2 \rightarrow CO_2 + H_2O. \qquad (5.8)$$

The autotrophic group of bacteria that degrade organic compounds containing nitrogen under aerobic conditions are called the nitrifying bacteria; the process is called ammonification and will be discussed below. Cooper *et al.* (1996) pointed out that both groups consume organics, but the greater metabolic rate of the heterotrophs means that mainly they are responsible for the decrease in the BOD of the system. An insufficient supply of oxygen to this group greatly decreases the performance of aerobic biological oxidation; however, if the oxygen supply is not limited, aerobic degradation is governed by the amount of active organic matter available to the organisms.

Biological degradation can take place within the bulk wastewater, although rates are usually low owing to the small numbers of bacteria present (Polprasert 1998). Nearly all degradation takes place within bacterial films present on solid surfaces, including sediments, soils, medium, litter and live submerged plant parts.

5.2.1.3 Anaerobic degradation

Anaerobic degradation is a multi-step process that occurs within constructed wetlands in the absence of dissolved oxygen (Cooper *et al.* 1996). The process can be performed by either facultative or obligate anaerobic heterotrophic bacteria. In the first step the primary end products of fermentation are fatty acids such as acetic acid (Equation 5.9), butyric acid and lactic acid (Equation 5.10), alcohols (5.11) and the gases CO_2 and H_2 (Vymazal 1995):

$$C_6H_{12}O_6 \rightarrow 3CH_3COOH + H_2, \quad (5.9)$$

$$C_6H_{12}O_6 \rightarrow 2CH_3CHOHCOOH \text{ (lactic acid)}, \quad (5.10)$$

$$C_6H_{12}O_6 \rightarrow 2CO_2 + 2CH_3CH_2OH \text{ (ethanol)}. \quad (5.11)$$

Acetic acid is the primary acid formed in most flooded soils and sediments. Strictly anaerobic sulphate-reducing (Equation 5.12) and methane-forming (Equations 5.13 and 5.14) bacteria then utilize the end-products of fermentation and, in fact, depend on the complex community of fermentative bacteria to supply substrate for their metabolic activities. Both groups are important in the decomposition of organic matter and carbon cycling in wetlands (Grant & Long 1985; Valiela 1984; Vymazal 1995):

$$CH_3COOH + H_2SO_4 \rightarrow 2CO_2 + 2H_2O + H_2S, \quad (5.12)$$

$$CH_3COOH + 4H_2 \rightarrow 2\,CH_4 + 2H_2O, \quad (5.13)$$

$$4H_2 + CO_2 \rightarrow 2CH_4 + 2H_2O. \quad (5.14)$$

The acid-forming bacteria are fairly adaptable, but the methane-formers are more sensitive and will operate only in the pH range 6.5–7.5. Overproduction of acid by the acid-

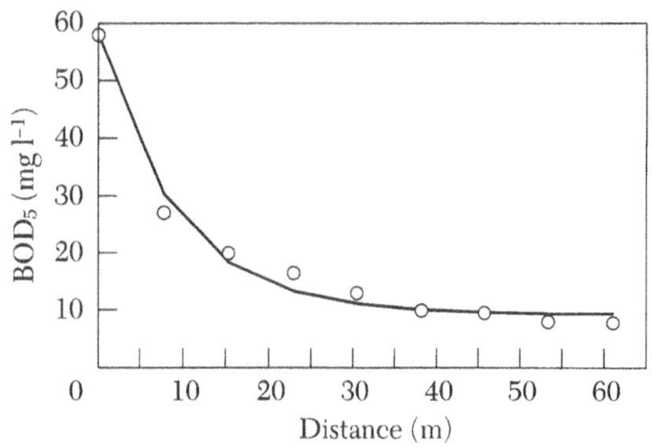

Figure 5.5. Longitudinal profile for BOD in FWS wetland in Arcata, California, USA. Each point is an 8-month average. the line is a plot of the areal model, with values k = 0.57 m d^{-1}, C^* = 9.5 mg l^{-1} and R^2 = 0.984.

formers can rapidly result in low pH, thus stopping the action of the methane-forming bacteria and resulting in the production of odorous compounds from the constructed wetland. Anaerobic degradation of organic compounds is much slower than aerobic degradation. However, when oxygen is limiting at high organic loadings, anaerobic degradation predominates (Cooper *et al.* 1996).

5.2.2 Performance

5.2.2.1 Surface flow

As a result of the combined processes, BOD_5 declines along the flow path from inlet to outlet, down to the background level (Figure 5.5). The k–C^* model provides a highly simplified description of the complex interactions of carbon in wetlands, and typically represents this progression quite well, accounting for *ca.* 90% of the intra-system variability ($R^2 \approx 0.9$). The central tendency of FWS rate constants for 38 FWS wetlands is about $k = 0.10$ m d^{-1} (Table 5.1). The central tendency of background concentrations is *ca.* 5.5 mg l^{-1} (Table 5.1). If C^* is presumed to be zero, the value is $k = 0.077$ m d^{-1} (N = 43 wetlands).

5.2.2.2 Subsurface flow

A regression equation for BOD in SSF wetlands in the NADB is:

$$C_o = 0.33C_i + 1.4, \quad (5.15)$$
$$R^2 = 0.48,\ N = 100,$$
$$\text{SE in } C_o = 5.0,$$
$$1 < C_i < 57 \text{ mg l}^{-1},$$
$$1 < C_o < 36 \text{ mg l}^{-1},$$
$$1.9 < q < 11.4 \text{ cm d}^{-1}.$$

On the basis of the data summary of Brix (1994), 70 Danish soil-based wetlands averaged values of $k = 0.16$ m d^{-1} for a presumed $C^* = 3.0$ mg l^{-1}. If C^* is presumed to be zero, the value is $k = 0.068$ m d^{-1}.

A regression equation for BOD in soil-based

Table 5.1. Rate constants for BOD reduction for some FWS wetland systems

Site		k value (m d^{-1})	Background C^* (mg l^{-1})
Listowel, Ontario, Canada	System 1	0.038	4.3
	System 2	0.018	3.3
	System 3	0.034	4.6
	System 4	0.101	10.4
	System 5	0.117	13.9
Arcata, California, USA	Pilot 1	0.126	4.3
	Pilot 2	0.207	9.7
	Pilot 3	0.112	6.6
	Pilot 4	0.132	6.4
	Pilot 5	0.135	7.6
	Pilot 6	0.134	4.1
	Pilot 7	0.072	0.0
	Pilot 8	0.092	11.3
	Pilot 9	0.055	0.0
	Pilot 10	0.058	1.4
	Pilot 11	0.058	0.0
	Pilot 12	0.136	7.4
Ouray, Colorado, USA	Marsh	0.073	10.8
Gustine, California, USA	Marsh 1A	0,050	11.6
	Marsh 1B	0.038	6.4
	Marsh 1C	0.026	13.0
	Marsh 1D	0.079	5.9
	Marsh 2A	0.060	7.8
	Marsh 2B	0.114	5.5
	Marsh 6A	0.091	3.5
	Pilot Marsh	0.059	4.7
Cobalt, Ontario, Canada	Marsh	0.148	4.7
Iron Bridge, Florida, USA	Marsh	0.062	2.1
Benton, Kentucky, USA	Marsh 1	0.257	5.4
	Marsh 2	0.163	7.9
Pembroke, Kentucky, USA	Marsh	0.141	3.3
West Jackson County, Mississippi, USA	Marsh	0.148	4.7
Lakeland, Florida, USA	Marsh 1	0.131	1.1
Wetwang, Yorkshire, UK	Marsh 2	0.143	3.9
	Marsh 3	0.141	5.2
Cannon Beach, Oregon, USA	Forested	0.048	3.8
Bear Bay, South Carolina, USA	Forested	0.019	1.9
Reedy Creek, Florida, USA,	Forested	0.094	1.7
Average		0.098	5.5
Standard deviation		0.053	3.6

wetlands in Denmark is:

$$C_o = 0.11C_i + 1.87, \tag{5.16}$$
$$R^2 = 0.74,\ N = 73 \text{ wetlands},$$
$$1 < C_i < 330 \text{ mg l}^{-1},$$
$$1 < C_o < 50 \text{ mg l}^{-1},$$
$$0.8 < q < 22 \text{ cm d}^{-1}.$$

A regression found in the Czech Republic (Vymazal 1998b) for vegetated beds was

$$C_o = 0.099C_i + 3.24, \tag{5.17}$$
$$R^2 = 0.33,\ N = 39,$$
$$5.8 < C_i < 328 \text{ mg l}^{-1},$$
$$1.3 < C_o < 51 \text{ mg l}^{-1},$$
$$0.6 < q < 14.2 \text{ cm d}^{-1};$$

for the whole system including pretreatment was

$$C_o = 0.029C_i + 9.22, \tag{5.18}$$
$$R^2 = 0.072,\ N = 41,$$
$$11.3 < C_i < 633 \text{ mg l}^{-1},$$
$$1 < C_o < 69 \text{ mg l}^{-1},$$
$$1.3 < q < 28.6 \text{ cm d}^{-1};$$

and for loadings of the vegetated beds was

$$L_o = 0.13L_i + 0.27, \tag{5.19}$$
$$R^2 = 0.57,\ N = 34,$$
$$2.6 < L_i < 99.6 \text{ kg ha}^{-1}\text{ d}^{-1},$$
$$0.32 < L_o < 21.7 \text{ kg ha}^{-1}\text{ d}^{-1},$$
$$0.6 < q < 14.2 \text{ cm d}^{-1}.$$

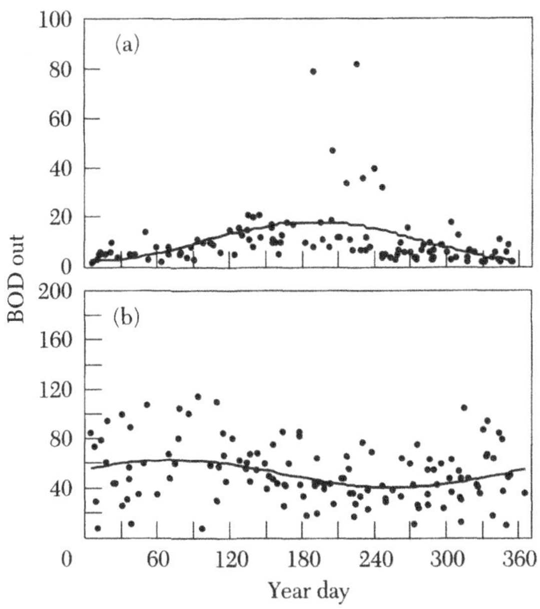

Figure 5.6. Annual patterns of effluent BOD from FWS (Airton, UK) and horizontal SSF (Westow, UK) wetlands providing secondary treatment in the UK (data from Cooper et al. 1996). There are data for 4 yr for Airton, and for 7 yr for Westow.

The central tendency of rate constants for secondary, horizontal SSF wetlands for $C^* = 0$ is approximately $k = 0.06$ m d^{-1} (Cooper *et al.* 1996). The average value of k for Czech Republic SSF wetlands is 0.133 m d^{-1}. In the Czech wetlands the longest period of operation and observation is 6 years. For that system the annual average BOD removal rate constants have generally improved over time from 0.07 to 0.097 to 0.13 to 0.18 to 0.31 to 0.17 m d^{-1}.

The rate constant for tertiary, horizontal SSF wetlands for $C^* = 0$ is about $k = 0.31$ m d^{-1} (Cooper *et al.* 1996). Fourteen SSF tertiary wetlands in the USA show $k = 0.17$ for $C^* = 0$.

Background concentrations seem to be quite low for tertiary SSF wetlands. Thirty-eight Severn Trent Water tertiary wetlands in UK produce average effluents with BOD in the range 1.0–2.5 mg l^{-1}.

5.2.2.3 Annual patterns

There is typically more BOD leaving FWS treatment wetlands in summer than in autumn and winter (Figure 5.6). The seasonal trend is discernible only if several years' data are available, because of the strong stochastic character of the data. A seasonal cyclic model can account for *ca.* 20% of the variability in the outlet BOD concentration ($R^2 = 0.20$). The underlying cause of seasonality is presumably the greater decomposition of organic matter and higher plant productivity during the warmer times of the year. This can offset a higher rate of BOD degradation during warm months.

As a result, FWS winter rate constants are

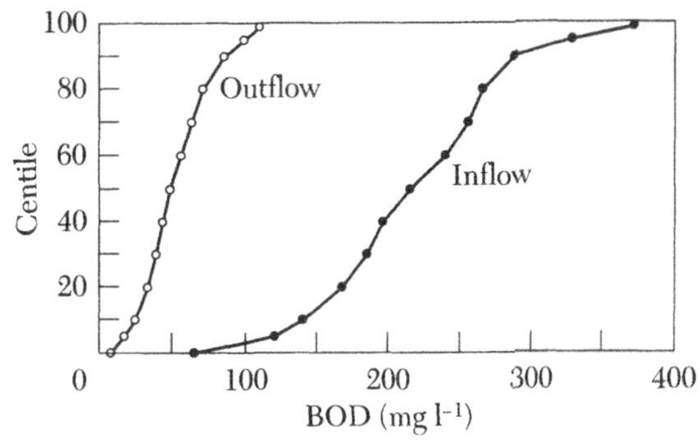

Figure 5.7. Centile patterns of influent and effluent BOD from a horizontal SSF (Westow, UK) wetland (data from Cooper et al. 1996). There are data for 7 yr for Westow.

higher than summer rate constants. For 23 FWS wetlands, the calibrated temperature coefficient was $\theta = 0.98 \pm -0.03$ (Table 5.2). However, temperature accounted for very little of the variance in rate constants. The mean fraction of the variance in rate constants described by the theta model was $R^2 = 0.066$, or less than 7% of the variance.

Measured temperature effects in SSF wetlands range from non-existent ($\theta = 1.00$) (Bavor 1988) to moderate ($\theta = 1.05$) (NRRI 1997).

5.2.2.4 Variability

The large data scatter in Figure 5.6 is characteristic of all treatment wetlands. The bandwidth of the scatter is approximately double the mean value of the trend line, which is in accordance with the performance ratio reported in Table 4.2. Both systems in Figure 5.6 provide a good BOD decrease on an annual average basis: 92% for Airton and 76% for Westow. However, individual measurements can vary substantially. The SSF system has poorer treatment in winter, on average over 7 years.

The stochastic component, which dominates the time series for wetland effluent BOD, can be represented as a centile graph (Figure 5.7). Note that the ratio of the maximum outlet BOD to the mean annual outlet BOD is 1.7 for this wetland.

5.3 Nitrogen

Nitrogen is a key element in wetland biogeochemical cycles. Nitrogen occurs in a number of different oxidation states in wastewaters and in treatment wetlands, and numerous biological and physicochemical processes can transform nitrogen between these different forms.

5.3.1 Processes

The major removal mechanism of organic nitrogen in treatment wetlands is the sequential processes of ammonification, nitrification and denitrification. Ammonia is oxidized to nitrate by nitrifying bacteria in aerobic zones. Organic N is mineralized to ammonia by hydrolysis and

Table 5.2. Temperature coefficients for one- and two-parameter BOD models

Wetland type	System	Cell	$C^* = 0; \theta_1$	$C^* \neq 0; \theta_2$
Surface flow	Orlando Easterly, Florida, USA	1	0.976	0.900
	Columbia, Missouri, USA	1	0.980	0.983
	Wetwang, Yorkshire, UK	1	0.973	
	Richmond, New South Wales, Australia	1	0.913	
	Brookhaven, New York, USA	1	0.991	
	Listowel, Ontario, Canada	1	0.930	
		2	1.011	
		3	0.972	0.964
		4	0.968	1.010
		5	0.997	
	Ouray, Colorado, USA	1	1.041	1.015
	Arcata, California, USA	1		0.993
		2		0.973
		3		0.993
		4		0.999
		5		0.978
		6		0.988
		7		0.989
		8		0.999
		9		0.980
		10		0.975
		11		0.992
		12		0.976
	Average (23 cells)		0.977	0.983
	Standard deviation		0.035	0.025
Overland flow	Smith & Schroeder (1985)		1.000	
	Hall *et al.* (1979)		1.019	
	Martell *et al.* (1982)		1.017	
Lagoons	Saqqar & Pescod (1995)		1.015	
	US Environmental Protection Agency (1983)		0.962	

bacterial degradation. Nitrates are converted to dinitrogen gas (N_2) and nitrous oxide (N_2O) by denitrifying bacteria in anoxic and anaerobic zones. The oxygen required for nitrification is supplied by diffusion from the atmosphere and leakage from macrophyte roots. Nitrogen is also taken up by plants, incorporated into the biomass and released back as organic nitrogen after decomposition. Other removal mechanisms include volatilization and adsorption. On average, these mechanisms are generally of less importance than nitrification-denitrification, but they can be seasonally important.

The nitrogen transformations in a treatment wetland are illustrated in Figure 5.8.

5.3.1.1 Ammonia volatilization

Ammonia volatilization is a physicochemical process in which NH_4-N is known to be in equilibrium between gaseous and hydroxy forms as indicated below:

$$NH_3(aq.) + H_2O \rightarrow NH_4^+ + OH^-. \quad (5.20)$$

Reddy & Patrick (1984) pointed out that losses of NH_3 through volatilization from flooded soils and sediments are insignificant if the pH is below 7.5 and very often losses are not serious if the pH is below 8.0. At pH of 9.3 the ratio of ammonia to ammonium ion is 1:1, and the losses via volatilization are significant. Algal photosynthesis in constructed wetlands as well as photosynthesis by free-floating and submerged macrophytes often creates high pH values.

In a broad literature review, Vymazal (1995) summarized that volatilization rate is controlled by the NH_4^+ concentration in water, temperature, wind velocity, solar radiation, the nature and number of aquatic plants, and the capacity of the system to change the pH in diurnal cycles (the absence of CO_2 increases volatilization).

5.3.1.2 Ammonification (mineralization)

Ammonification (mineralization) is the process in which Org-N is converted into inorganic N, especially NH_4-N. Mineralization rates are fastest in the oxygenated zone and decrease as mineralization switches from aerobic to facultative anaerobic and obligate anaerobic microflora. The rate of ammonification in wetlands is dependent on temperature, pH, the C:N ratio

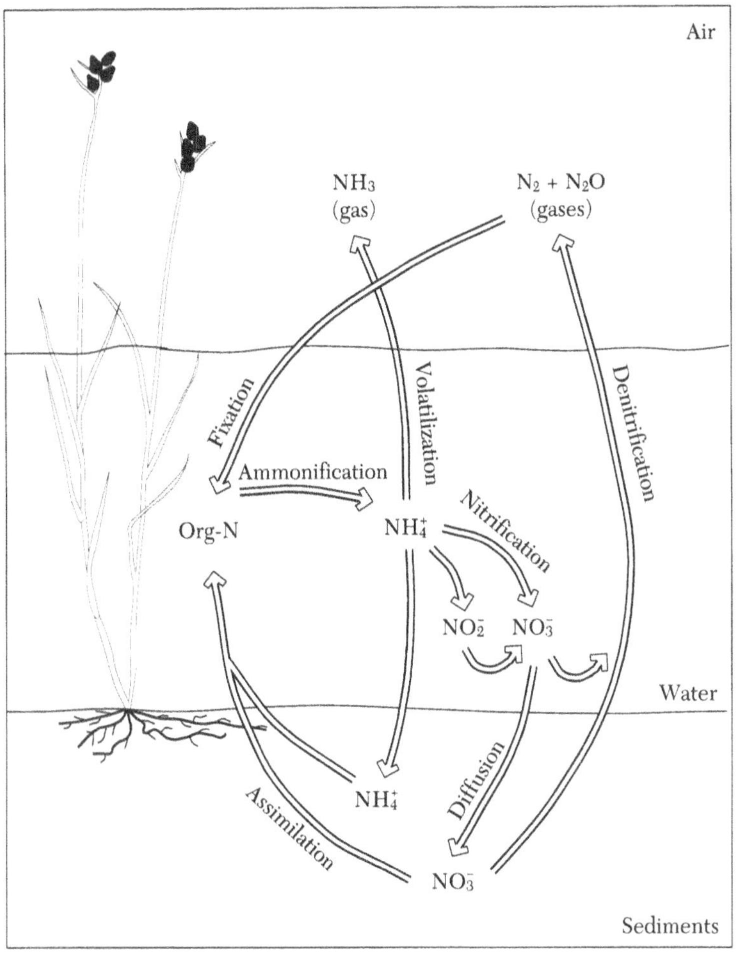

Figure 5.8. Simplified wetland nitrogen cycle (Kadlec & Knight 1996).

of the residue, available nutrients in the system, and soil conditions such as texture and structure (Reddy & Patrick 1984). The optimum pH range for the ammonification process is between 6.5 and 8.5. In saturated soils, pH is buffered around neutrality, whereas under well-drained conditions the pH value of the soil decreases as a result of nitrate accumulation and the production of H^+ ions (Equation 5.23) during mineralization (Patrick & Wyatt 1964). Reddy *et al.* (1979) concluded from published data that the rate of aerobic ammonification doubles with a temperature increase of 10 °C.

5.3.1.3 Nitrification/denitrification

Nitrification

Nitrification is usually defined as the biological oxidation of ammonium to nitrate with nitrite as an intermediate in the reaction sequence. Nitrification is a chemoautotrophic process. The nitrifying bacteria derive energy from the oxidation of ammonia and/or nitrite, and carbon dioxide is used as a carbon source for the synthesis of new cells. These organisms require O_2 during NH_4-N oxidation to nitrite-N and nitrite-N oxidation to nitrate-N (Equations 5.21, 5.22, and 5.23). The oxidation of ammonium to nitrate is a two-step process (Wallace & Nicholas 1969; Hauck 1984):

$$NH_4^+ + 1.5O_2 \rightarrow NO_2^- + 2H^+ + H_2O \qquad (5.21)$$

$$NO_2^- + 0.5O_2 \rightarrow NO_3 \qquad (5.22)$$

$$NH_4^+ + 2O_2 \rightarrow NO_3^- + 2H^+ + H_2O \qquad (5.23)$$

The first step, the oxidation of ammonium to nitrite, is executed by strictly chemolithotrophic (strictly aerobic) bacteria, which are entirely dependent on the oxidation of ammonia for the generation of energy for growth. In soil, species belonging to the genera *Nitrosospira* (*Nitrosospira briensis*), *Nitrosovibrio* (*Nitrosovibrio tenuis*), *Nitrosolobus* (*Nitrosolobus multiformis*), *Nitrosococcus* (*Nitrosococcus nitrosus*) and *Nitrosomonas* (*Nitrosomonas europaea*) have been identified. *Nitrosomonas europaea* is also found in fresh waters (Grant & Long 1981, 1985; Schmidt 1982). The probable reaction sequence for the oxidation of ammonia to nitrite by Nitroso group bacteria is (Hauck 1984):

$$\underset{\text{ammonia}}{NH_3/NH_4^+} \rightarrow \underset{\text{hydroxylamine}}{NH_2OH} \rightarrow \underset{\text{nitroxyl}}{NOH} \rightarrow$$

$$\underset{\text{nitrohydroxylamine}}{NO_2.NH_2OH} \rightarrow \underset{\text{nitrite}}{NO_2^-}. \qquad (5.24)$$

The postulated intermediate compounds NOH and $NO_2.NH_2OH$ have never been isolated, but their participation in the reaction sequence is consistent with the assumption that two electrons are transferred for each oxidation step between NH_4^+ and NO_2^- (Hauck 1984, and references cited therein).

The second step in the process of nitrification, the oxidation of nitrite to nitrate, is performed by facultative chemolitrotrophic bacteria, which can also use organic compounds in addition to nitrite for the generation of energy for growth. In contrast with the ammonia-oxidizing bacteria, only one species of nitrite-oxidizing bacteria is found in the soil and fresh water, i.e. *Nitrobacter winogradskyi* (Grant & Long 1981). Schmidt (1982), however, reported that a genus *Nitrospira* was found in addition to *Nitrobacter* in soil and fresh water as well as in marine environments. In addition, in contrast to ammonia-oxidizing bacteria, at least some species of nitrite-oxidizing bacteria can grow mixotrophically on nitrite and a carbon source, or are even able to grow in the absence of oxygen (Bock *et al.* 1986).

Vymazal (1995) summarizes that nitrification is influenced by temperature, pH, alkalinity, inorganic C source, the microbial population and concentrations of NH_4-N and dissolved oxygen. The optimum temperature for nitrification in pure cultures ranges from 25 to 35 °C and in soils from 30 to 40 °C. Lower temperatures (below 15 °C) have a greater effect on nitrification rate than temperatures between 15 and 35 °C. Cooper *et al.* (1996) pointed out that the minimum temperatures for growth of *Nitrosomonas* and *Nitrobacter* are 5 and 4 °C, respectively.

Nitrifying bacteria are sensitive organisms and are extremely susceptible to a wide range of inhibitors, including high concentrations of ammoniacal nitrogen. A narrow pH optimum range (7.5–8.6) also exists; however, acclimatized systems can be operated to nitrify at a much lower pH value. Approximately 4.3 mg of O_2 per mg of ammoniacal nitrogen oxidized to nitrate nitrogen is needed. In the conversion process, a large amount of alkalinity is consumed, *ca.* 8.64 mg of HCO_3^- per mg of ammoniacal nitrogen oxidized (Cooper*et al.* 1996).

Denitrification

The first anoxic oxidation process to occur after oxygen depletion is the reduction of nitrate to molecular nitrogen or nitrogen gases. This process is called denitrification. From a biochemical viewpoint, denitrification is a bacterial process in which nitrogen oxides (in ionic and gaseous forms) serve as terminal electron acceptors for respiratory electron transport. Electrons are carried from an electron-donating substrate (usually, but not exclusively, organic compounds) through several carrier systems to a more oxidized N form. The resultant free energy is conserved in ATP, after phosphorylation, and is used by the denitrifying organisms to support respiration. Denitrification is illustrated by the following equation (Hauck 1984):

$$6(CH_2O) + 4NO_3^- \rightarrow 6CO_2 + 2N_2 + 6H_2O. \quad (5.25)$$

This reaction is irreversible and occurs in the presence of available organic substrate only under anaerobic or anoxic conditions (E_h = +350 to +100 mV), in which nitrogen is used as an electron acceptor in place of oxygen. More and more evidence is being provided from pure-culture studies that nitrate reduction can occur in the presence of oxygen. Hence, in waterlogged soils nitrate reduction might also start before the oxygen is depleted (Laanbroek 1990).

Gaseous N production during denitrification can also be depicted as follows (Hauck 1984):

$$4(CH_2O) + 4NO_3^- \rightarrow 4HCO_3^- + 2N_2O + 2H_2O, \quad (5.26)$$

$$5(CH_2O) + 4NO_3^- \rightarrow H_2CO_3 + 4HCO_3^- + 2N_2 + 2H_2O. \quad (5.27)$$

Denitrifying ability has been demonstrated in 17 genera of bacteria. Most denitrifying bacteria are chemoheterotrophs, obtaining energy solely through chemical reactions and use organic compounds as electron donors and as a source of cellular carbon (Hauck 1984). The genera *Bacillus*, *Micrococcus* and *Pseudomonas* are probably the most important in soils; in the aquatic environment the most important are *Pseudomonas*, *Aeromonas* and *Vibrio* (Grant & Long 1981). Other denitrifiers include members of the genera *Achromobacter*, *Aerobacter*, *Alcaligenes*, *Azospirillum*, *Brevibacterium*, *Flavobacterium*, *Spirillum* and *Thiobacillus*. A list of genera involved in the denitrification process has been given by Focht & Verstraete (1977). When oxygen is available, these organisms oxidize a carbohydrate substrate to carbon dioxide and water (Reddy & Patrick 1984). Aerobic respiration with oxygen as an electron acceptor, or anaerobic respiration using nitrogen for this purpose, is accomplished by the denitrifier with the same series of electron transport system. This facility to function both as an aerobe and as an anaerobe is of great practical importance because it enables denitrification to proceed at a significant rate soon after the onset of anoxic conditions (a redox potential of *ca.* 300 mV) without change in microbial population (Hauck 1984). Because denitrification is performed almost exclusively by facultative anaerobic heterotrophs that

Table 5.3. Nutrient removal of potential wetland plant species used in wetland treatment systems (Reddy & Debusk 1987; Vymazal 1995, 1999)

	Biomass		Nitrogen		Phosphorus	
Genus	Stock (kg ha^{-1})	Growth (kg ha yr^{-1})	Stock (kg ha^{-1})	Growth (kg ha yr^{-1})	Stock (kg ha^{-1})	Growth (kg ha yr^{-1})
Typha	480–68,030	5740–93,390	6–1560	111–2630	16–375	8–400
Juncus	130–22,000	7960–53,300	200–300	800	30–40	110
Scripus	280–42,000	7850–46,000	66–550	125–775	14–110	18–150
Phragmites	1820–127,000	1830–60,000	85–2200	750–2450	3–191	25–199
Phalaris	100–24,600	8000–35,000	20–470	80–1200	10–105	23–140
Glyceria	6590–26,900	9000–28,600	66–1340	75–1500	9–230	50–425
Eichhornia	7980–30,000	80 000–110,000	300–2340	420–5850	32–180	105–1260
Hydrocotyle	1880–11,000	4000–60,000	90–300	540–3200	23–75	130–770
Lemna	90–29,500	6000–26,000	4–50	350–6110	1–345	116–800

substitute oxidized N forms for O_2 as electron acceptors in respiratory processes, and because these processes follow aerobic biochemical routes it can be misleading to refer to denitrification as an anaerobic process; rather, it is one that takes place under anoxic conditions (Hauck 1984).

It is generally agreed that the actual sequence of biochemical changes from nitrate to elemental gaseous nitrogen is (Vymazal 1995)

$$2NO_3 \rightarrow 2NO_2 \rightarrow 2NO \rightarrow N_2O \rightarrow N_2. \quad (5.28)$$

Vymazal (1995) summarizes the environmental factors known to influence denitrification rates, including the absence of O_2, redox potential, soil moisture, temperature, pH, the presence of denitrifiers, soil type, organic matter and the presence of overlying water. The quantity of N_2O evolved during denitrification depends on the amount of nitrogen denitrified and the ratio of N_2 to N_2O produced. The ratio is also affected by aeration, pH, temperature and the ratio of nitrate to ammonia in the denitrifying system.

Cooper *et al.* (1996) pointed out that the presence of dissolved oxygen suppresses the enzyme needed for denitrification and is a critical parameter. The optimum pH range lies between 7 and 8; however, alkalinity produced during denitrification can result in an increase in pH. Denitrification is also strongly temperature-dependent and proceeds only very slowly at temperatures below 5 °C. The process of denitrification and its consequences have been reviewed extensively by Payne (1981).

Nitrification and denitrification are known to occur simultaneously in flooded soils in which both aerobic and anaerobic zones exist, such as would occur in a flooded soil or water bottom containing an aerobic surface layer over an anaerobic layer, or in the aerobic rhizosphere microsites in otherwise anaerobic soil. In combination these two reactions, a balanced equation occurring in aerobic and anaerobic layers, can be written as (Reddy & Patrick 1984)

$$24NH_4^+ + 48O_2 \rightarrow 24NO_3^- + 24H_2O + 48H^+ \quad (5.29)$$

$$24NO_3^- + 5C_6H_{12}O_6 + 24H^+ \rightarrow 12N_2 + 30CO_2 + 42H_2O \quad (5.30)$$

$$24NH_4^+ + 5C_6H_{12}O_6 + 48O_2 \rightarrow 12N_2 + 30CO_2 + 66H_2O + 24H^+. \quad (5.31)$$

5.3.1.4 Plant uptake

The potential rate of nutrient uptake by a plant is limited by its net productivity (growth rate) and the concentration of nutrients in the plant tissue. Nutrient storage is similarly dependent on plant tissue nutrient concentrations and also on the ultimate potential for biomass accumulation, that is, the maximum standing crop. Desirable traits of a plant used for nutrient assimilation and storage would therefore include rapid growth, high tissue nutrient content and the capability of attain a high standing crop (biomass per unit area) (Reddy & DeBusk 1987).

In the literature there are many reviews on nitrogen concentrations in plant tissue as well as nitrogen standing stocks for plants found in natural stands (Reddy & DeBusk 1987; Vymazal 1995). The uptake capacity of emergent macrophytes, and thus the amount that can be removed if the biomass is harvested, is roughly in the range 1000–2500 kg N ha^{-1} yr^{-1} (Table 5.3). The highly productive water hyacinth (*Eichhornia crassipes*) has a higher uptake capacity (up to nearly 6000 kg N ha^{-1} yr^{-1}), whereas the capacity of submerged macrophytes is lower (*ca.* 700 kg N ha^{-1} yr^{-1}) (Brix 1994; Vymazal 1995).

However, only a few data have been reported for plants from constructed wetlands treating wastewaters. In addition, it is important to note that the amounts of nutrients that can be removed by harvesting in secondary treatment systems are generally insignificant in comparison with the loadings into the constructed wetlands with the wastewater (Brix 1994). This is especially true of constructed wetlands with emergent macrophytes. It has been reported

Table 5.4. Typical nitrogen species composition of various wastewaters

Wastewater	Nitrogen concentration (mg l^{-1})			
	Organic	Ammonium	Oxidized	Total
Crop runoff	0	0	3–100	3–100
Nitrified/denitrified secondary	2	2	3–6	7–10
Domestic facultative lagoon	10	10	0	20
Nitrified secondary	2	2–10	18–20	24–30
Primary	15	40	2	57
Secondary	10–30	10–20	0	30–40
Vegetable processing	100	50	0	150
Nitrified meat processing	15	60	120	195
Animal lagoon	100–200	240–300	0	340–500

that under optimum conditions the amount of nitrogen removed with the biomass does not exceed 10% of the total removed nitrogen (Gersberg *et al.* 1985; Herskowitz 1986; Vymazal *et al.* 1999). The removal of nutrients through harvesting might be more important in treatment systems designed for polishing.

If the wetland is not harvested, the vast majority of the nutrients that have been incorporated into the plant tissue will be returned to the water by decomposition processes. Long-term storage of nutrients in the wetland system results from the undecomposed fraction of the litter produced by the various elements of the biogeochemical cycles as well as the deposition of refractory nutrient-containing compounds (Brix 1996). Seasonality of plant harvesting can also be important in the amount of nutrient mass that can be harvested. For example, it is not possible to harvest common reed during the period of peak standing stock because the plant is easily killed by such activity. *Phragmites* does not translocate storage products to its rhizomes during the growing season; it moves them to the rhizomes just before the end of the growing season. The best time to harvest common reed without damaging plant growth is in the early spring; however, above-ground plant tissue nutrient concentrations are about 30–50% of those during the peak growing season.

5.3.1.5 Matrix adsorption

In a reduced state, NH_4-N is stable and can be adsorbed on active sites of an SSF bed matrix or on the sediments of a FWS wetland. However, the ion exchange of NH_4-N on cation-exchange sites of the matrix is not considered to be a long-term sink for NH_4-N removal. Rather, sorption of NH_4-N is presumed to be rapidly reversible. As the NH_4-N is lost from the system via nitrification, the exchange equilibrium is expected to redistribute itself. The sorbed NH_4-N in a continuous-flow system will therefore be in equilibrium with the NH_4-N in solution. In the course of seasonal variations in ammonium content in the water, there can be alternate loading and unloading of sorption sites. Intermittent loading of a system will show rapid removals of NH_4-N by adsorption mechanisms owing to the depletion of NH_4-N on the sorption sites during rest periods.

The Freundlich equation can be used to model NH_4-N sorption (Kadlec & Knight 1996).

5.3.2 Performance

Because of the complex transformations affecting nitrogen forms in wetlands, the sequential series of reactions must be considered to describe treatment performance adequately, even on the most elementary level. Figure 5.8 illustrates these major interconversions. Mass balance equations for these interrelated reactions have been published for plug flow hydraulics in treatment wetlands (Kadlec & Knight 1996).

The nature of the influent wastewater is a very important determinant of nitrogen processing in the wetland (Table 5.4). If organic nitrogen dominates the influent, then mineralization can increase the ammonium content until nitrification, uptake and sorption can decrease it. Data from Listowel, Ontario, Canada, illustrate this progression as water travels through the wetland (Figure 5.9). If NH_4-N dominates the influent, oxidized nitrogen might peak before denitrification can decrease its concentration.

5.3.2.1 Surface flow

Regression equations

Treatment wetland input-output data can be summarized by the use of a regression model. Comparing linear and log-normal regressions of the SF nitrogen data in the NADB and incorporating HLR and concentration produces the regression equations in Table 5.5 (Kadlec & Knight 1996). Low correlation coefficients for these data indicate the importance of other factors not included in these simple regression models.

Table 5.5. Regression equations for nitrogen outlet concentration in FWS treatment wetlands (modified from Kadlec & Knight (1996))

Parameter		R^2	N	SE in C_2	Data range (median)		
					q (cm d^{-1})	C_1 (mg l^{-1})	C_2 (mg l^{-1})
Org-N	$C_2 = 1.0C_1^{0.476}$	0.52	243	1.80	0.02–27 (2.9)	0.09–20 (2.8)	0.16–16 (1.4)
NH_4-N	$C_2 = 0.336C_1^{0.728}q^{0.456}$	0.44	542	4.40	0.1–33 (2.9)	0.4–59 (2.2)	0.01–58 (0.6)
NO_3-N	$C_2 = 0.093C_1^{0.474}q^{0.745}$	0.35	553	4.90	0.02–27 (2.7)	0.01–25 (1.7)	0.01–22 (0.2)
TKN	$C_2 = 0.569C_1^{0.840}q^{0.282}$	0.74	419	1.90	0.1–24 (2.9)	0.2–97 (87)	0.15–48 (3.0)
TN	$C_2 = 0.409C_1 + 1.22q$	0.48	408	3.50	0.2–29 (2.5)	13 180.0 (9.1)	0.4–29 (2.2)

Abbreviations: SE, standard error; TKN, total Kjeldahl nitrogen.

First-order rate constants

The sequential k–C^* model represents these processes reasonably well. Table 5.6 presents average global rate constants, background concentrations and temperature correction values for nitrogen forms (Kadlec & Knight 1996). These constants can be used for single species feeds in a single disappearance equation or in the sequential multi-reaction sequence set of equations. They should not be used in a single equation for a single species in a mixture.

A background concentration of *ca.* 1.5 mg l^{-1} of organic nitrogen is common to all FWS wetlands.

5.3.2.2 Horizontal subsurface flow

Regression equations

Treatment wetland input-output data from systems in the NADB produces the regression equations in Table 5.7 (Kadlec & Knight 1996). These wetlands had low amounts of ammonium and virtually no oxidized nitrogen in the influent. Low correlation coefficients for these data indicate the importance of other factors not included in these simple regression models.

A regression equation for TN in soil based wetlands in Denmark is

$$C_o = 0.52C_i + 3.1, \quad (5.32)$$
$$R^2 = 0.63, N = 58 \text{ wetlands},$$
$$4 < C_i < 142 \text{ mg l}^{-1},$$
$$5 < C_o < 69 \text{ mg l}^{-1},$$
$$0.8 < q < 22 \text{ cm d}^{-1}.$$

Regressions for the Czech Republic were as follows (Vymazal 1998*b*):

TN for vegetated beds:

$$C_o = 0.42C_i + 7.68, \quad (5.33)$$
$$R^2 = 0.72, N = 25,$$
$$16.4 < C_i < 93 \text{ mg l}^{-1},$$
$$10.7 < C_o < 49 \text{ mg l}^{-1},$$
$$1.7 < q < 14.2 \text{ cm d}^{-1};$$

TN for the whole system including pretreatment:

$$C_o = 0.36C_i + 7.54, \quad (5.34)$$
$$R^2 = 0.59, N = 25,$$
$$11.1 < C_i < 100 \text{ mg l}^{-1},$$
$$0.5 < C_o < 49 \text{ mg l}^{-1},$$
$$1.7 < q < 14.2 \text{ cm d}^{-1};$$

Table 5.6. Preliminary global rate constants, background concentrations and temperature factors in FWS wetlands (Kadlec & Knight 1996)

Parameter	k (m yr^{-1})	C^* (mg l^{-1})	θ
Org-N	17	1.50	1.05
NH_4-N	18	0.00	1.05
NO_3-N	60	0.00	1.05
TN	22	1.50	1.08

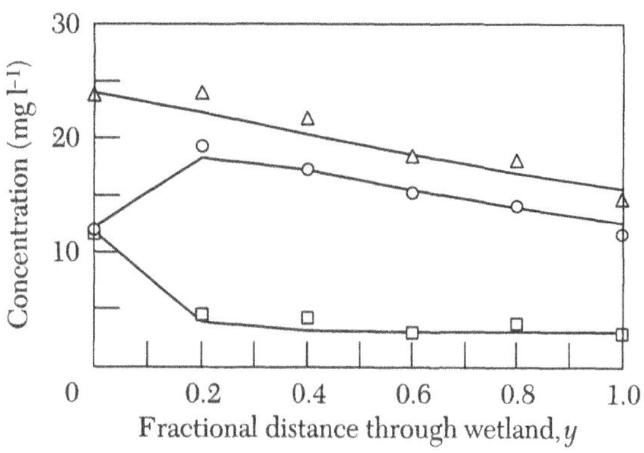

Figure 5.9. Profiles of major dissolved nitrogen species in Listowel, Ontario, Canada, wetland 4, in summer 1984. Lines are model calibrations; symbols denote data points that are averages of data collected every two weeks over 3 months (from Kadlec & Knight 1996). Symbols: △, total Kjeldahl nitrogen; ○, NH_4-N; □, Org-N.

TN for loadings of the vegetated beds:

$$L_o = 0.68 L_i + 0.27, \quad (5.35)$$
$$R^2 = 0.96, N = 24,$$
$$145 < L_i < 1894 \text{ g/m}^2 \text{ yr}^{-1},$$
$$134 < L_o < 1330 \text{ g/m}^2 \text{ yr}^{-1},$$
$$1.7 < q < 14.2 \text{ cm d}^{-1}.$$

NH_4-N for vegetated beds:

$$C_o = 0.42C_i + 4.37, \quad (5.36)$$
$$R^2 = 0.65, N = 26,$$
$$3.4 < C_i < 66 \text{ mg l}^{-1},$$
$$1.7 < C_o < 37 \text{ mg l}^{-1},$$
$$1.7 < q < 14.2 \text{ cm d}^{-1};$$

Table 5.7. Regression equations for nitrogen outlet concentration in horizontal SSF treatment wetlands (modified from Kadlec & Knight (1996))

Parameter		R^2	N	SE in C_2	Data range (median)		
					q (cm d^{-1})	C_1 (mg l^{-1})	C_2 (mg l^{-1})
Org-N	$C_2 = 0.1C_1 + 1.0$	0.07	289	1.90	0.7–49 (6.2)	0.6–22 (6.9)	0.1–11 (1.1)
NH_4-N	$C_2 = 3.3 + 0.46C_1$	0.63	92	4.40	0.7–49 (5.5)	0.1–44 (6.7)	0.1–27 (6.1)
NO_3-N	$C_2 = 0.62C_1$	0.80	95	2.40	0.7–49 (5.5)	0.1–27 (0.3)	0.1–21 (0.4)
TKN	$C_2 = 0.569C_1^{0.840}q^{0.282}$	0.74	92	1.70	0.7–49 (5.5)	0.7–58.2 (15.2)	0.6–36 (8.2)
TN	$C_2 = 0.409C_1 + 1.22q$	0.45	135	6.10	0.7–49 (7.1)	5–59 (21.0)	2–38 (14.0)

Abbreviations: SE, standard error; TKN, total Kjeldahl nitrogen.

Table 5.8. Rate constants, background concentrations and temperature factors in horizontal SSF wetlands

Parameter	k (m yr^{-1})	C^* (mg l^{-1})	θ
Org-N	35	1.50	1.05
NH_4-N	34	0.00	1.05
NO_3-N	50	0.00	1.05
TN	27	1.50	1.05

These preliminary global rate constants are from wetlands with low to moderate N loadings (Kadlec & Knight 1996).

NH_4-N for the whole system including pre-treatment:

$$C_o = 0.36C_i + 7.54, \quad (5.37)$$
$R^2 = 0.54, N = 31,$
$2.5 < C_i < 53$ mg l^{-1},
$0.1 < C_o < 28$ mg l^{-1},
$1.7 < q < 20.6$ cm d^{-1};

NH_4-N for loadings of the vegetated beds:

$$L_o = 0.81L_i - 72.86, \quad (5.38)$$
$R^2 = 0.86, N = 26,$
$83 < L_i < 867$ g m^{-2} yr^{-1},
$53 < L_o < 807$ g m^{-2} yr^{-1},
$1.7 < q < 14.2$ cm d^{-1}.

Org-N for vegetated beds:

$$C_o = 0.23C_i + 1.39, \quad (5.39)$$
$R^2 = 0.39, N = 14,$
$0.9 < C_i < 18$ mg l^{-1},
$0.55 < C_o < 5.5$ mg l^{-1},
$1.7 < q < 14.2$ cm d^{-1};

Org-N for loadings of the vegetated beds:

$$L_o = 0.49\,L_i + 7.56, \quad (5.40)$$
$R^2 = 0.72, N = 13,$
$22 < L_i < 309$ g m^{-2} yr^{-1},
$8.8 < L_o < 210$ g m^{-2} yr^{-1},
$1.7 < q < 14.2$ cm d^{-1}.

NO_3-N for vegetated beds:

$$C_o = 0.55\,C_i + 3.10, \quad (5.41)$$
$R^2 = 0.41, N = 16,$
$0.79 < C_i < 22$ mg l^{-1},
$0.7 < C_o < 16$ mg l^{-1},
$1.7 < q < 14.2$ cm d^{-1};

NO_3-N for loadings of the vegetated beds:

$$L_o = 0.28\,L_i + 47.25, \quad (5.42)$$
$R^2 = 0.26, N = 14,$
$6.4 < L_i < 1141$ g m^{-2} yr^{-1},
$5.3 < L_o < 830$ g m^{-2} yr^{-1},
$1.7 < q < 14.2$ cm d^{-1}.

First-order rate constants

The sequential k–C^* model represents these processes reasonably well. Table 5.8 presents average global rate constants, background concentrations, and temperature correction values for nitrogen forms (Kadlec & Knight 1996). For the Czech SSF systems the average k value for TN is 0.028 m d^{-1} (10.2 m yr^{-1}).

Rate constants for strong influents are lower. For instance, for TN in an SSF wetland treating septic-tank effluent (TN ≈ 80 mg l^{-1}), k_{20} = 10 m yr^{-1}.

5.3.2.3 Vertical subsurface flow

Intermittent operation of VF wetlands, a variant of the intermittent sand filter, results in a greater oxygen supply and thus a higher rate of nitrification (Cooper *et al.* 1996). The sources of the enhanced oxygen supply are (1) the convection of air into bed void spaces caused by bed draining, and (2) the diffusion of oxygen into voids caused by oxygen depletion to the interstitial water; both of which augment the net plant aeration flux, if any. This oxygen, together with any nitrate in the water, acts to decrease BOD/COD and to nitrify NH_4-nitrogen.

Because of the periodic character of the process, the use of a simple first-order model, as for other types of constructed wetland, is inappropriate (Platzer 1998). It is possible to modify the application of the first-order model to a more complex form involving the dynamics of the filling and draining cycle (Sun *et al.* 1998). So far, there have been no calibrations of such models to domestic sewage treatment systems.

One design approach is to rely on empirical

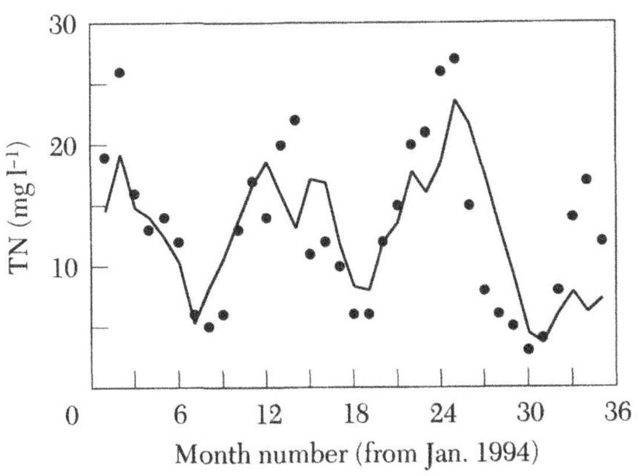

Figure 5.10. Annual pattern of TN leaving the Linköping (Sweden) wetlands. The line is a smoothed average. Values for the k–C model:* k_{AN20} = *16 m* yr^{-1} (θ = *1.042);* k_{NN20} = *78 m* yr^{-1} (θ = *1.042). (*k_{AN20} *is the rate constant for* NH_4*-N at 20 °C;* k_{NN20} *is nitrate nitrogen.)*

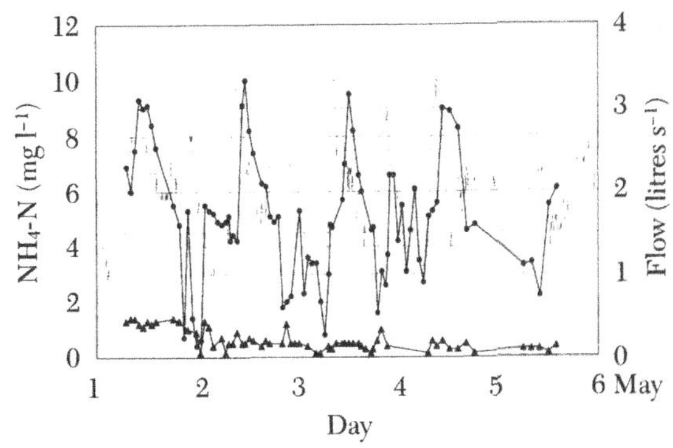

Figure 5.11. Close interval sampling at Leek Wootton, UK (Cooper et al. 1996). Symbols: ●, inflow NH_4*-N; ▲, effluent* NH_4*-N; thin line, flow rate.*

observations of the apparent oxygen transfer rates from operating systems. Platzer (1998) has measured values of oxygen transfer rate (OTR) in the range 23–64 g O_2 m^{-2} d^{-1}; Green *et al.* (1997) report 56–60 g O_2 m^{-2} d^{-1}, and Cooper *et al.* (1998) report 50-90 g O_2 m^{-2} d^{-1}. Oxygen demand (OD) is due to nitrification and BOD/COD decrease, according to Cooper (1998):

$$OD = Q[4.3\Delta(NH_4\text{-}N) + \Delta(BOD)] \text{ g } O_2 \text{ d}^{-1}, \quad (5.43)$$

where Q = flow rate (m^3 d^{-1}).

This demand can be lessened by BOD settlement, ammonia volatilization, plant uptake and denitrification. The observed ranges of oxygen supply per unit area can be combined with the estimate of OD to determine the bed area:

$$A = OD/OTR \ (m^2). \quad (5.44)$$

The areas so determined are relatively small. For instance, suppose that the incoming flow is 200 l/PE, and $\Delta(NH_4\text{-}N)$ = 20 mg l^{-1} and Δ(BOD) = 40 mg l^{-1}. Then the OD is calculated to be 25.2 g O_2 m^{-2} d^{-1}. If the OTR is taken to be 50 g O_2 m^{-2} d^{-1}, then only 0.5 m^2/PE is required.

5.3.2.4 Annual patterns

There is typically much greater nitrogen mass reduction in treatment wetlands in spring and summer than in autumn and winter (Figure 5.10). The seasonal trend is masked somewhat by the strong stochastic character of the data. A seasonal cyclic model can account for *ca.* 33% of the variability in the outlet nitrogen concentrations (R^2 = 0.34). The underlying cause for seasonality is presumably the larger biological (microbiological and macrobiological) activity during the warmer times of the year.

As a result, summer rate constants are higher than winter rate constants. For many FWS and SSF wetlands, the calibrated temperature coefficients are θ = 1.05 for all three transformation reactions. For the Linköping (Sweden) system (Figure 5.10), temperature accounted for most of the variance in rate constants. The mean fraction of the variance in rate constants described by the theta model was R^2 = 0.75, or *ca.* 75% of the variance. However, if rate constants are regressed to season instead of temperature, 92% of the variance is explained. The k values peak earlier than temperature.

5.3.2.5 Variability

The data scatter in Figure 5.10 is characteristic of all treatment wetlands. The bandwidth of the scatter is *ca.* 25% of the mean value of the trend line. Consequently, the performance ratio reported in Table 4.2 represents both stochastic and seasonal variation in performance.

The simple first-order models work well on annual and seasonal average bases, but cannot be used on an instantaneous basis. First, there is nominal delay of one detention time between the entrance and the exit of a water parcel. Secondly, temporary increases and decreases in the wetland storages can easily affect instantaneous performance. The outlet concentrations therefore do not 'track' the inlet concentrations and flows according to a first-order model (Figure 5.11).

5.4 Phosphorus

Constructed and natural wetlands are capable of absorbing new phosphorus (P) loadings and in appropriate circumstances can provide a low-cost alternative to chemical and biological treatment. Phosphorus interacts strongly with wetland soils and biota, which provide both short-term and sustainable long-term storage of this nutrient.

5.4.1 Processes

In SSF wetlands, the sorption capacity of the media can be designed to provide significant P removal (Maehlum *et al.* 1995). This storage

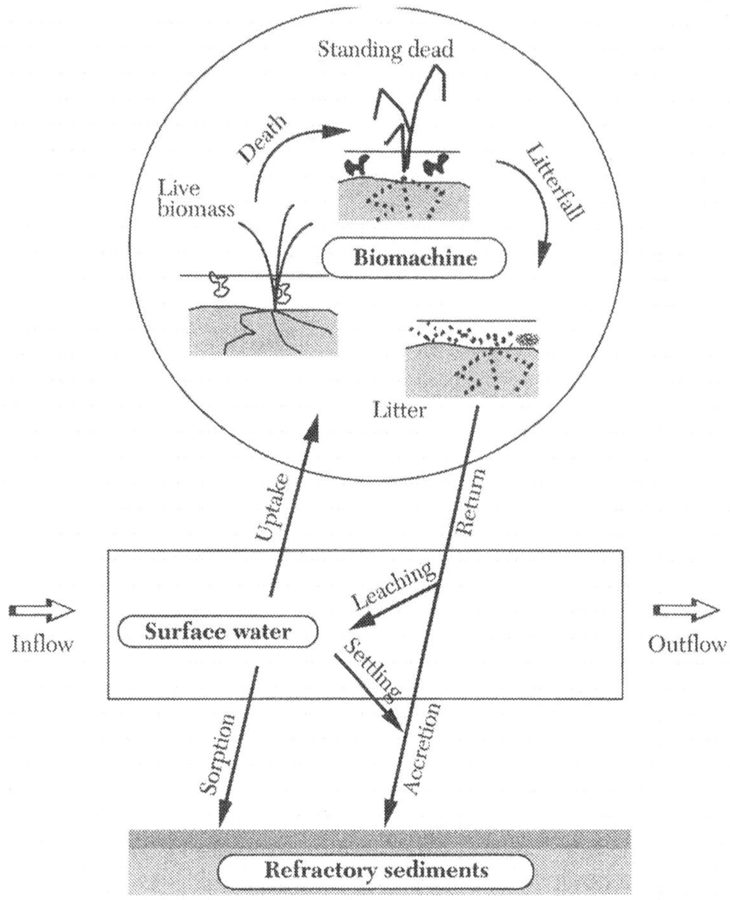

Figure 5.12. Wetland biogeochemical processing of phosphorus. The two temporary sinks for P are sorption on antecedent soils and expansion of the biomachine. Macrophytes (plants) and microphytes (algae) can both be important, depending on wetland type. The sustainable removal pathway is the accretion of new soils and sediments via the deposition of organic and inorganic forms of P in FWS. (Kadlec (1996).)

eventually becomes saturated, necessitating the replacement of the medium and the re-establishment of the wetland.

In FWS wetlands, soil sorption can provide initial removal, but this partly reversible storage eventually becomes saturated. For some antecedent soil conditions, there can even be an initial release of P. A new source of P acts to fertilize the wetland, and some P is used in the establishment of a new or larger standing crop of vegetation.

The sustainable removal processes involve the accretion of new wetland sediments. Uptake by small organisms, including bacteria, algae and duckweed, forms a rapid-action, partly reversible removal mechanism (Figure 5.12). Cycling through growth, death and decomposition returns most of the microbiotic uptake via leaching, but an important residual contributes to long-term accretion in newly formed sediments and soils. Macrophytes, such as cattails and bulrushes, follow a similar cycle but on a slower time scale of months or years. The detrital residual from the macrophyte cycle also contributes to the long-term storage in accreted solids. Direct settling and trapping of particulate P can contribute to the accretion process. There can also be biological enhancement of mineralogical processes, such as iron and aluminium uptake and subsequent P binding in detritus and the algae-driven precipitation of P with calcium.

5.4.2 Performance

5.4.2.1 Surface flow

P removal in wetlands is in a class by itself: the amount of data and analysis is enormously greater than for other pollutants. There are hundreds of wetland-years of performance data, spanning two decades. There are hundreds of scientific papers, and modelling has been in progress for over two decades. Investigations are currently funded in aggregate at many millions of dollars per year.

The regression of input-output data provides one means of description of the general trends in intersystem performance. Within the set of linear and log-linear regressions on loading and concentration, the best fit of marsh data is produced by

$$C_o = 0.195q^{0.53}C_i^{0.91}, \quad (5.43)$$
$R^2 = 0.77, N = 373,$
SE in $\ln C_o = 1.00,$
$0.02 < C_i < 20 \text{ mg l}^{-1},$
$0.009 < C_o < 20 \text{ mg l}^{-1},$
$0.1 < q_{av} < 33 \text{ cm d}^{-1}.$

SF wetlands provide sustainable removal of P but at relatively low rates. The internal progression of removal causes concentrations to decrease exponentially to a background value along the water flow path (Figure 5.13). The first-order areal mass balance model is currently the most supportable level of detail for describing long-term sustainable performance. It typically explains about 80% of the variability in transect data and explains internal profiles as well as input–output data for individual wetlands. This model must be applied over more than 3 to 5 detention times to avoid transit time effects.

The background concentration C^* is not a well-known quantity but seems to be in the range 10–50 µg l^{-1}, on the basis of information from large natural and constructed wetlands. It therefore does not exert a strong influence on model predictions until outlet concentrations reach this low range. The first-order rate constants for a number of non-forested wetlands show a central tendency of $k \approx 10$ m yr^{-1}. Forested systems have lower rate constants of *ca.* 3 m yr^{-1} (Kadlec & Knight 1996).

5.4.2.2 Subsurface flow

Data from SSF wetlands on P removal are very sparse.

A regression equation for TP in soil-based wetlands in Denmark is (Brix 1994*b*):

$$C_o = 0.65C_i + 0.71, \quad (5.44)$$
$R^2 = 0.75, N = 61$ wetlands,
$0.5 < C_i < 19 \text{ mg l}^{-1},$
$0.1 < C_o < 14 \text{ mg l}^{-1},$
$0.8 < q < 22 \text{ cm d}^{-1}.$

A regression for TP found in the Czech Republic (Vymazal 1998*b*) for vegetated beds was

$$C_o = 0.26C_i + 1.52, \quad (5.45)$$
$R^2 = 0.23, N = 27,$
$0.77 < C_i < 14.3 \text{ mg l}^{-1},$
$0.4 < C_o < 8.4 \text{ mg l}^{-1}.$
$1.7 < q < 14.2 \text{ cm d}^{-1};$

for the whole system including pretreatment was

$$C_o = 0.29C_i + 1.12, \quad (5.46)$$
$R^2 = 0.27, N = 27,$
$1 < C_i < 13.5 \text{ mg l}^{-1},$
$0.4 < C_o < 8.4 \text{ mg l}^{-1},$
$1.7 < q < 14.2 \text{ cm d}^{-1};$

and for loadings of the vegetated beds was

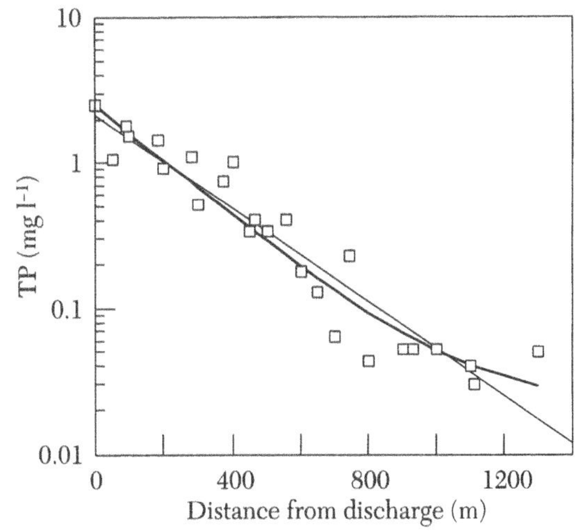

Figure 5.13. Transect P data for the wetland treatment system at Houghton Lake, Michigan, USA. Each data point is the average of values for the period 1987–95 for each distance. The straight regression line is for $C^* = 0$ ($y = 2.14^{-0.0037x}$; $R^2 = 0.90$); the curve is for $C^* = 0.022$ mg l^{-1} ($k = 9.6$ m yr^{-1}; $R^2 = 0.92$). Because of the low values of C^*, both produce good fits.

$$L_o = 0.67L_i - 9.03, \quad (5.47)$$
$R^2 = 0.58, N = 24,$
$28 < L_i < 307 \text{ g m}^{-2}\text{ yr}^{-1},$
$11.4 < L_o < 175 \text{ g m}^{-2}\text{ yr}^{-1},$
$1.7 < q < 14.2 \text{ cm d}^{-1}.$

The TP removal rate constant for Czech SSF wetlands averaged 0.025 m d^{-1}.

5.4.2.3 Annual patterns

Start-up processes differ from the long-term sustainable processes. Sorption and biomass growth enhance early results; leaching of antecedent loads decreases performance. Data show a period of 1–4 years for start-up transients to disappear (Kadlec & Knight 1996).

Seasonal and temperature effects are of minor importance in FWS wetlands for P removal. The theta factors are close to unity: $\theta = 0.999$, $R^2 = 0.006$ for Listowel system 4; and $\theta = 1.005$, $R^2 = 0.003$ for Listowel system 5. The low R^2 values indicate that the use of a temperature correction accounts for only 0.3–0.6% of the variability in rate constants for these systems.

5.4.2.4 Variability

The first-order model is a surrogate for a slow biogeochemical cycle, with a turnover time of many months for macrophyte-dominated systems. Consequently, it is not applicable on a short time scale such as daily or weekly. There is typically considerable stochastic scatter in the time sequence of output concentrations (Figure 5.14), which is the result of variability in influent flow rate and concentration, meteorology and biological processes. There is not yet a calibrated general model available to describe

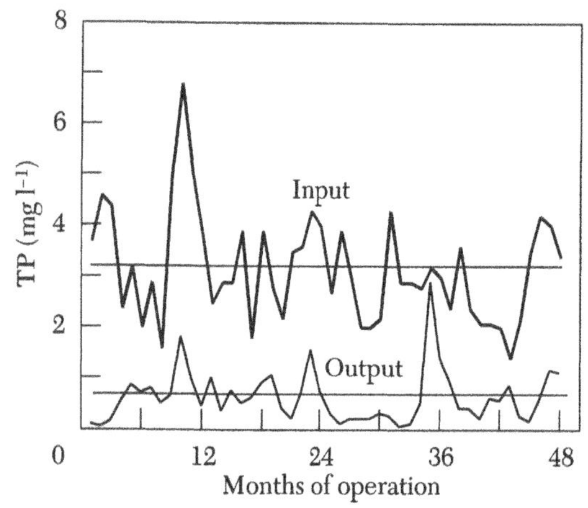

Figure 5.14. Input-output P concentration data for wetland number 4 at Listowel, Ontario, Canada, for its 4 yr period of operation (1980–84). The mean inlet concentration of 3.17 mg l^{-1} was decreased to 0.62 mg l^{-1}. k is 12.2 m yr^{-1}; C^* does not affect the data fit at these high concentrations. The output does not track the input at all times, indicating significant stochastic influences.

the daily, weekly and monthly scatter; consequently it is necessary to be aware of the probability distribution associated with the mean long-term performance. The occasional random spikes and valleys in output are reflected in the tails of these distributions and are not predictable from models. The maximum monthly outlet P concentration is typically 1.8 times higher than the long-term mean (Kadlec & Knight 1996).

5.5 Pathogens

Domestic wastewaters contain human pathogens that can survive pretreatment and enter treatment wetlands. These include bacteria, viruses, protozoans and helminths. The commonly used regulatory measure is for faecal coliforms, but faecal streptococci, *Salmonella*, *Yersinia*, *Pseudomonas* and *Clostridium* have all been studied in treatment wetlands (Herskowitz 1986). These human enteric organisms are typically decreased in numbers in passage through SF wetlands. Viruses are also attenuated in wetlands (Gersberg *et al.* 1989). Less is known about protozoans and their cysts, such as *Giardia* and *Cryptosporidium*, but these also are decreased in wetlands (Rivera *et al.* 1994). Helminths, including eggs of the nematode *Ascaris*, and various species of amoebae, were also decreased in soil-based systems (Rivera *et al.* 1994).

5.5.1 Processes

The ecology of microorganisms in a constructed wetland, as in any biological wastewater treatment system, is extremely complex. The important organisms from a public health point of view are the pathogenic bacteria and viruses. Protozoan pathogens and helminth worms are also of particular importance in tropical and subtropical countries (Rivera *et al.* 1995). In the aerobic environment of a VF wetland and the colder partly aerobic environment of an HF SSF system, they have minimal growth. Pathogens are removed during the passage of wastewater through the system mainly by sedimentation, filtration and adsorption on sediments. Once these organisms are entrapped within the system their numbers decrease rapidly, mainly by the processes of natural die-off and predation.

Wetlands are known to offer a suitable combination of physical, chemical and biological factors for the removal of pathogenic organisms. Physical factors include mechanical filtration, exposure to ultraviolet, and sedimentation. Chemical factors include oxidation, exposure to biocides excreted by some plants, and absorption by organic matter. Biological removal mechanisms include antibiosis (Seidel *et al.* 1978), predation by nematodes and protists, attack by lytic bacteria and viruses, and natural die-off (Gersberg *et al.* 1989).

Lower temperatures are known to affect the survival of sewage bacteria adversely. However, higher temperatures favour not only the pathogens but also their predators. Physical processes, such as sorption or settling, are not particularly temperature-sensitive. Annual irradiation patterns mimic the annual temperature cycle, and hence ultraviolet-induced mortality should be higher at higher temperatures.

Some pathogens are associated with warm-blooded animals other than humans, most especially faecal coliforms, streptococci and *Salmonella*. It is therefore possible for birds and mammals to contribute to the occurrence of these organisms in the wetland environment. A negative effect of treatment has been observed for treatment wetlands with high bird populations (PBSJ 1989). Non-zero background concentrations of faecal coliforms are typically present in natural, unimpacted wetlands.

5.5.2 Performance

Performance data are characterized by rapid declines to background concentrations.

5.5.2.1 Surface flow

The k–C^* pattern is present for water in continuous flow-through treatment wetlands: incoming concentrations of pathogenic organisms are decreased as the water moves through the system. Data from an Australian treatment facility demonstrate this progression (Figure 5.15). The background level in that wetland was relatively high (*ca.* 650 per 100 ml).

The central tendency is for $k = 72$ m yr^{-1} (Table 5.9). For a water depth of 30 cm, this

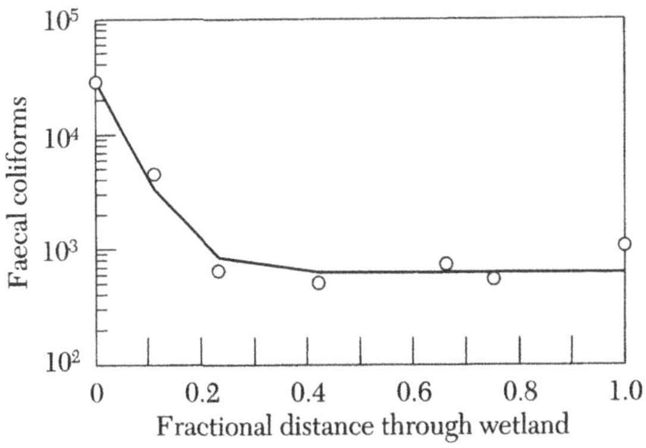

Figure 5.15. Decrease in faecal coliforms during water passage through full-scale wetlands at Byron Bay, Australia, in 1990–93 (Byron Shire Council, unpublished data). The decline in organisms takes place in wetlands dominated by a mixture of flow regimes and vegetated by cattails, bulrushes and Phragmites. The last section of this wetland was composed of open water in a tea tree (Melaleuca quinqinervia) forest. Heavy use by birds was presumed to have caused an increase in bacterial counts in that forest. Parameters for the line graph: k = 78 m yr^{-1}; C^* = 646/100 ml; R^2 = 0.96.

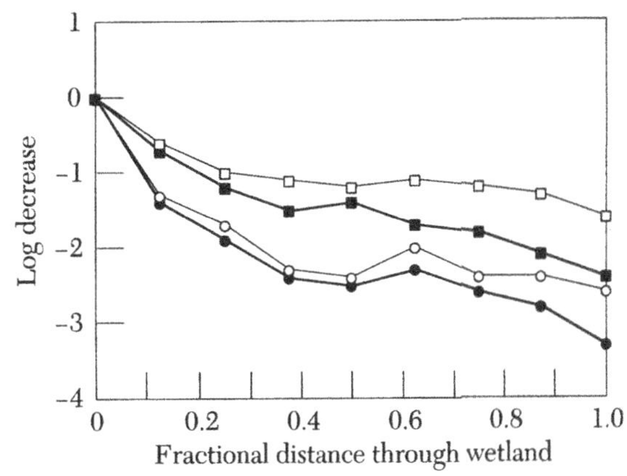

Figure 5.16. Decrease in virus during water passage through pilot-scale wetlands at Arcata, California, USA, in 1986 (redrawn from Gersberg et al. (1989)). The virus, bacteriophage MS-2, was continuously seeded into the wetland, which was dominated by cattail and bulrush. Symbols: ■, April; □, May; ○, June; ●, September.

corresponds to a 99% decrease to background in 7 d of detention. Background numbers are quite variable, ranging to over 2000 per 100 ml at Denham Springs, Louisiana, USA. That wetland, designed for SSF, actually operates in SF and harbours large populations of water birds.

Viruses follow a pattern similar to that for bacteria: a decline as water progresses through the wetland. Figure 5.16 shows this progression for a virus that affects bacteria (bacteriophage), but human enteric viruses followed the same pattern in both experimental channels and a receiving wetland (Scheuerman *et al.* 1989). Background levels of human enteric viruses should theoretically be absent from wetlands.

5.5.2.2 Subsurface flow

Removal of coliform bacteria in SSF wetlands has been described by several authors, including Gersberg *et al* (1989*a*, *b*), Bavor *et al.* (1989) and Williams *et al.* (1985). Gersberg *et al.* (1989*a*) demonstrated 97% (1.52 log) removal in a gravel-filled artificial wetland, planted with *Scirpus* in Santee, California, USA, with a theoretical retention time of 1.5 d. Bavor *et al.* (1989) looked at, among other things, the removal of coliforms in long, gravel-filled, trenches in Richmond, Australia, comparing vegetated (*Typha*) and unvegetated systems. By sampling along the length of the channels, they produced data that fitted a first-order reaction equation and calculated removal rate constants for the different systems. With their model it is possible to predict the requirement of 2 d retention to achieve 90% removal in a gravel trench at 20 °C, whereas the *Typha*-planted trench required more than 3 d to achieve 90% removal at 20 °C. Williams *et al.* (1985) also sampled for coliform bacteria along the length of gravel-filled wetland systems. Their tertiary treatment beds achieved 99% removal of faecal coliforms with a retention time of 1 d. The secondary bed required approx. 2.5 d retention time to achieve 90% removal of faecal coliforms.

These results are similar to those reported for SSF systems in the USA (Kadlec & Knight 1996). The corresponding k values range from 50 to 300 m yr^{-1}, with a central tendency of approx. 100 m yr^{-1}.

5.5.2.3 Annual patterns

Temperature does not strongly affect the rate constant for faecal coliform reduction; θ = 1.003 ± -0.024 for five FWS wetlands. Further, the inclusion of this factor accounts for a little of the variability (2.1%) in the rate constants and is therefore not of great importance.

5.5.2.4 Variability

There is considerable variability in the time series of organism counts produced by treatment wetlands. This phenomenon is also present for other treatment technologies and leads to the use of geometric averaging to determine monthly mean values from daily or weekly measurements. A factor of more than 10 still remains between individual monthly values and the long-term mean for the Listowel system and also for other treatment wetlands. As indicated by the foregoing discussion, exiting organisms did not necessarily originate with the incoming wastewater.

Table 5.9. Reduction rate constants for faecal coliforms in SF wetlands (NADB 1993; Kadlec & Knight 1996)

Site	System	HLR (cm d^{-1})	FC in (cfu/100 ml)	FC out (cfu/100 ml)	k_1 ($C^* = 0$) (cm d^{-1})	k (cm d^{-1})	C^* (cfu/100 ml)
Arcata, California, USA	Pilot 1	13.33	3,183	416	27.1		
	Pilot 2	7.89	12,500	316	29.0		
	Transect	27.72	15,850	1608	39.0	45	118
	Woolgrass	4.72	4,747	135	16.8	38	28
	Cattail	4.72	4,747	458	11.0	34	151
Boggy Gut, South Carolina, USA		3.01	2	236			
Brookhaven, New York, USA		2.02	4,175	378	4.8		
Byron Bay, Australia		5.53	28,918	667	20.8	21.4	646
Carolina Bays, South Carolina, USA		0.15	66,000	56	1.1		
Central Slough, South Carolina, USA		0.51	857	50	1.4		
Cobalt, Ontario, Canada		1.7	159,300	1087	8.5		
Denham Springs, Louisiana, USA	1	12.18	39,620	4115	27.6	66	2 325
	2	12.18	42,030	3810	29.2	59	2 080
	3	12.18	39,866	2854	32.1	66	2 034
Harriman, Pennsylvania, USA	1	3.75	1,953,329	14,180	18.5		
	2	3.75	29,278	538	15.0		
Iron Bridge, Florida, USA	1990	2.97	1	33			
	1991	2.85	1	91			
Lakeland, Florida, USA	1	4.37	25,536	55	26.9	124	26
Listowel, Ontario, Canada	1	2.80	1,773	72	9.0	27	2
	2	2.92	1,773	573	3.3	32	86
	3	2.10	1,773	56	7.3	47	4
	4	1.95	228,292	141	14.0	17	4
	5	2.60	228,292	2251	11.0	12	98
Neshaminy, Pennsylvania, USA		5.28	1,290,600	5600	28.7		
Pembroke, Kentucky, USA		4.38	165,959	266	28.2	288	60
Richmond, NSW, Australia	Open water	6.40	1,698,244	50,119	22.5		
	Myriophyllum	7.35	1,698,244	56,234	25.0		
Waldo, Florida, USA	Pilot	17.64	7,700,000	270,000	59.1		
West Jackson Co., Mississippi, USA		3.18	239	674			
Whangarei, New Zealand	Trial	6.00	400,000	2300	31.0		
	Full-scale	7.50	1,085	481	6.1		
Average					19.79 (72 m yr^{-1})		

Abbreviations: cfu, colony-forming units; FC, faecal coliforms.

5.6 Metals

Trace metals have a high affinity for adsorption and complexation with organic material and are accumulated in a wetland ecosystem. The processes of metal removal in wetlands have been reviewed by Richards *et al* (1992) and are shown in Figure 5.17. Although some metals are required for plant and animal growth in trace quantities (such as barium, beryllium, boron, chromium, cobalt, copper, iron, magnesium, manganese, molybdenum, nickel, selenium, sulphur and zinc), these same metals can be toxic at higher concentrations. Other metals (such as arsenic, cadmium, lead, mercury and silver) have no known biological role, and can be toxic at even lower concentrations.

5.6.1 Processes

Metals can occur in either the soluble or particulate associated forms, with the former representing the most bioavailable form, particularly when the metal is present as either an ionic or weakly complexed species. The distribution between particulate and dissolved phases is determined by physicochemical processes such as sorption, precipitation, complexation, sedimentation, erosion and diffusion. Certain metals, such as Cd and Zn, have been shown to have a stronger affinity for the dissolved phase, whereas Pb tends to be predominantly particulate-associated (Morrison *et al.* 1984). Specific parameters that control the sediment–water partitioning of metals include the ratio of flow to suspended solids, oxic/anoxic conditions, ionic strength, pH, dissolved and particulate organic carbon contents, organic and inorganic ligand concentrations, and metal mobilization by biochemically mediated reactions.

5.6.1.1 Adsorption and cation exchange

Adsorption involves the binding of particles or dissolved substances in solution to sites on the plant or matrix surface. In a cation exchange reaction, positively charged metal ions in solution bind to negatively charged sites on the surface of the adsorption material. The attractive force for cation exchange is electrostatic; the size of this force depends on a wide range of factors. A cation in solution will displace a cation bound to a site on the surface of a material if the electrostatic attraction of the site for the dissolved cation exceeds that of the bound cation. The cation exchange capacity

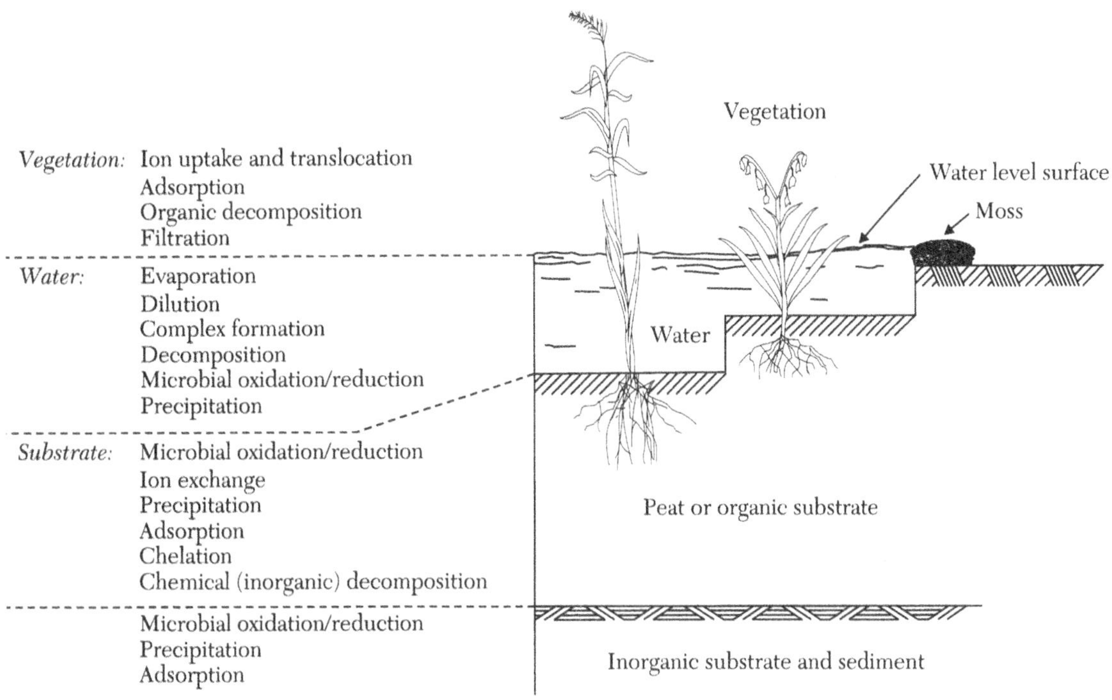

Figure 5.17. Processes of metal removal in wetlands. (Kleinman & Girts (1987).)

(CEC) of a material is a measure of the number of binding sites per mass or volume.

The cation exchange properties of wetland substrates have been attributed to carboxy functional groups (-COOH) in the humic acids of plant cellular tissue. Studies have been undertaken to calculate the CEC value of many macrophytes and other plant materials (Howard *et al.* 1988). The CEC value has also been shown to be the same whether the plant is alive or dead. Wetland sediments and soils also have large CEC values. The adsorption of metals on the surface of soils is therefore a process that can be significant in treatment wetlands. The CEC of SSF matrices depends on the material of construction selected, but most gravel or soil matrices tend to become saturated with metals in time.

5.6.1.2 Microbially mediated processes

The wetland can be differentiated into two zones: aerobic and anaerobic. The presence of metal-oxidizing bacteria in the aerobic zones and sulphate-reducing bacteria in the anaerobic zones, which cause the precipitation of metal oxides and sulphates respectively, has been established by Batal *et al.* (1987).

For instance, microbially mediated iron oxidation by *Thiobacillus ferrooxidans*, followed by the subsequent precipitation of iron oxyhydroxide, is considered the most important iron removal mechanism in wetlands treating metal-rich mine wastewaters. In unbalanced equation form:

$$Fe^{2+} + O_2 + H_2O \rightarrow Fe(OH)_3 + H^+. \qquad (5.48)$$

Similar chemistries and limited investigation suggest similar oxidations for many other metals including nickel, copper, lead, zinc, silver and gold. Wetland plants can potentially stimulate the growth of metal-oxidizing bacteria by oxygen transfer into the rhizosphere.

Microbially mediated sulphate reduction consumes sulphate ions and produces hydrogen sulphide and alkalinity in the form of the bicarbonate ion. In unbalanced equation form, where 'CH_2O' represents a simple organic molecule:

$$SO_4^{2-} + CH_2O \rightarrow H_2S + HCO_2. \qquad (5.49)$$

The H_2S dissolves and ionizes to give sulphide ions, which react with a range of metal ions to produce metal sulphide precipitates.

Precipitation of metals as sulphides rather than oxides has the following advantages:

- alkalinity produced by sulphate reduction helps to neutralize acidity
- sulphate precipitates are denser than oxide precipitates
- sulphides are precipitated within the organic sediments and so are less vulnerable to disruption by sudden surges in flow.

5.6.1.3 Filtration

Vegetation can assist in metal removal by aiding the direct filtration of particulate matter. Macrophyte species with high plant surface areas have been shown to be very effective at retaining metal hydroxide particles that have precipitated out of solution.

5.6.1.4 Plant uptake

Some wetland species have a well-established

Table 5.10. Metal removal data from surface flow treatment wetlands

Metal	Wetland type	Concentration (µg ml^{-1}) In	Out	Mass removal (kg ha^{-1} yr^{-1})	Reference
Cadmium	Constructed	43	0.6	2.4	Hendrey *et al.* (1979)
Chromium	Constructed	160	20	7.9	Hendrey *et al.* (1979)
	Constructed	3.4	1.5	4.5	Crites *et al.* (1995)
Copper	Constructed	1510	60	82	Hendrey *et al.* (1979)
	Constructed	8	3	11	Crites *et al.* (1995)
	Natural	20.4	6.1	0.21	CH2M HILL (1992)
Iron	Constructed	6430	2140	243	Hendrey *et al.* (1979)
	Constructed	205,000	6300	29,900	Edwards (1993)
	Natural	241	766	−4.3	CH2M HILL (1992)
Lead	Constructed	1.7	0.4	3.1	Crites *et al.* (1995)
	Constructed	2.2	1.63	0.085	Edwards (1993)
	Natural	2.0	5.5	−0.03	CH2M HILL (1992)
Manganese	Constructed	210	120	5.1	Hendrey *et al.* (1979)
	Constructed	7400	3900	526	Edwards (1993)
Mercury	Natural	<0.2	0.21	0.0001	CH2M HILL (1992)
Nickel	Constructed	35	10	1.4	Hendrey *et al.* (1979)
	Constructed	7.5	3.8	0.8	Crites *et al.* (1995)
	Natural	17.0	9.1	0.14	CH2M HILL (1992)
Silver	Natural	0.36	0.53	-0.0005	CH2M HILL (1992)
Zinc	Constructed	2200	230	112	Hendrey *et al.* (1979)
	Constructed	36	11	60	Crites *et al.* (1995)
	Natural	20.6	5.6	0.22	CH2M HILL (1992)

ability for direct uptake of heavy metals. Unfortunately, accumulation can become sufficient to kill the plant within just one growing season. Fortunately, some species such as *Typha latifolia* have a species-wide constitutional tolerance for heavy metals and do not accumulate metals to toxic levels. The presence of an iron plaque in the plant root system decreases the uptake of metals by the root hairs (Ye *et al.* 1994). Present information suggests that other species such as *Phragmites* have acid-tolerant and metal-tolerant ecotypes (populations within a species) that do not significantly accumulate metals. A number of mechanisms have been proposed, including the prevention of uptake of metals and their storage in a non-toxic form within the cells. Acclimatization over prolonged periods has also been observed but not yet quantified.

In a study of *Typha latifolia* in urban wetlands, the metal load distributions for lead, copper, zinc and cadmium were 50-62% in the rhizome, 30–33% in the leaf and 6–10% in the roots (Shutes *et al.* 1993). The maximum recorded metal loads of lead (414.9 g ha^{-1}), copper (502.9 g ha^{-1}), zinc (766.9 g ha^{-1}) and cadmium (62.9 g ha^{-1}) in this study indicate that metal uptake and storage can be significant in this macrophyte species.

It should be noted that direct uptake is an active process, requiring the plant to be alive. Plant matter liberates its metal content on decomposing. Harvesting of the foliage would only minimally assist metal removal because of the low concentration of metals in the aboveground parts of the plants. It is preferable to allow litter to form, as this can provide new sites for metal removal and thermal insulation.

5.6.2 Performance

5.6.2.1 Surface flow

Information from SF treatment wetlands indicates that a fraction of the incoming metal load will be trapped and effectively removed through sequestration in plants and soils. Table 5.10 provides a summary of published concentrations of metals at treatment wetland inlets and outlets from a variety of sites. For many metals, the limited data indicate that EFF and RED are correlated with inflow concentration and mass loading rate (Kadlec & Knight 1996). Wetland background metal concentrations and internal profiles are not well known. Metal k values have not been estimated for SF treatment wetlands.

5.6.2.2 Mine drainage

Information from mine-drainage treatment wetlands indicates that a fraction of the incoming metal load is trapped and effectively removed through sequestration in plants and the bed medium. Table 5.11 provides a sum-

Table 5.11. Summary statistics for treatment efficiencies from constructed wetlands treating acid mine drainage (AMD)

Chemical constituent of AMD	Treatment efficiencies for a centile of:					Centile for zero treatment efficiency	Mean±SE	*n*
	100	75	50	25	0			
H^+	100.0	98.2	68.4	0.0	−15,749.0	29	−311 ± 167	125
Acidity	100.0	100.0	66.8	32.8	−200.0	11	56.6 ± 6.1	74
Fe	99.9	93.8	80.9	46.6	−567.0	13	58.2 ± 6.5	126
Al	90.9	78.9	47.7	5.6	−52.9	25	39.0 ± 10.0	20
Mn	99.9	64.1	34.1	9A	−1100.0	17	16.8 ± 13.0	124
SO_4^{2-}	88.6	25.6	8.1	−3.5	−812.0	38	0.6 ± 8.5	106

For each chemical constituent of AMD, the tabular value in the 100 centile column represents the maximum treatment efficiency and the tabular value in the 0 centile column represents the minimum treatment efficiency. The intermediate centiles denote the treatment efficiency below which a certain percentage of all observations fell. Also provided are the arithmetic mean and standard error (SE) for the treatment efficiency for each chemical parameter; *n* is the number of wetlands (Weider 1989).

mary of metal removal efficiencies for over 100 treatment wetland systems treating acid mine drainage.

5.7 Ancillary water chemistry

5.7.1 Oxygen

Several treatment wetland processes, such as oxidation, respiration and nitrification, depend on dissolved oxygen. Plant roots require oxygen, which is normally transported downwards through passages (aerenchyma) in stems and roots. Some surplus of oxygen can be released from small roots into their immediate environs, but it is quickly consumed in the decrease of local OD (Brix 1994). Wetland soils are typically anoxic or anaerobic (Reddy & D'Angelo 1994).

Wetland surface waters are aerated by oxygen transfer from air, through the air-water interface. Reaeration mechanisms include dissolution and diffusion (O'Connor & Dobbins 1958), as well as turbulent transfer associated with rainfall surface mixing (South Florida Water Management District, unpublished data). In unshaded (open water) areas, photosynthesis by algae within the water column causes oxygen production, sometimes creating dissolved oxygen in excess of the saturation limit (Schwegler 1978). Photosynthesis stops at night, and respiratory use dominates. The result is a strong diurnal variation in water-column DO for lightly loaded, algal open water wetlands.

In an HF treatment wetland, the supply of oxygen to the microbial population within the bed matrix comprises that transferred by the macrophyes plus that which diffuses from the surface of the bed matrix. A total flux of gaseous oxygen into the bed substrate of 5.9 g O_2 m^{-2} d^{-1}, of which 2.08 g O_2 m^{-2} d^{-1} was through the hollow culms of standing-dead culms of *Phragmites australis*, has been measured during winter when the plants were senescent (Brix *et al.* 1992). Unfortunately, the respiratory oxygen consumption of the roots and rhizomes was found to balance almost perfectly the oxygen influx through the culms, leaving only 0.02 g O_2 m^{-2} d^{-1} to be released to the surrounding matrix. This study suggests that macrophyte-induced rhizosphere oxygenation is of no quantitative importance during winter months when the reeds are dead. It should be noted that these oxygen fluxes are far lower than those quoted for VF systems and also much lower than the OD of 8–10 g O_2 m^{-2} d^{-1} for a bed designed at 5 m^2/PE for BOD removal only.

The wetland carbon cycle creates micro-detritus and dissolved carbonaceous materials, which create an OD. These materials are often located near the wetland bottom but can also be distributed in the water column as flocculent deposits on litter or plant stems or in a floating or suspended form. Wetlands receiving higher nutrient loads have larger amounts of living and dead biomass. These, in turn, depress the oxygen levels in the water and dampen the diurnal cycle (Mitsch & Wu 1995). Higher nutrient loadings also promote dense macrophyte growth, which leads to shading of the water column and the suppression of algal activity.

Water entering the treatment wetland has carbonaceous and nitrogenous oxygen demand (NOD). After entering the wetland, several competing processes affect the concentrations of oxygen, BOD and nitrogen species. Dissolved oxygen is depleted to meet wetland oxygen requirements in four major categories: sediment or litter OD, respiration requirements, dissolved carbonaceous BOD and

Table 5.12. Dissolved oxygen entering and leaving several SSF wetlands

Type	Wetland	Inlet DO	Outlet DO
Steady flow	Benton 3, Kentucky, USA	8.20	1.00
	Hardin 1, Kentucky, USA	5.20	1.20
	Hardin 2, Kentucky, USA	5.20	0.70
	Rector, Arkansas, USA	7.23	0.97
	Waldo, Florida, USA	8.87	0.10
	Richmond *Typha*, NSW, Australia	1.01	0.04
	Richmond cattail, NSW, Australia	1.01	0.00
	Richmond gravel 1, NSW, Australia	1.01	0.25
	Richmond gravel 2, NSW, Australia	1.20	0.13
Intermittent flow	Portsmouth, UK	3.00	6.40
	Phillips High School, Alabama, USA	6.10	5.37

dissolved NOD. The sediment OD is the result of decomposing detritus generated by carbon fixation in the wetland, as well as the decomposition of precipitated organic solids that entered with the water. The NOD is exerted primarily by ammonium nitrogen, but ammonium can be supplemented by the mineralization of dissolved organic nitrogen. Decomposition processes in the wetland also contribute to NOD and BOD. Microorganisms, primarily attached to solid immersed surfaces, mediate the reactions between DO and the oxygen-consuming chemicals.

In approximate terms, FWS wetlands with open water receiving highly pretreated water possess moderately well oxygenated waters. FWS wetlands with dense emergent vegetation receiving secondary wastewater have low DO levels, in the range 1–2 mg l^{-1}. Free dissolved oxygen is rarely found below the water/sediment interface.

Horizontal SSF wetlands contain virtually no free oxygen in the steady-flow mode (Table 5.12). However, intermittent flow in either the vertical or horizontal direction greatly improves oxygen availability.

5.7.2 Hydrogen ions

Natural wetlands exhibit pH values ranging from slightly basic in alkaline fens (pH 7–8) to quite acidic in sphagnum bogs (pH 3–4) (Mitsch & Gosselink 1993). Natural freshwater marsh pH values are generally slightly acidic (pH 6–7). Treatment wetland effluent hydrogen ion concentrations are typically around neutral to slightly acidic. Open water zones within wetlands can develop high levels of algal activity, which in turn create a high-pH environment. Data on an open-water, unvegetated treatment 'wetland' displayed high pH during some summer periods (pH > 9), with circumneutral influent (7.0 < pH < 7.4) (Bavor *et al.* 1988). Algal photosynthetic processes peak during the daytime hours, creating a high pH during the day, followed by a night-time sag with a low pH as respiration replaces photosynthesis.

The organic substances generated within a wetland via growth, death and decomposition cycles are the source of natural acidity. The resulting humic substances are large complex molecules with multiple carboxy and phenolate groups. The protonated forms have a tendency to be less soluble in water and precipitate under acidic conditions. As a consequence, wetland soil–water systems are buffered against incoming basic substances. They are less well buffered against incoming acidic substances because the water column contains a limited quantity of soluble humics.

The result is that treatment wetlands act to adjust the pH of entering water to *ca.* 7. Listowel system 3 received lagoon water, which periodically exhibited high pH owing to algal activity in the lagoon. During the first year of operation, little or no buffer capacity was evident. This was evidently due to the start-up conditions, during which the vegetation spread to cover the wetland, and litter formation and decomposition became operative. In later years, high incoming pH values were effectively damped out by the wetland.

6 Design

6.1 Sizing the system

The design sizing and description of treatment wetlands involves two principal features: hydraulics and pollutant removal. Previous literature used concepts of Darcy-type flow in SSF wetlands (Hobson 1989; Fisher 1990) and of vegetated open channel flow in FWS wetlands (Hokosawa & Horie 1992). For treatment wetlands, previous literature has suggested first-order removal models of irreversible pollutant decrease. First-order models can be either area-specific, and thus determine the necessary wetland acreage (Schierup *et al.* 1990; Mitsch *et al.* 1995; Upton & Green 1995) or volume-specific, and thus determine the wetland water volume (US Environmental Protection Agency 1993). These current models are deterministic, meaning that the equations purport to represent the wetland output concentrations in response to inlet concentrations, flow rate, and area or volume. However, wetland performance also includes a good measure of variability that is not predicted by the average values of these forcing variables. That variability is caused by unpredictable events, such as the fluctuations in input flows and concentrations, and by changes in internal storages, as well as by weather, animal activity and other ecosystem factors.

More complex models can be contrived to include the dynamic behaviour of the various ecosystem compartments and processes (Kadlec 1996), but these require very large quantities of data for proper calibration. Input/output (I/O) data on flows and concentrations are generally insufficient for calibration, and consequently little is known in general about the numerous model parameters. It can be presumed that calibrated compartmental models will provide more details of internal allocations of chemicals, but it is not clear that more detailed deterministic models will provide more accurate descriptions of overall wetland performance. At this point in the evolution of treatment wetland technology, only simple models can be calibrated for most operational systems. However, expanding intersystem databases provide the opportunity to examine the assumptions inherent in current models. This section examines the current design frameworks to provide further insights into their uses and limitations.

6.1.1 Treatment sizing

There are two distinct types of sizing procedure: one for steady-flow wetlands; and another for stormwater wetlands, which are driven by rainfall events. So far the two have evolved separately over the history of the technology.

6.1.1.1 Computations for steady flow

The rate and temperature equations are of the form given in Chapter 4.

Design often contemplates a stable period of operation, over which input and outputs are averaged. The rate and temperature equations are of the form:

$$J = k(C - C^*), \tag{6.1}$$

$$R = k_V(C - C^*), \tag{6.2}$$

$$k = k_{20}\theta^{(T-20)}, \tag{6.3}$$

$$k_V = k_{V20}\theta^{(T-20)}, \tag{6.4}$$

where

C = concentration (mg l^{-1}),
C^* = background concentration (mg l^{-1}),
J = areal removal rate (g m^{-2} yr^{-1}),
k = areal removal rate constant at T °C (m yr^{-1}),
k_{20} = areal removal rate constant at 20T °C (m yr^{-1}),
ε = porosity,
k_V = volumetric removal rate constant at T °C (d^{-1}),
k_{V20} = volumetric removal rate constant at 20 °C (d^{-1}),
R = volumetric removal rate (g m^{-3} d^{-1}),
T = temperature (C),
θ = temperature coefficient.

The two alternative rate constants are related by the free water depth, with $k = (\varepsilon h)k_V$, usually together with a change in time scale from years to days. The rates (Equation 6.1 or 6.2) are used in combination with the water mass balance (Equation 4.7) to obtain pollutant concentration profiles. If flow is plug flow, with constant water volume ($P = ET$), exponential

Table 6.1. Variation of Arcata (California, USA) BOD rate 'constants' with depth

Flow (m^3 d^{-1})	Depth(m)	Increase in HRT (%)	BOD rate constant (d^{-1})	Decrease in k_v (%)
93	0.40		0.29	
94	0.55	37	0.17	42
86	0.36		0.25	
83	0.61	76	0.13	49
45	0.30		0.28	
49	0.49	49	0.14	48
29	0.33		0.14	
29	0.53	78	0.08	40
23	0.35		0.14	
24	0.50	39	0.09	36

profiles are predicted (reaching a plateau of $C = C^*$):

$$\frac{C - C^*}{C_i - C^*} = \exp\left(-\frac{ky}{q}\right) \tag{6.5}$$

$$= \exp(-k_V \tau y), \tag{6.6}$$

where

q = hydraulic loading rate (m yr^{-1}),
y = fractional distance through wetland ($=x/L$),
τ = nominal detention time (d).

However, if flow varies owing to $a = P - ET \neq 0$, then power-law profiles are predicted:

$$\frac{C - C''}{C_i - C'} = \left(1 + \frac{ay}{q}\right)^{-(1+k/a)}, \tag{6.7}$$

$$C' = C^*\left(\frac{k}{k+a}\right). \tag{6.8}$$

Evapotranspiration (rain) has two effects: lengthening (shortening) of detention time, and concentration (dilution) of dissolved constituents. The use of Equations 6.5 and 6.6 with an average flow rate compensates for altered detention time, but not for dilution or concentration. The fractional error due to flow averaging is approximately a/q, for $a/q > -0.5$. Thus, if 25% of the inflow evaporates, the use of Equations 6.5 and 6.6 with average flow predicts concentrations 25% lower than required by the mass balance. If rain adds 25% to the flow, the use of Equations 6.5 and 6.6 predicts concentrations 25% higher.

In the design mode, these equations are used to calculate the hydraulic loading (q) or detention time (τ) that will produce the required concentration (C_o) at the wetland outlet ($y = 1$). Equation 4.1 or 4.3 is then used to compute the wetland surface area.

6.1.1.2 Parametric variability

The parameters of the first-order models are referred to as 'rate constants', but there is no *a priori* reason to believe that these 'constants' do not in fact depend on other operational characteristics of the wetland. Intersystem variability and intrasystem stochastic effects are to be expected. However, the design variables of depth and hydraulic loading, which combine to determine nominal detention time, are directly involved in sizing computations. If k changes with depth and flow rate, then those effects must be accounted for in design.

Depth

The relation $k = (\varepsilon h)k_V$ requires that both k and k_V cannot be independent of depth. If k_V is constant with respect to depth, then k is proportional to depth. That condition requires chemical reaction to be uniformly distributed vertically throughout the water column. If k is constant, k_V is inversely proportional to depth. That condition corresponds to chemical reaction apportioned to wetland surface area. Neither ideal extreme is likely to be present in a treatment wetland, but data show FWS wetlands to behave with constant k, not constant k_V.

The implications for design are very important. If the volumetric model is used in FWS calculations, there seems to be the option of increasing performance by increasing the water depth, and hence increasing the nominal detention time. That advantage is lost if the volumetric rate 'constant' decreases with increasing depth. Table 6.1 illustrates this effect for the side-by-side tests at Arcata, California, USA (Gearhart *et al.* 1983). The rate constants in Table 6.1 are determined for $C^* = 0$, and are designated k_{V1}, indicating a one-parameter rate model. Data from side-by-side tests of horizontal SSF wetlands show the same effect: deeper wetlands have lower rate constants (George *et al.* 1994).

The depth of the wetland is selected from a narrow range of possibilities for both FWS and SSF wetlands. In FWS systems, the water must be deep enough to cover all portions of the basin bottom and to cover the litter layer, which

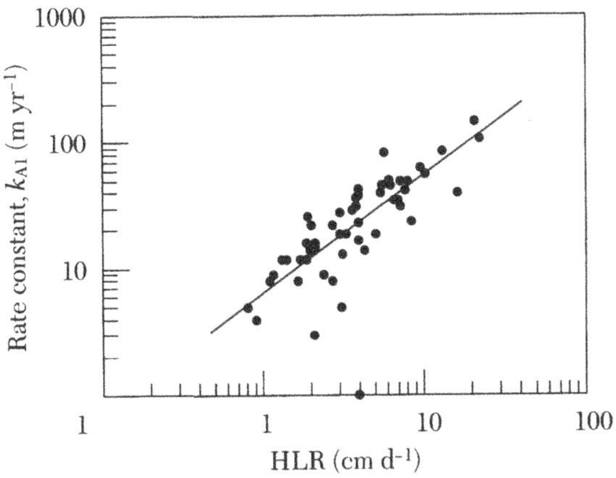

Figure 6.1. HLR effect on BOD rate constant; results from Danish soil-based wetlands, for the one-parameter, area-based rate constant $k_{A1} = 6.9q^{0.92}$.

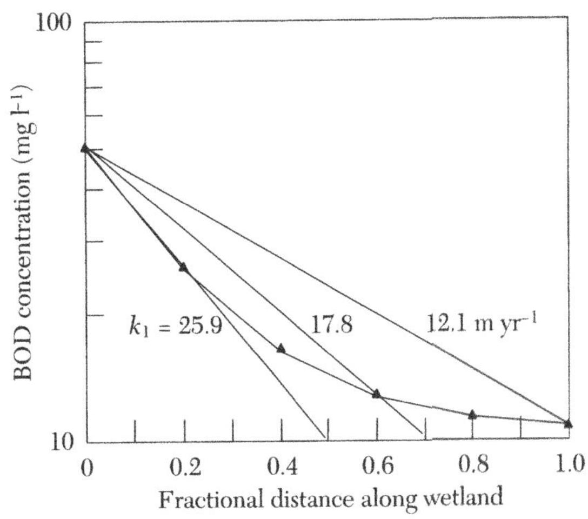

Figure 6.2. BOD rate constants from I/O data from Listowel channel 4 in 1983/84. Points show average annual data, and the lines show the k–C model for different values of k_1. k = 38 m yr^{-1}; C^* = 10.5 mg l^{-1}; R^2 = 0.999.*

contains a large fraction of the active surface area for microbes. In horizontal SSF systems, the water should occupy most of the depth of the medium. If plants are to have an effect on treatment, they must be able to contact the flowing water. Therefore the depth of medium in an SSF is restricted to approximately the rooting depth of the plants. This is a rather narrow range, typically from 20 to 80 cm and in most cases less than 40 cm (Cooper*et al.* 1996).

Hydraulic loading

One-parameter k values are strongly dependent on hydraulic loading rate, as illustrated by the BOD data from Danish soil-based wetlands (Schierup *et al.* 1990) (Figure 6.1). This is due in part to the existence of background concentrations, which create this effect in I/O data analysis, and in part to mechanistic influences within the ecosystem, such as the velocity dependence of mass transfer and non-plug-flow detention time distributions. As the effluent point of the treatment wetland moves farther out along the C^* plateau, the k_1 value, determined as the slope of the logarithmic model, becomes smaller (Figure 6.2).

As a result of these effects, k_1 or k_{V1} is proportional to hydraulic loading rate to a power, $k \propto q^m$. For example, data from 55 SSF wetlands for BOD show $k_{V1} \propto q^{0.63}$; data from 47 FWS wetlands for BOD show $k_{V1} \propto q^{0.79}$. This effect is not as large for pollutants with very low values of C^* or if the decrease is not close to C^*. This loading dependence, together with the depth dependence discussed above, indicates that the one-parameter version of Equations 6.5 and 6.6 should more properly read:

$$\frac{C}{C_1} = \exp\left(-\frac{k'_1 y}{q^{1-m}}\right) = \exp(-hk'_{V1}\tau^{1-m}y). \quad (6.9)$$

If the hydraulic efficiency of the wetland in design is less than that in the data sets that generated the rate constants, as indicated by more mixing or short-circuiting, then corrections for the degree of non-ideality should be applied (Kadlec & Knight 1996).

Temperature

Many individual biological processes have temperature-sensitive rates, and consequently the rate constants that represent the consortia of wetland processes might also be temperature-sensitive. However, the overall decrease in a pollutant's concentration typically involves an intricate web of transfers and transformations, which involve physical processes such as sedimentation and sorption, microbially mediated storages and conversions, uptake and storage in biota of varying sizes and life histories, and transfers of other reactants, such as oxygen and carbon dioxide. Some processes, and some incoming flows and concentrations, are seasonally variable, and those influences can become confused with temperature effects. This complexity indicates that ecosystem data are the only sure source of information on the influence of temperature on decreases in wetland pollutant concentrations.

Chapter 5 showed that ecosystem nitrogen species reaction rates are slower at lower temperatures but that there is no temperature effect on TSS removal. In contrast with these intuitive results, wetland data indicate no temperature effect on BOD, TP and faecal coliform decreases. Of these, the non-effect for BOD is somewhat counterintuitive and deserves further discussion. Some wastewater treatment technologies show a significant lowering of BOD k values as temperature is lowered, notably suspended growth processes

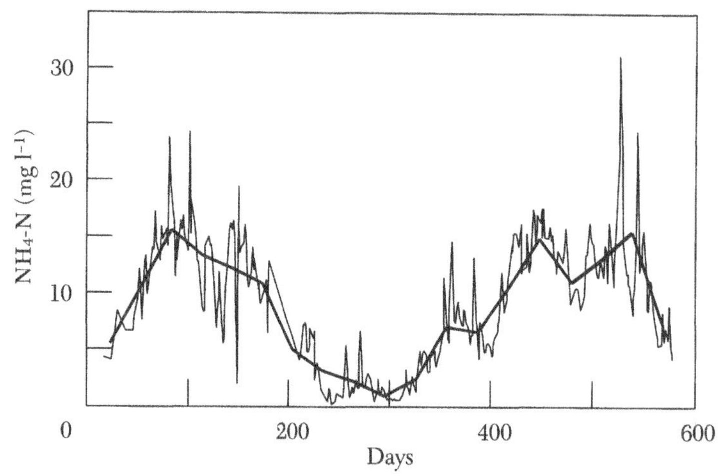

Figure 6.3. Stochastic chatter in a time sequence of effluent ammonium nitrogen in Columbia, Missouri, USA.

and aerated lagoons (Metcalf & Eddy 1991). The most relevant companion technologies are overland flow and facultative lagoons, which show little temperature change in BOD decrease. Of equal importance is the fact that the inclusion of a temperature coefficient in data analysis accounts for very little of the variance in data. For instance, including a theta factor in the FWS data accounts for only 6.6% of the variance.

6.1.1.3 Stochastic variability

Stochastic effects are a large part of treatment wetland performance. There are many causes, such as short-term dynamics, input variations in flow and concentration, meteorological events of rain, drought and heatwaves, and biological influences due to algae, insects, fish, birds and animals. The result is a large degree of 'chatter' about the mean performance, as illustrated in Figure 6.3. In this instance, there seems to be I/O tracking of the seasonal trend, but the daily measurements occupy a wide band about the mean.

In addition to the mean behaviour described by the equations given above, measures of the variance about this mean are required. The frequency distributions of inlet and outlet concentrations provide this additional description. Current regulatory requirements in the USA dictate a maximum monthly value; other countries place a maximum on a given centile, typically the 80th or 90th. Design must acknowledge regulatory requirements in most cases, and so it must account for stochastic as well as deterministic effects. Where seasonal patterns are known to be significant, the design equations can be applied on that seasonal basis, but random variability still remains. Design can include this chatter if the design target is adjusted downwards by a factor of approx. 2.0 to meet a monthly maximum cap.

6.1.1.4 Setting the area

In the simplest case there is a known flow and inlet concentration of a single pollutant, such as BOD. A desired outlet concentration is to be met at a specified measurement frequency. For example, suppose a FWS wetland is to decrease BOD from 100 mg l^{-1} to an annual average of 20 mg l^{-1}, and with a monthly effluent limit of 30 mg l^{-1}. The flow rate is to be a steady 2000 m^3 d^{-1}. The first-order area-based model is used with k = 35 m yr^{-1} (0.096 m d^{-1}), C^* = 6 mg l^{-1}. At the bed outlet (y = 1),

$$\ln\left(\frac{C_e - C^*}{C_i - C^*}\right) = -\frac{k}{q}, \qquad (6.10)$$

$$\ln\left(\frac{20-6}{100-6}\right) = -\frac{35}{q}.$$

The annual average, C_e = 15, can be achieved with q = 18.4 m yr^{-1} = 5.0 cm d^{-1}. The required area is $A = Q/q$ = 2000/0.05 = 39 700 m^2 = 4.0 ha. Alternately, the designer can calculate the area directly by using Equation 4.28 or, if C^* 0, Equation 4.32.

The maximum monthly value, C_e = 30, reflects a departure from a lower annual mean. From Table 4.2, the ratio of maximum monthly value to annual average is 1.7. The required annual average is therefore lower, C_e = 30/1.7 = 17.6 mg l^{-1}. This requires q = 16.7 m yr^{-1} = 4.6 cm d^{-1}. The required area is $A = Q/q$ = 43 600 m^2 = 4.4 ha. Meeting the maximum monthly requirement is more stringent and needs 10% more area.

In smaller domestic systems, there is unlikely to be flow information on which to base the design. Rather, only the population is known. Each person contributes a volume of water and a loading of various pollutants. A PE becomes the design basis. In USA, the water volume is approximately Q = 200 l d^{-1} per PE. Specific estimating tables have been published (see, for example, Metcalf & Eddy (1991)). In Europe the typical flow value per person is 150–180 l d^{-1} per PE for larger cities and about 80–120 l d^{-1} per PE for smaller communities (populations up to 500 PE). The area specification for the wetland then takes the form of the area (in m^2) required for one PE, A' (m^2 per PE). For instance, take the following values:

for performance model,
k = 36.5 m yr^{-1} (0.10 m d^{-1}),
C^* = 6 mg l^{-1};

for flow characteristics,
Q = 200 l d^{-1} per PE (0.20 m^3 d^{-1} per PE),
C_i = 240 mg l^{-1} (UK range 150–300 mg l^{-1});

for design goal,
C_e = 25 mg l^{-1};

for area requirement,
A' = 5.0 m^2 per PE.

WRc recommends 5 m^2 per PE for secondary treatment in SSF wetlands (Cooper *et al.* 1996). In Denmark a lower value of k_{BOD5} has been used, resulting in A' = 10.0 m^2 per PE. Severn Trent recommends 0.7 m^2 per PE for tertiary treatment.

6.1.1.5 Computations for stormwater wetlands

The amount of water and the amounts of pollutants that reach stormwater treatment wetlands from contributing watersheds are not typically known in advance. The number and duration of the events that produce input, together with the spacing between events, affect the efficiency of the stormwater wetland (Wong & Somes 1996). Because of this variability of actual flow rates, the detention time and hydraulic loading are difficult to define, and other single-number sizing rules have evolved.

One rule of thumb states that the wetland size should be a specified fraction of the contributing watershed, usually in the range 1.0–5.0%. A little arithmetic shows that this is equivalent to the range of hydraulic loading rates cited above for point-source treatment wetlands. In a moderate climate,

annual rainfall = 60 cm,
average annual rain rate = 60/365 = 0.164 cm d^{-1},
watershed runoff coefficient = 0.75,
runoff = 0.75 × 0.164 = 0.123,
watershed:wetland area ratio (WWAR) = 1/0.04 = 25 (for 4% of the watershed),
average annual wetland HLR = 25 × 0.123 = 3.1 cm d^{-1}.

Because the average annual HLR is close to the mode of the distribution of HLRs for point-source-driven marshes (Kadlec & Knight 1996), it is reasonable to expect that stormwater wetlands designed in this way would perform somewhere near the average for the emergent marsh database set. For instance, the mean decrease in P in 50 NADB marsh cells was 57% at an average HLR of 4.2 cm d^{-1}; the mean decrease for several constructed stormwater marshes, with an average WWAR of 4.3%, was also 57% (Strecker *et al.* 1992).

A stormwater wetland can also be sized to contain a specific volume of water, usually the volume associated with a rain event of a specified return frequency (probability of occurrence). For instance, Schueler (1992) suggests that the wetland should have sufficient volume to contain fully any rain event up to the 90th centile of the rainstorm quantity distribution. Again, this can be shown to match the loading and detention design ranges used for point-source wetland systems.

In the vicinity of Washington, DC, USA, there is an annual rainfall of 104 cm; the 90th-centile storm is 3.18 cm:

watershed area = 40 ha = 400,000 m^2,
watershed runoff coefficient = 0.75,
design storm runoff volume = 0.75 × 0.0318 × 400,000 = 9540 m^3,
wetland area at 0.3 m depth = 9540/0.3 = 31,800 m^2 = 3.18 ha,
WWAR = 100 × (3.18/40) = 8%,
annual flow = 0.75 × 1.04 × 400,000 = 312,000 m^3,
average annual detention time = 9540/312,000 = 0.03 yr = 11 d,
average annual wetland HLR = 312,000/31,800 = 9.8 m yr^{-1} = 2.7 cm d^{-1}.

This single-number design technique has the advantage of permitting a variable percentage of watershed, depending on the annual rainfall pattern and annual rainfall total. As in the case of WWAR design, the loading and detention times correspond to the mean values for point-source treatment wetlands. It is therefore not surprising that Schueler (1992) lists pollutant decreases that are in the mid range for other treatment wetlands. For instance, the TP removal is projected to be 45%, in comparison with the 57% mean for marshes in the NADB.

6.1.2 Hydraulic sizing

The calculations of the previous section set the area of the constructed wetland. It is further necessary to select the length (L) and width (W) for that area, as given by the aspect ratio, L/W. The presumption that high aspect ratios would favour a more efficient (close to plug flow) mode has proved to be untrue in many tracer tests of constructed wetlands. Consequently, any aspect ratio with good inlet distribution can be considered.

Aspect ratio is also a principal determinant of the hydraulic profile in the wetland. Flow through vegetation, or bed medium, creates a decreasing elevation of water surface from inlet to outlet. The decrease in water-surface elevation from inlet to outlet is the head loss for the system. The hydraulic profile must be contained properly in the wetland.

6.1.2.1 Surface-flow wetlands

The determinants of the hydraulic profile of a FWS wetland are:

- flow rate
- outlet weir setting
- aspect ratio (more generally, system planar geometry)
- bottom slope (more generally, vertical morphology)
- vegetation resistance.

These combine via Equations 4.7, 4.8 and 4.9 to produce the calculated water surface profile. Unfortunately, there is no closed-form easy solution to these equations; numerical methods are needed. Simple cases have been reduced to graphical representations (Kadlec & Knight 1996). The maximum head loss will occur for the maximum expected flow.

Recommended friction coefficients for Equation 4.8 are (notation in §4.1.2.1):

- $a = 10^7\ \mathrm{m^1\ d^{-1}}$ (dense vegetation)
- $a = 5 \times 10^7\ \mathrm{m^{-1}\ d^{-1}}$ (sparse vegetation)
- $b = 3$
- $c = 1$.

A simple, approximate calculation can be made for rectangular wetlands with flat, horizontal bottoms, with the resulting approximate criterion for a 20% depth increase from outlet to inlet:

if $$qL^2/ah_o^4 < 0.2, \tag{6.11}$$

then $$h_i/h_o < 1.2. \tag{6.12}$$

A second useful approximation for this simple situation is based on *normal depth* of flow, which separates distance-thinning and distance thickening flows:

$$h_n = \frac{Q/W}{a(\mathrm{d}B/\mathrm{d}x)^c}, \tag{6.13}$$

where h_n is the normal depth for the given bottom slope. If the exit weir is set above this depth, the flow will be distance thickening; if below, distance thinning.

6.1.2.2 Subsurface horizontal flow wetlands

The requirements for stable and controllable water flow and for proper vegetation conditions serve to restrict the geometry of the bed and the size of the medium.

These requirements are:

(a) Expected flows must pass through the bed without overland flow or flooding.
(b) Expected flows must pass through the bed without stranding the plants above water; i.e. there must not be protracted, excessive headspace.
(c) Operation should remain acceptable in the likely event of changing hydraulic conductivity. As the bed clogs with roots and sediments, it should not flood.
(d) The bed should be drainable.
(e) The bed should be floodable.
(f) Water levels within the system should be fully controllable through the use of inlet and outlet structures.
(g) The configuration must fit the site, in terms of project boundaries and in terms of hydraulic profiles.

Such constraints must be met for all expected operating conditions, including initial and clogged conductivity, and the range of expected operating flows, including daily maximum and minimum values.

Bed depth (δ) is usually in a narrow range and is set by conditions other than hydraulics. There is no theoretical need for a slope to the top of the bed. If control of water level is to include the ability to inundate the bed totally for vegetation management, then a top slope is detrimental. The bed depth is usually selected to be in the range of 30–60 cm, based on assumptions on plant rooting depth and its effect on treatment potential. Such depth 'criteria' remain speculative. However, ice formation can use some of the water depth, and there needs to be some room for sediment accretion in the bottom of the bed. The upper half of the range, 45–60 cm, therefore seems to be the best choice.

The determinants of the hydraulic profile of a horizontal SSF wetland are:

- flow rate
- outlet weir or standpipe setting
- aspect ratio (more generally, system planar geometry)
- bottom slope (more generally, vertical morphology)
- media resistance (hydraulic conductivity).

These combine via Equations 4.7, 4.9 and 4.11–4.14 to produce the calculated water surface profile. Unfortunately, there is no closed-form easy solution to these equations; numerical methods are needed.

Bed slope criterion

The bottom slope should be set to provide for complete bed drainage. Normally, a few centimetres of elevation differential allow for this requirement. *Bottom slope should not be considered as the design driving force for water movement.* The reason is that designs based on bed slope are excessively sensitive to changing conditions of flow and hydraulic conductivity; dryout or flooding is virtually certain to occur with such designs.

The bottom of the bed should be slightly sloped from inlet to outlet to provide for drainage. However, if there is too great a distance from the bed top to the water, plant roots cannot reach the water. This excessive headspace is prevented by limiting the bed slope. For instance, a value of 10% of the bed depth might be chosen, on the basis of the desire to use most of the bed depth for treatment and to encourage proper rooting. The worst-case water surface is the level pool created by low flows. This criterion is purely geometrical:

$$\Delta B = B_i - B_o, \quad (6.14)$$

$$\Delta B < 0.1\delta. \quad (6.15)$$

Loading criterion

Excessive loading or low hydraulic conductivity leads to excessive gradients of the water surface and flooding. A limit should be placed on the head loss through the bed, for instance a value of 10% of the bed depth. This is also a geometrical constraint:

$$\Delta H = H_i - H_o, \quad (6.16)$$

$$\Delta H < 0.1\delta. \quad (6.17)$$

Design should not be in the regime of severe distance thickening or thinning flows. Therefore, to a first approximation, the linear version of the friction equation (Equation 4.13) applies for average water depth and hydraulic conductivity:

$$\Delta H = \frac{QL}{\tilde{k}\tilde{h}W} < 0.1\delta, \quad (6.18)$$

where tildes indicate spatially averaged values. To prevent flooding, under this criterion, the outlet level should be controlled below the bed surface.

Operational hydraulic conductivity

Hydraulic conductivity can easily be measured in field tests (Kadlec & Knight 1996). Otherwise, Equation 4.14 provides a rough estimate of the hydraulic conductivity for clean, unrooted medium. The operational conductivity of the front end of the bed has been found to decrease with time to about one-tenth of that value because of biomass and other clogging (Fisher 1990). As a consequence, it is prudent to presume that the operational conductivity is:

$$k = 0.1k_{clean}. \quad (6.19)$$

The variability of k with time and distance plus the uncertainty of an estimate combine to underscore the unreliability of designs based on bed slope and hydraulic conductivity.

Control of the water surface should be designed to be determined nearly entirely by the outlet depth setting at the standpipe or weir.

6.1.2.3 Subsurface vertical flow wetlands

Because the flow is vertical, the aspect ratio is no longer a determinant of hydraulics, but the conductivity of the medium becomes more important, especially during the drainage portion of the cycle. The considerations of flow and saturation are complicated, and they have not yet been reduced to design guidelines.

6.2 System layout

Pre-existing topographic, geological and soil chemistry conditions can greatly affect the cost and performance of a wetland. Excessive site topography creates large earthwork volumes for a given wetland area, significantly increasing the construction cost of a wetland. Surface and subsurface geological conditions can also increase costs by requiring the removal of rock or by resulting in the need for liner materials to decrease groundwater exchanges.

Given the total required wetland area and the concepts of system configuration, there still remains the placement of the wetland on the site. The principal considerations are adaptation to the boundaries and contours of the site, minimization of inter-cell conveyance and minimization of earthmoving.

Site boundaries often determine the external shape of the overall system because there is often neither extra land nor the ability to choose the shape of the available land. In that event, the various pieces of the overall system must conform to the space available. The topology of the conceptual layout is retained, but shapes and perhaps areas are sacrificed. The available lands are likely to be bounded by streams, roads, railways and ownership boundaries. As a consequence, the actual layout might not be completely rectangular; neither might all the cells be in close proximity.

6.3 Compartmentation

At this point in the design procedure, the size and shape limits for the wetland system have been determined. There is next a need to set the compartmentation of the FWS or SSF system. The number of wetland cells in the design of treatment wetlands is based on considerations of redundancy, maintenance and topography. Constructed wetland treatment systems should have at least two cells that can operate in parallel to permit operational flexibility. Having at least two parallel cells is especially important because of unexpected

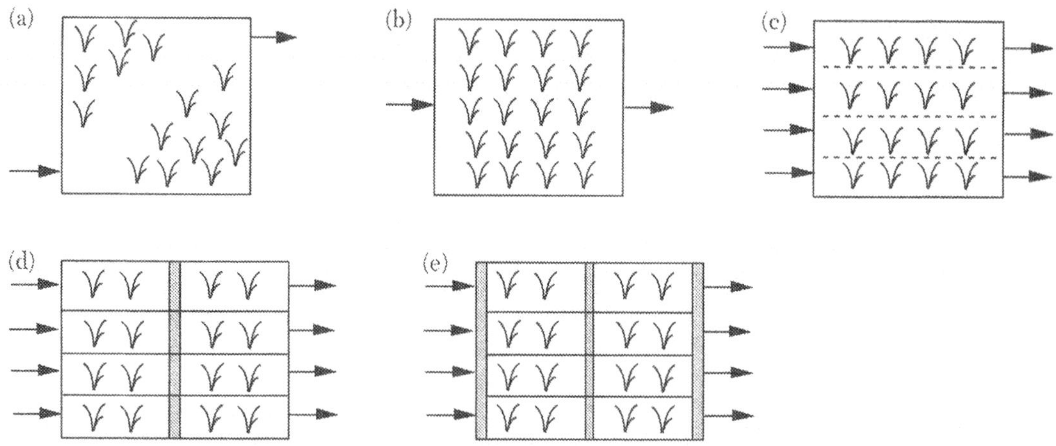

Figure 6.4. Configurations of wetland system elements (modified from Kadlec & Knight 1996): (a) a bad configuration (preferential flow channel from inlet to outlet); (b) a poor configuration (large corner zones not in flow path); (c) a better configuration (multiple inlets and flow-control berms); (d) a still better configuration (separation dikes and redistribution); (e) another very good configuration (deep zones for distribution, redistribution and collection). (Modified from Kadlec & Knight (1996).)

events such as vegetation die-off, pretreatment failures and subsequent wetland contamination, and failures of berms or other structures. Multiple flow paths allow the loading rate to be manipulated to meet varying inflow water quality. In addition, parallel flow paths allow cells to be drained for replanting, rodent control, harvesting, burning, leak patching or other possible operational controls. In the extreme long term, the replacement of structures and piping becomes necessary. Alternative conceptual compartmentations are shown in Figures 3.1 and 6.4.

The number of cells required must be determined by evaluating the cost of more cells (the ratio of berm area to wetland surface area increases with more cells), site constraints where sloped ground mandates a terraced multi-cell design, and operational flexibility to isolate various fractions of the total wetland treatment area. For example, with two cells, half of the treatment area must be shut off to conduct any maintenance, but with five cells, as little as 20% of the treatment area must be turned off. Large systems can profitably incorporate more than two flow paths, for purposes of internal flow control. However, a multiplicity of inlet and outlet control structures can add significant cost to the overall project.

6.4 Pond zones

Deep-water zones are advantageous for the collection of large amounts of sediments because they provide extra space for collection and are easier to clean out. As a result, forebays are recommended when the incoming water has a high TSS load.

Deep-water zones can become dominated by planktonic algae, which contribute to TSS. Consequently, large open water zones should not be the final element in the constructed wetland complex.

Deep cross-zones in SF constructed wetlands serve several purposes (Knight & Iverson 1990). These deeper areas extend below the bottom of the vegetated basin areas by at least 1 m to exclude the development of rooted macrophytes. Unvegetated cross-ditches provide a low-resistance path for water to move laterally and re-establish a constant head across the wetland. They also provide for extra detention time, but in a deep-water zone. Such ditches often become covered with duckweed (*Lemna* spp.) and can be used by wetland birds and fish as reliable habitat. These redistribution ditches materially change the overall degree of mixing within the wetland because high-speed rivulets are intercepted and mixed with slower-moving water. However, the redistribution ditch adds a potential for wind mixing that compensates for the decreased short-circuiting. Water is more effectively distributed over the wetland, improving the gross areal efficiency (Knight *et al.* 1994).

6.5 Sealing the basin

Constructed wetlands can require sealing to prevent the contamination of groundwater or to prevent groundwater from infiltrating into the wetland. In general, FWS treatment wetlands providing advanced wastewater treatment do not pose a threat to groundwaters and do not need to be lined. SSF wetlands providing secondary treatment are generally lined to prevent direct contact between the wastewater and groundwater.

The effects of leaky SF treatment wetlands on groundwater have been documented for a few systems (Kadlec & Knight 1996; Knight & Ferda 1989). General findings for secondary wastewaters discharged to sandy soils are that concentrations of NO_3-N and faecal coliforms reaching the shallow groundwater are very low and are not likely to be a problem. Additional information about this issue needs to be collected and summarized to determine when liners are necessary to protect groundwater resources.

Where on-site soils or clay provide an adequate seal, compaction of these materials can be sufficient to line the wetland. Sites underlain by karst, fractured bedrock, or gravelly or sandy soils will have to be sealed by some other method. It might be necessary to have a laboratory analyse the construction material before choosing a sealing method. Soils that contain more than 15% clay are generally suitable. Bentonite, as well as other clays, provide adsorption/reaction sites and contribute alkalinity. The SCS (now the NRCS) South National Technical Center (SNTC) Technical Note 716, 'Design and construction guidelines for considering seepage from agricultural waste storage ponds and treatment lagoons' (1993) and its companion SNTC Technical Note 717, 'Measurement and estimation of permeability of soils for animal waste storage facilities' (1991), provide guidance in determining when soils *in situ* will meet seepage control needs.

Synthetic liners include asphalt, synthetic butyl rubber, and plastic membranes (for example, 0.5–10.0 mm high-density polyethylene). The liner must be strong, thick and smooth to prevent root attachment or penetration. If the site soils contain angular stones, then sand bedding or geotextile cushions should be placed under the liner to prevent punc-tures. A synthetic liner in an SSF system will also typically be covered by a geofabric to prevent punctures.

Liners in FWS wetlands should, if necessary, be covered with 6–12 inches of soil to prevent the roots of the vegetation from penetrating the liner. If the wetland is to be used for mine-drainage treatment, the reaction of the clay or synthetic liner should be tested before it is used, because some clays and synthetics are affected by some acid-mine drainages.

The bottom of the wetland, as well as the core of containment dikes, can be formed of compacted clays or bentonite. Locally available clays are preferred from the standpoint of cost reduction. Plastic liners might be feasible for smaller wetlands. This clay layer, or other sealant, should not be penetrated by plant roots, so that it retains its integrity. Wetland cells might need to be lined with clay or plastic if regulatory requirements prohibit mixing with groundwater or if natural infiltration rates will make it difficult to maintain surface-water wetland conditions.

Most of the systems in the UK have used a plastic liner or membrane such as high-density polyethylene or low-density polyethylene. The liner most often used has been Monarflex (low-density polyethylene with glass fibre reinforcement) 0.5–0.75 mm thick.

Recently a number of systems have been built with liners made from bentonite and a geotextile such as Fibertex. The advice given in the European Guidelines (Cooper 1990) was that if the local soil had a hydraulic conductivity of 10^{-8} m s^{-1} or less, it was likely that it contained a high clay content and could be 'puddled' to provide adequate sealing for the bed.

6.6 Substrate selection

6.6.1 Surface flow systems

The topsoil from the site should be stockpiled and replaced within the wetland to form a rooting medium. The roots and rhizomes of emergent macrophytes such as cattails, bulrushes and common reed usually occupy the top 30–40 cm of the soil column. A layer of that thickness should therefore be used. The original topsoil from the site might be usable; if so, it should be stockpiled separately from the other soils during construction. This topsoil will contain seeds of the wetland plants of the region, which can assist in vegetating the wetland. If topsoil is not available at the site, it might need to be imported to optimize plant survival and growth.

Wetland plant growth and survival is also dependent on environmental factors other than water regime. Two of these factors include soil texture and soil chemistry. Many plants grow most rapidly in soils of sandy to loamy texture. Excessive rock or clay can retard plant growth and result in mortality. Excessively acidic or basic conditions can limit the availability of plant growth nutrients. In some cases, soil concentrations of macronutrients or micronutrients might not be available in the native soil to allow adequate initial establishment of plants.

6.6.2 Subsurface-flow systems

SSF wetlands have been designed and built with substrates ranging from fine textured soil to 30 cm fieldstone. Very small particles have very low hydraulic conductivity and create surface flow. Very large rocks have high conductivity, but have little wetted surface area per unit volume for microbial habitat. Large and angular medium is inimical to root propagation. The compromise is for intermediate-sized materials, generally characterized as gravels.

6.6.2.1 Horizontal-flow beds

The advice given by root-zone designers in 1985 to the UK group that visited Germany was that fully developed RZM beds built with soil would have a hydraulic conductivity of 3×10^{-3} m s^{-1} (260 m d^{-1}) (Boon 1985). This has not been borne out by experience in the UK or other European countries, leading to the advice given in the European Guidelines of 1990 (Cooper 1990) 'not to assume a hydraulic conductivity greater than that of the original media'. This advice has been followed over the past 5 years and it is still very important. Unfortunately a number of systems built from 1985 to 1989 used soil for which it was assumed that the hydraulic conductivity would increase. Some of these beds suffered from surface flow; this led to channelling and scouring of the surface, which resulted in areas of the bed being starved of water, leading in turn to poor reed growth. It also led to by-passing and poor treatment. Similar problems occurred with plants built in Germany and Denmark. As indicated in Section 6.1.2.2, the reverse is likely to occur: clogging of the clean medium.

As a result of these problems, WRc decided in 1986/87 to recommend the use of gravels in UK systems at Little Stretton (Severn Trent Water) and Gravesend (Southern Water) because this would allow through-flow of water from the start. It was postulated that if the gravel beds filtered out solids and started to block the voids, this might be counter-balanced by the roots and rhizomes opening up the bed. This change has been very successful: a large number of gravel beds have been built and the operators are happy with the way in which they have performed. The oldest gravel beds are now 10–11 yr old, and none of them has yet become blocked.

Typical gravel sizes are 3–6 mm, 5–10 mm and 6–12 mm. The most frequently used size fraction in Europe is 8–16 mm. It is recommended that the gravels are washed because this removes fines that could block the void spaces. Most gravels used have been washed river gravels, usually silica quartz, but broken limestone has also been used successfully. Crushed rock is also used in some European countries.

A number of specialty media have been tested. One system used a waste product pulverized fuel ash from a coal-fired power station (Dickson 1995). Light expanded clay aggregates have been tested in Norway because of their high sorption capacity for P. Broken glass is in place in a secondary system in Washington State, USA (Kadlec 1998).

6.6.2.2 Vertical-flow beds

Conventional wisdom on intermittent sand filters suggests a clean washed sand with an effective size of 0.25–0.5 mm with a uniformity coefficient of approx. 3.5 (Metcalf & Eddy 1991). VF beds use layers of graded gravel, usually with a top layer of 'washed sharp sand'. In the UK the specification of graded gravel has been that used by Burka at Oaklands Park (Burka & Lawrence 1990). This is:

	Depth	
top layer	8 cm	'sharp sand'
	15 cm	6 mm washed pea-gravel
	10 cm	12 mm round washed gravel
bottom layer	15 cm	30-60 mm round washed gravel

In addition, large stones were placed around the agricultural drainage pipe that formed the underdrain system.

6.7 Inlet and outlet structures

6.7.1 Inlets

Inlets at FWS wetlands are usually simple: an open-ended pipe, channel or gated pipe that releases water into the wetland (Figure 6.5). As the *L/W* ratio decreases, equal flow distribution becomes more important. Accessible and easily adjustable inlets are mandatory for systems with small *L/W* ratios. For systems intended to operate under ice in winter, the inlet distribution system must be placed below the ice line.

Inlet structures at SSF systems include surface and subsurface manifolds (such as a perforated pipe 150 mm in diameter), open trenches perpendicular to the direction of flow, and simple single-point weir boxes (Figure 6.6). A subsurface manifold avoids the build-up of algal slimes and the consequent clogging that can occur next to surface manifolds, but it is difficult to adjust and maintain. Subsurface distribution is mandatory in northern environments, to accommodate the formation of frost and ice. A surface manifold, with adjustable outlets, provides the maximum flexibility for future adjustments and maintenance, and is recommended if the system is to operate only under ice-free conditions. A surface manifold also avoids back-pressure problems. The distance above the water surface of the wet-land is typically 12–24 cm. The use of coarse rock (8–15 cm) in the entry zone ensures rapid infiltration and prevents ponding and algal growth. To discourage the growth of algae, open water areas near the outlet should be avoided. Shading with either vegetation or a structure in the summer and some thermal protection in the winter will probably be necessary.

A flow splitter will be needed for parallel cells. A typical design consists of a pipe, flume,

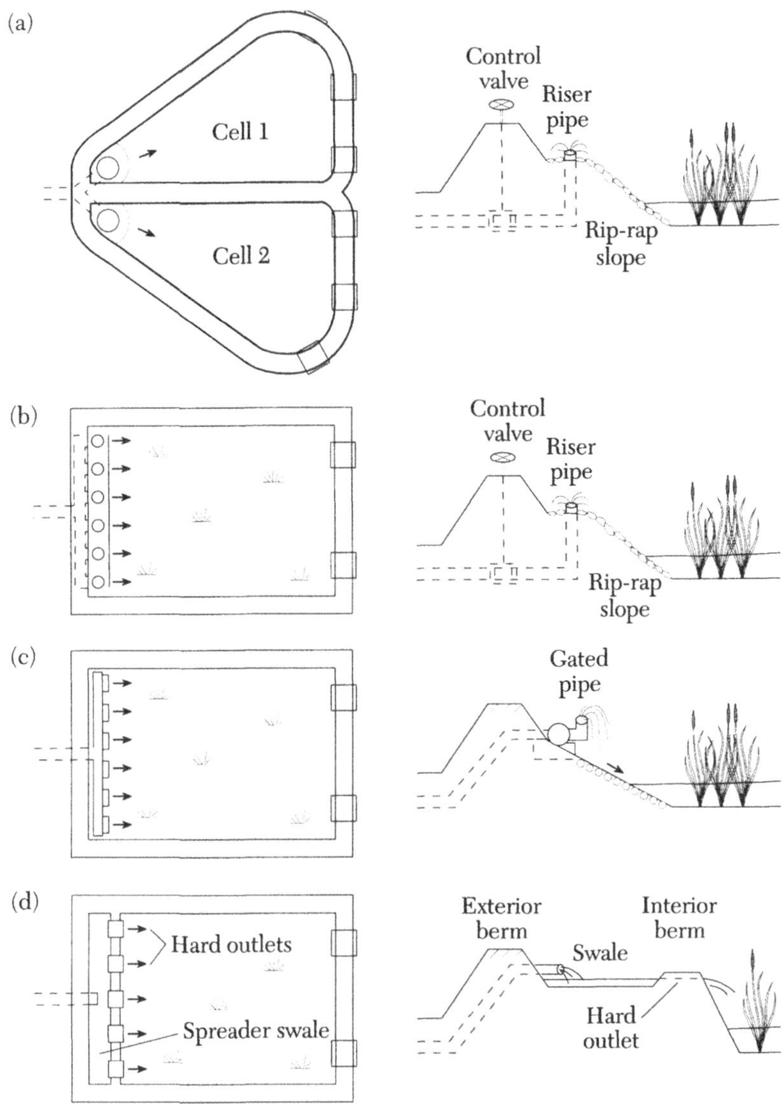

Figure 6.5. Configurations of FWS treatment wetland inlet: (a) point discharge; (b) multiple-point discharge; (c) gated pipe; (d) level spreader swale. (From Kadlec & Knight (1996).)

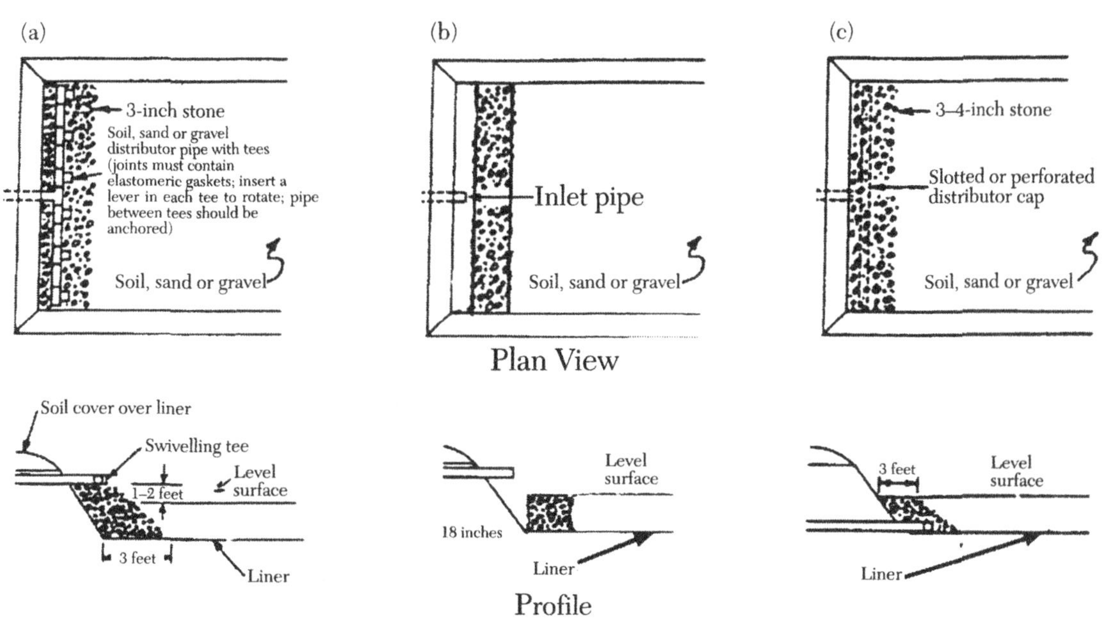

Figure 6.6. Inlet configurations for SSF wetlands: (a) inlet with swivelling tees; (b) inlet with gabion; (c) inlet with buried distributor pipe. (Davis (1995).)

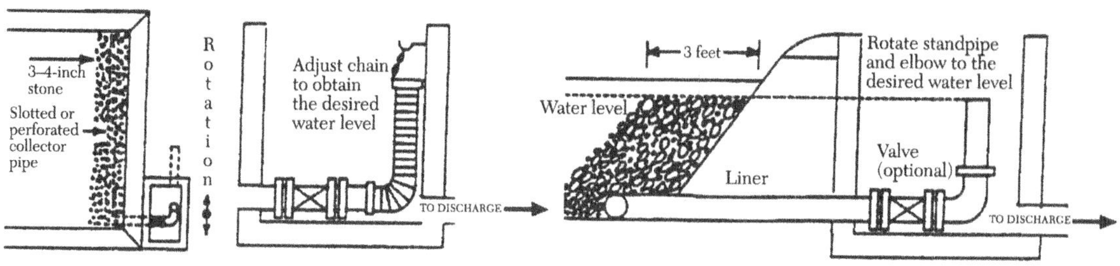

Figure 6.8. Outlet control structures for SSF wetlands: (a) plan view of outlet structure with swivelling standpipe; (b) central structure with collapsible tubing; (c) central structure with swivelling standpipe. (Davis (1995).)

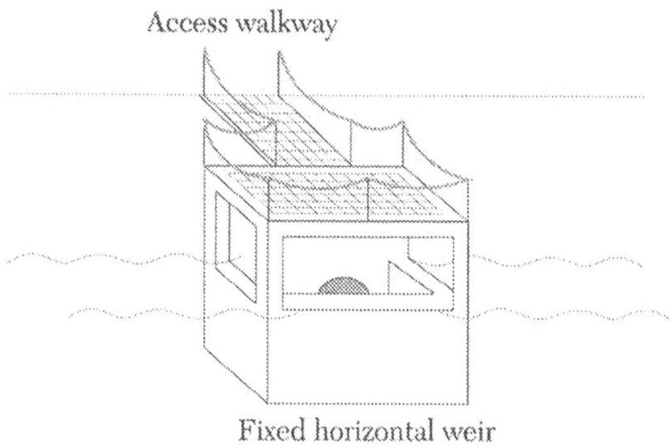

Figure 6.7. Example of a constructed FWS wetland outlet weir. (From Kadlec & Knight (1996).)

or weir with parallel orifices of equal sizes at the same elevation. Valves are impracticable because they require daily adjustment. Weirs are relatively inexpensive and can easily be replaced or modified. Flumes minimize clogging in applications with high solids but are more expensive than weirs.

6.7.2 Outlets

At FWS wetlands, the water level is controlled by the outlet structure, which can be a weir, spillway or adjustable riser pipe. A variable-height weir, such as a box with removable stoplogs, allows the water levels to be adjusted easily. More sophisticated structures might be desirable on large systems (Figure 6.7). Skimmer boards and debris fences are required to prevent floating litter from clogging the outlet.

Large short-duration flows can occur in treatment wetlands owing to extreme rainfall events. Weirs and spillways must be designed to pass the maximum probable flow. Spillways should consist of wide cuts in the dike with gentle side slopes, no steeper than 2:1 (horizontal:vertical) and lined with non-biodegradable erosion control fabric. If high flows are expected, coarse rip-rap should be used. Vegetated spillways overlying erosion control fabric provide the most natural-looking and stable spillways. Weirs or spillways should be used for mine-drainage wetlands because pipes tend to clog with deposits of iron precipitates.

At SSF wetlands, outlets include subsurface manifold and weir boxes or similar gated structures. The manifold should be located just above the bottom of the bed to provide complete control of water level, including draining. The use of an adjustable outlet, which is recommended to maintain an adequate hydraulic gradient in the bed, can also have significant benefits in operating and maintaining the wetland. Adjustable riser pipes or flexible hoses anchored by a chain offer simple control of water level (Figure 6.8). A PVC elbow attached to a swivel offers easy control of the water level. If pipes are used, small-diameter pipes should be avoided because they clog with litter.

The surface of the bed can be flooded to encourage the development of newly planted vegetation and to suppress undesirable weeds, and the water level can be lowered in anticipation of major storms and to provide additional thermal protection against freezing in the winter. The design of SSF beds should allow controlled flooding to 6 inches (15 cm) to foster desirable plant growth and to control weeds. A perforated subsurface manifold connected to an adjustable outlet offers the maximum flexibility and reliability as the outlet device for SSF systems. Because the manifold is buried and inaccessible after construction, careful grading and sub-base compaction are required during construction, and clean-out risers in the line must be provided.

The final discharge point from the wetland system should be placed high enough above the receiving water for a rise in the water level in the receiving water, for instance after a storm, not to interfere with the flow of water through the wetland.

7 Plants and planting

7.1 Varieties of vegetation

Constructed wetlands can be planted with a number of adapted, emergent wetland plant species (Table 7.1). Wetlands created as part of compensatory mitigation or for wildlife habitat typically include a large number of planted species. However, in constructed wetland treatment systems, diversity is typically quite low.

7.1.1 Surface-flow wetland vegetation

The selection of plant species for wetlands should consider the following variables: expected water quality, normal and extreme water depths, climate and latitude, maintenance requirements and project goals. At present there is no clear evidence that treatment performance is superior or different between the common emergent wetland plant species used in treatment wetlands. The best selection criteria are growth potential, survivability and cost of planting and maintenance. It is clear that densely vegetated areas are more effective at treating pollutants than are sparsely vegetated areas. A corollary to this observation is that plant species that provide structure year-round perform better than species that die below the water line after the onset of cold temperatures. For these reasons, fast-growing emergent species that have high lignin contents and that are adapted to variable water depths are the most appropriate for constructed wetland treatment systems. Wetland plant genera that most successfully meet these criteria include *Typha*, *Scirpus* and *Phragmites* (Figure 7.1).

Only a small fraction of the ultimate plant density is planted in the new wetland. Planting densities range from 1000 to 25,000 plants ha^{-1}. Through vegetative reproduction, these plants spread to shoot densities of more than 1,000,000 plants ha^{-1}. As the first round of plants mature and die, rhizomes send up new shoots, thus maintaining the wetland plant community. Most constructed wetlands also have colonizing plant species around the shallow edges and in unvegetated areas inside the cells. Although these colonizers typically do not provide much cover, they do provide some buffer against plant pathogens, provide habitat diversity important to wildlife, and fill niches that the dominant plant species might otherwise not occupy.

The performance of an FWS treatment wetland is not sensitive to the particular plant species that populate the wetland. It is difficult to sort this effect from other phenomena in most wetland treatment systems, but there are some side-by-side wetland comparisons that give strong indications of this lack of sensitivity. The facility in Tarrant County, Texas, USA, is one such source of evidence. Three separate trains of three wetlands each were geometric and hydraulic replicates. However, different plant species were established in the three trains: train 1 was bulrush, cattail, arrowhead and smartweed; train 2 was softrush, pondweed and water primrose; train 3 was natural regrowth, including Colorado River hemp, arrowhead, reed canarygrass and smartweed. After a one-year startup period, which produced full vegetative cover, there were no measured differences in performance.

It is likely that small performance differences exist that are due to vegetation type in FWS wetlands, but these are often masked by other unavoidable differences in comparison wetlands. At the time of writing, the superiority of a particular plant species has not been proved or disproved; the evidence points towards minimal differences.

From the standpoint of system resiliency, the wetland should probably contain a diverse mix of macrophyte species and thus be in a position to accommodate changes in water quality and timing that might occur. In other words, a polyculture is preferable to a monoculture. Most FWS treatment wetlands undergo a process of alteration after an initial planting, with the more robust species gaining dominance, typically cattails, *Phragmites* and bulrushes. However, in ultra-polishing systems, with very high water quality, a very diverse species composition can develop (Schwartz 1992).

If the wetland is to be planted, the cost and availability of plant materials must be add-

Table 7.1. Approximate hydroperiod and depth tolerance for emergent, herbaceous wetland plants capable of continuous inundation

Scientific name	Common name	Maximum water depth (m)	Flooding duration (%)
Alternanthera philoxeroides	Alligator weed	0.1–1.0	70–100
Canna spp.	Canna lilies	<0.05–0.25	50–100
Carex spp.	Sedges	<0.05–0.25	50–100
Ceratophyllum spp.	Coontail	>3	75–100
Cladium jamaicense	Sawgrass	0.1–0.25	50–100
Colocasia esculenta	Wild taro	0.1–0.5	25–100
Cyperus spp.	Sedges	<0.05–0.50	50–100
Eleocharis spp.	Spikerushes	<0.05–0.5	50–100
Elodea spp.	Waterweed	>3	75–100
Glyceria spp.	Mannagrass	<0.05–0.30	0–100
Hydrocloa caroliniensis	Watergrass	<0.05–1.0	75–100
Iris spp.	Iris or blue flag iris	<0.05–0.2	50–100
Juncus spp.	Rushes	<0.05–0.25	50–100
Lemna spp.	Duckweed	None	75–100
Ludwigia spp.	Water primroses	0.1–0.5	70–100
Panicum hemitomon	Maidencane	0.1–0.3	50–100
Panicum repens	Torpedo grass	<0.05–0.5	50–100
Peltandra spp.	Spoon flowers	<0.05–0.25	50–100
Phalaris arundinacea	Reed canarygrass	<0.05-0.30	13–100
Phragmites australis	Common reed	<0.05–0.5	70–100
Polygonum spp.	Smartweeds	<0.05–0.25	50–100
Pontederia spp.	Pickerelweeds	0.1–0.25	70–100
Rhynchospora spp.	Beak-rush	<0.05–0.5	50–100
Sagittaria spp.	Arrowheads	0.2–0.5	50–100
Saururus cernuus	Lizard's-tail	<0.05–0.2	50–100
Scirpus spp. (*Schoenoplectus*)	Bulrush	0.1–1.5	75–100
Sparganium spp.	Bur-reed	0.1–0.5	70–100
Sphagnum spp.	Sphagnum mosses	<0.05–0.1	75–100
Thalia geniculata	Arrowroot	0.1–0.75	70–100
Typha spp.	Cattail, reedmace, bulrush	0.1–0.75	70–100
Zizania aquatica	Wild rice	0.1–1.0	70–100
Zizaniopsis milacea	Southern wild rice	0.1–1.0	70–100

ressed early in the design process. The option of establishing an on-site wetland plant nursery must be decided very early because mature, 1-2-year-old plants are preferred. These have the energy reserves to survive the transplanting operation. Consequently, the establishment of the nursery must be complete well in advance of other construction.

Another option is to allow natural regrowth of the wetland basins. In southern climates, this process is complete within one growing season, but it can require two or more seasons in northern climates. In either climate, the option of transplanting will accelerate the establishment of vegetation. The design decision is based on economics and the regulatory requirements on start-up. A one-year delay in the imposition of permit requirements allows for natural regrowth and can save a large amount of money.

7.1.2 Subsurface-flow wetland vegetation

The three genera of wetland plants that are most frequently used in SF wetland treatment systems are also used in SSF wetlands. Commonly used plants are *Phalaris arundinacea* (reed canarygrass), *Typha* spp. (cattails), *Scirpus* spp. (bulrushes) and *Glyceria maxima* (sweet mannagrass). However, the most frequently used plant species worldwide is *Phragmites australis* (common reed). This species has remarkable growth rates, root development and tolerance to saturated soil conditions. They are known to provide some ancillary benefits in terms of wildlife habitat in the UK (Merritt 1994).

Phragmites is planted by using rhizomes, seedlings or field-harvested reeds. All of these techniques are effective if the plants are healthy and if adequate (but not excessive) soil moisture is maintained during plant establish-

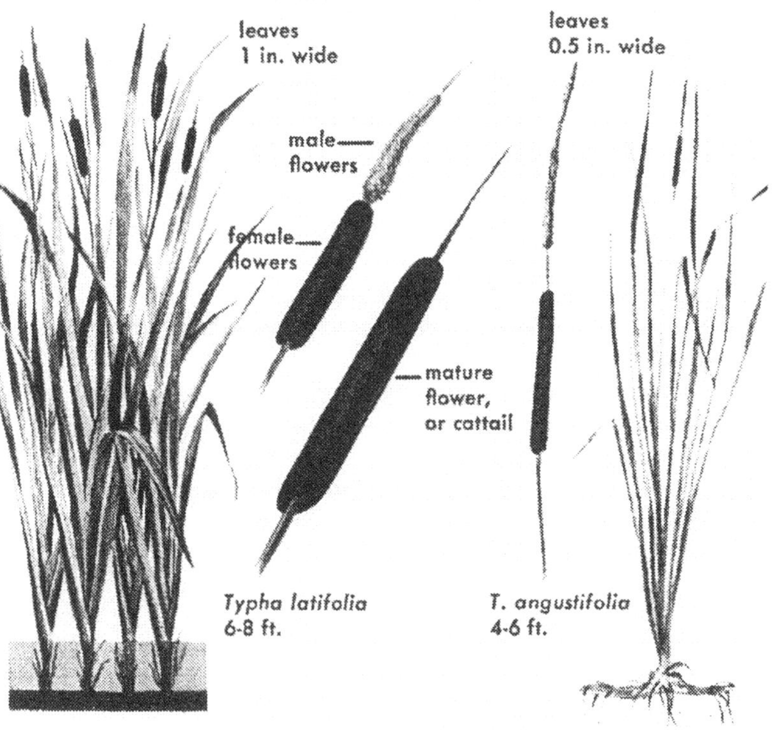

Figure 7.1. The above-ground structure of Typha. Plants are typically 2–4 m tall. (Reid (1987).)

ment. Planting densities between 2 and 6 m^{-2} (20,000–60,000 ha^{-1}) are normally recommended for *Phragmites* (Cooper 1990; ATV 1989).

A gravel bed will require planting because seed banks are typically lacking and the medium is not optimal for germination. If a portion of the bed remains flooded, a litter layer can develop that is conducive to the germination of wetland plant seeds, thus permitting invasion. More frequently, a portion of the bed can remain too dry, permitting invasion by terrestrial species (weeds).

The presence of macrophytes is important for many, if not all, pollutant-removal functions in SSF wetlands also. However, the question of which plant might be best has not yet been resolved. The results of various side-by-side investigations are inconclusive, as will be discussed for ammonium removal. The project at Santee, California, USA, ranked *Scirpus* best, *Phragmites* second and *Typha* a distant third, close to no plants (fourth) (Gersberg *et al.* 1984). The project at Lake Buena Vista, Florida, USA, ranked *Sagittaria* better than *Scirpus* (DeBusk *et al.* 1989). The project at Hamilton, New Zealand, ranked *Glyceria* better than *Schoenoplectus* (*Scirpus*) better than no plants (van Oostrom & Cooper 1990). The project at Pretoria, South Africa, ranked *Phragmites* better than *Scirpus* better than *Typha* for lagoon effluent, but *Scirpus* better than *Typha* better than *Phragmites* for settled sewage (Batchelor *et al.* 1990). Bavor *et al.* (1988) found very little difference between *Schoenoplectus* (*Scirpus*) and *Typha* and no plants at Richmond, NSW, Australia. At Hardin, Kentucky, USA, *Phragmites* was better than *Scirpus* (NADB 1993).

All of these results read like a set of football game results: the reader can use a sequence of his or her choice to prove that a particular plant is better than another, just as game scores can be used to establish one team's superiority.

7.2 Role of macrophytes in the treatment of wetlands

The macrophytes growing in constructed treatment wetlands have several properties in relation to the treatment processes that make them an essential component of the design. The most important effects of the macrophytes in relation to the wastewater treatment processes are the physical effects that the plant tissues give rise to (such as erosion control, filtration effect and the provision of surface area for attached microorganisms). The metabolism of the macrophytes (such as plant uptake and oxygen release) affects the treatment processes to different extents depending on design. The macrophytes have other site-specific valuable functions, such as providing a suitable habitat for wildlife and giving systems an aesthetic appearance. The major roles of macrophytes in constructed treatment wetlands are summarized in Table 7.2.

Table 7.2. Summary of the major roles of macrophytes in constructed treatment wetlands (from Brix 1997)

Macrophyte property	Role in treatment process
Aerial plant tissue	Light attenuation → reduced growth of phytoplankton Influence on microclimate → insulation during winter Reduced wind velocity → reduced risk of resuspension Aesthetically pleasing appearance of system Storage of nutrients
Plant tissue in water	Filtering effect → filter out large debris Reduce current velocity → increase rate of sedimentation, reduces risk of resuspension Provide surface area for attached biofilms Excretion of photosynthetic oxygen → increases aerobic degradation Uptake of nutrients
Roots and rhizomes in the sediment	Provide surface for attached bacteria and other microorganisms Stabilizing the sediment surface → less erosion Prevents the medium from clogging in VF systems Release of oxygen increase degradation (and nitrification) Uptake of nutrients Release of antibiotics

7.2.1 Physical effects

The presence of vegetation in wetlands distributes and decreases the current velocities of the water (Pettecrew & Kalff 1992; Somes*et al.* 1996). This creates better conditions for the sedimentation of suspended solids, decreases the risk of erosion and resuspension, and increases the contact time between the water and the plant surface areas. The macrophytes are also important for stablizing the soil surface in treatment wetlands, because their dense root systems impede the formation of erosion channels. In vertical flow systems the presence of the macrophytes, together with an intermittent loading regime, helps to prevent clogging of the medium (Bahlo & Wach 1990). The movements of the plants, as a consequence of wind and other factors, keep the surface open, and the growth of roots within the filter medium helps to decompose organic matter and prevents clogging.

The vegetation cover in a wetland can be regarded as a thick biofilm located between the atmosphere and the wetland soil or water surface in which significant gradients in different environmental parameters occur (Figure 7.2). Wind velocities are decreased near the soil or water surface in comparison with the velocities above the vegetation, which decreases the resuspension of settled material and thereby improves the removal of suspended solids by sedimentation. A drawback of decreased wind velocities near the water surface is, however, the decreased aeration of the water column (Figure 7.2a).

Light is attenuated, hindering the production of algae in the water below the vegetation cover (Figure 7.2b). This property is used in duckweed-based systems, as algae die and settle out beneath the dense cover of duckweed (Ngo 1987). Another important effect of the plants is the insulation that the cover provides during winter, especially in temperate areas (Smith *et al.* 1996). When the standing litter is covered by snow it provides a perfect insulation and helps to keep the soil free of frost (Figure 7.2d). The litter layer also helps to protect the soil from freezing during winter; however, it also keeps the soil cooler during spring (Haslam 1971a, b; Brix 1994).

7.2.2 Effects on soil hydraulic conductivity

In constructed wetlands with subsurface horizontal water flow, the flow of water in the bed is intended to be largely subsurface through channels created by the living and dead roots and rhizomes, as well as through soil pores. As the roots and rhizomes grow they disturb and loosen the soil. Furthermore, when roots and rhizomes die and decay, they can leave behind tubular pores and channels (macropores), which are thought by some to increase and stabilize the hydraulic conductivity of the soil (Boon 1985). The structure of the macropore system is dependent on the plant species and the conditions of growth, and it can be very effective in channelling water through a soil bed (Beven & Germann 1982). Claims have been made (Boon 1985) that after a period of 5 years (five full growing seasons) any soil will develop a hydraulic conductivity of 10^3 m s^{-1}. Therefore, the hydraulic dimensioning of con-

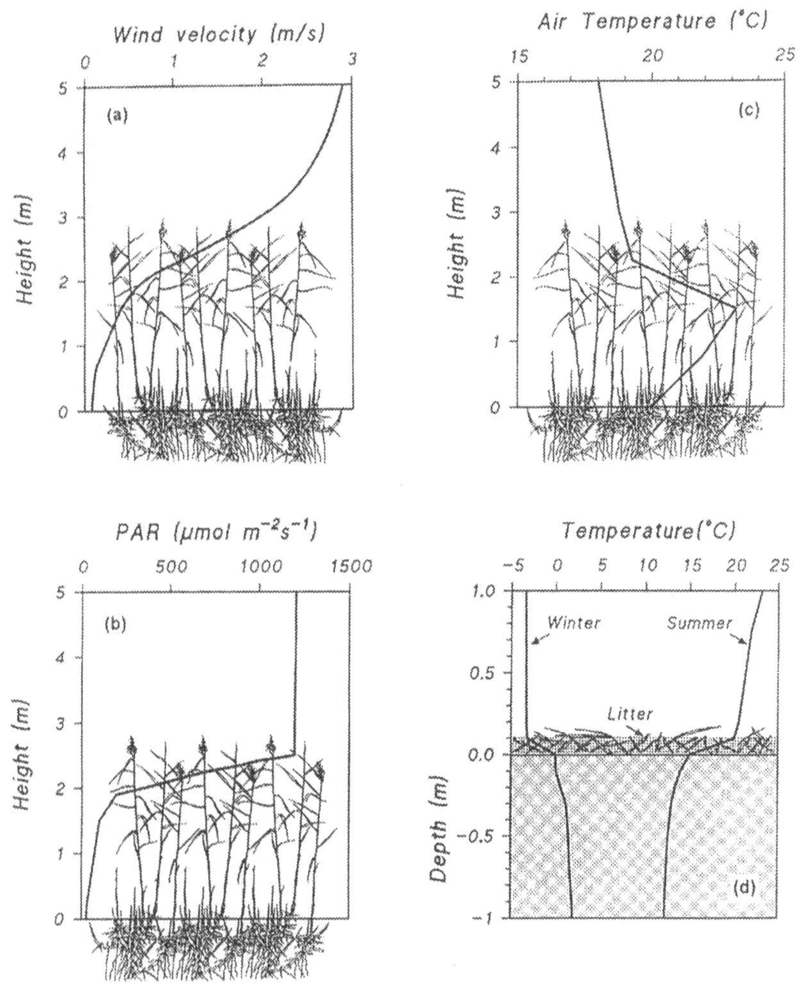

Figure 7.2. Effects of a dense canopy of Phragmites australis on (a) the wind velocity, (b) the incident light intensity and (c) air temperature during summer, and (d) effects of the litter layer on the soil temperature during winter and summer, respectively. (Modified from Brix (1994).)

structed wetlands with subsurface flow should not be based on the assumption that the hydraulic conductivity will increase as a consequence of root and rhizome growth.

7.2.3 Surface area for attached microbial growth

The stems and leaves of macrophytes that are submerged in the water column provide a huge surface area for biofilms (Gumbricht 1993*a*, *b*; Chappell & Goulder 1994). The plant tissues are colonized by dense communities of photosynthetic algae as well as by bacteria and protozoa. Similarly, the roots and rhizomes that are buried in the wetland soil provide a substrate for the attached growth of microorganisms (Hofmann 1986). Thus, biofilms are present on both the above-ground and below-ground tissue of the macrophytes. These biofilms, as well as the biofims on all other immersed solid surfaces in the wetland system, including dead macrophyte tissues, are responsible for most of the microbial processing that occurs in wetlands.

7.2.4 Nutrient uptake

Wetland plants require nutrients for growth and reproduction, and the rooted macrophytes take up nutrients primarily through their root systems. Some uptake also occurs through immersed stems and leaves from the surrounding water. Because wetland plants are very productive, considerable quantities of nutrients can be bound in the biomass. The uptake capacity of emergent macrophytes, and thus the amount that can be removed if the biomass is harvested, is roughly in the range 30–150 kg P ha^{-1} yr^{-1} and 200–2500 kg N ha^{-1} yr^{-1} (Brix & Schierup 1989; Gumbricht 1993*a*, *b*; Brix 1994). The highly productive *Eichhornia crassipes* (water hyacinth) has a higher uptake capacity (approx. 350 kg P and 2000 kg N ha^{-1} yr^{-1}), whereas the capacity of submerged macrophytes is lower (less than 100 kg P and 700 kg N ha^{-1} yr^{-1}). However, the quantities of nutrients that can be removed by harvesting is generally insignificant in comparison with the loading into the constructed wetlands with the wastewater (Brix 1994; Geller 1996). If the wetlands are not harvested, the vast majority of the nutrients that have been incorporated into the plant tissue will be returned to the water by decomposition processes. Long-term storage of nutrients in the wetland systems results from the undecomposed fraction of the litter produced by the various elements of the biogeochemical cycles as well as the deposition of refractory nutrient-containing compounds (Kadlec & Knight 1996).

7.2.5 Root release

It is well documented that aquatic macrophytes release oxygen from roots into the rhizosphere and that this release influences the biogeochemical cycles in the sediments through the effects on the redox status of the sediments (Barko *et al.* 1991; Sorrell & Boon 1992). Qualitatively this is easily detected by the reddish colour associated with oxidized forms of iron on the surface of the roots. However, the quantitative magnitude of the oxygen release under conditions *in situ* remains a matter of controversy (Bedford *et al.* 1991; Sorrell & Armstrong 1994).

Oxygen release rates from the roots depend on the internal oxygen concentration, the oxygen demand of the surrounding medium and the permeability of the root walls (Sorrell & Armstrong 1994). Wetland plants conserve internal oxygen because of suberized and lignified layers in the hypodermis and outer cortex (Armstrong & Armstrong 1988). These stop radial leakage outwards, allowing more oxygen to reach the apical meristem. Thus, wetland plants attempt to minimize their oxygen losses to the rhizosphere. Wetland plants do, however, leak oxygen from their roots. Rates of oxygen leakage are generally highest in the sub-apical region of roots and decrease with distance from the root apex (Armstrong 1979). The oxygen leakage at the root tips serves to oxidize and detoxify potentially harmful reducing substances in the rhizosphere. Species possessing an internal convective through-flow ventilation system have higher internal oxygen concentrations in the rhizomes and roots than species relying exclusively on the diffusive transfer of oxygen (Armstrong & Armstrong 1990), and the convective through-flow of gas significantly increases the root length that can be aerated, in comparison with the length by diffusion alone (Brix 1994). Wetland plants with a convective through-flow mechanism therefore have the potential to release more oxygen from their roots than species without convective through-flow.

Studies on individual roots have been made with oxygen microelectrodes to measure radial oxygen losses in oxygen-depleted solutions (Armstrong 1967; Laan *et al.* 1989). The oxygen release rates obtained by this technique vary from less than 10 to 160 ng O_2 min^{-1} cm^{-2} of root surface, depending on species. Oxygen release from fine laterals at the base of roots can be significant, but in general no release of oxygen from old roots and rhizomes is detected (Armstrong & Armstrong 1988). The inhomogeneity of the oxygen release pattern of wetland roots makes it difficult or impossible to extrapolate from results obtained by the oxygen microelectrode technique release rates *in situ*. By using different assumptions of root oxygen release rates, root dimensions, numbers and permeability, Lawson (1985) calculated a possible oxygen flux from roots of *Phragmites* of up to 4.3 g d^{-1} m^{-2}. Others, with different techniques, have estimated root oxygen release rates from *Phragmites* to be 0.02 g d^{-1} m^{-2} (Brix 1990; Brix & Schierup 1990), 1–2 g d^{-1} m^{-2} (Gries *et al.* 1990) and 5–12 g d^{-1} m^{-2} (Armstrong *et al.* 1990). Root oxygen release rates from a number of submerged aquatic plants are reported to be in the range 0.5–5.2 g d^{-1} m^{-2} (Sand-Jensen *et al.* 1982; Kemp & Murray 1986; Caffrey & Kemp 1991) and from free-floating plants 0.25–9.6 g d^{-1} m^{-2} (Moorhead & Reddy 1988; Perdomo *et al.* 1996). The wide range in these values is caused by species-specific differences, by the seasonal variation in oxygen release rates and by the different experimental techniques used in the studies. The importance of providing an external oxygen sink during experiments attempting to quantify the oxygen release from entire root systems has been demonstrated by Sorrell & Armstrong (1994). The study concluded that oxygen release rates reported in earlier studies might have been underestimated.

Root systems also release other substances besides oxygen. In some early studies Dr K. Seidel from the Max Planck Institute in Germany showed that the bulrush *Schoenoplectus* released antibiotics from its roots (Seidel 1964, 1966). A range of bacteria (coliforms, *Salmonella* and enterococci) obviously disappeared from polluted water by passing through a vegetation of bulrushes. It is also well known that a range of submerged macrophytes releases compounds that affect the growth of other species. However, the role of this attribute in treatment wetlands has not yet been experimentally verified. Plants also release a wide range of organic compounds by the roots (Rovira 1965, 1969; Barber & Martin 1976).

The magnitude of this release is still unclear, but reported values are generally 5–25% of the photosynthetically fixed carbon. This organic carbon exuded by the roots might act as a carbon source for denitrifiers and thus increase nitrate removal in some types of treatment wetland (Platzer 1996).

7.2.6 Other roles

The macrophytes in constructed treatment wetlands can have functions that are not directly related to the water treatment processes. In large systems, the wetland vegetation can support a diverse wildlife, including birds and reptiles (Knight 1996; Worrall *et al.* 1996). This can be of importance, as natural wetlands and thereby wetland habitats have been destroyed at a high rate in many places. Another point that is perhaps most important in small systems serving, for example, single houses and hotels is the aesthetic value of the macrophytes. It is possible to select attractive wetland plants such as *Iris pseudacorus* (yellow flag) or *Canna* spp. (canna lilies) and in this way give the sewage treatment system a pleasant appearance.

7.3 Establishment

A healthy stand of emergent macrophytic vegetation is the most important feature affecting the consistent performance of wetland treatment systems. Attaining and maintaining that vegetative cover can be a challenging obstacle for many contractors. The science of effectively establishing wetlands vegetation on the first attempt is relatively simple, but the knowledge necessary to accomplish this goal has been laboriously relearnt on dozens of projects. These trial-and-error attempts to grow wetlands successfully can delay project implementation and displease clients and regulators.

7.3.1 Plant propagules and sources

The propagation and sale of wetland plant species has become a big business in several areas of the USA, in the UK and in Europe. Wetland plant nurseries supply thousands of plants that are used to renovate altered landscapes, such as phosphate-mining and coal-mining areas, and to create landscaped wetlands and ponds (aquascaping) for wildlife habitat. Although most of these plants are currently being propagated for use in habitat creation projects, this market has attracted suppliers that can propagate the types and quantities of plants required for constructing large wetland treatment systems.

Plant propagules that are frequently used to establish constructed wetlands for wastewater treatment include seeds, bare-root seedlings (sprigs), rhizomes, greenhouse-grown potted seedlings and field-harvested plants. Each of these plant propagule types has different qualities for wetlands planting.

7.3.1.1 Bare-root seedlings

Seedlings are young plants that have been established from fertile seeds that were field-collected or collected from nursery brood stock. Both herbaceous and woody wetland plants can be propagated as seedlings. Wetland seed germination is a highly variable process that depends on species-specific dormancy conditions and on some phenotypic differences for the same plant species gathered from different geographical areas. Bare-root seedlings are easily planted in the field in shallow individual holes prepared with a shovel, trowel, spike or dibble. The survival rate of bare-root seedlings is significantly higher than for the field germination of seeds and can generally be maintained at 80% or higher with healthy plant stock and an adequate moisture regime.

7.3.1.2 Seeds

Wetland plants can be established directly from seeds, given suitable seed stock and suitable soil moisture, light and temperature conditions. Some species have seeds that can be field-harvested in very large numbers, whereas other species have few seeds. For example, a typical cattail seed head contains thousands of individual seeds, whereas mature bulrush culms might contain only 20–30 seeds each. Most wetland seeds can be broadcast by using rotary seeders or by hand, and lightly harrowed into the surface soil layer. An estimated 1.2×10^6 seeds per hectare are required for establishment of a *Spartina* saltmarsh (Broome *et al.* 1988).

A second approach to establishing a wetland plant community from seed is to harvest the seed bank from a neighbouring wetland (natural or constructed) that has a plant community similar to that desired for the new constructed wetland. This seed bank is harvested by scraping the top 10–20 cm of topsoil or muck from the donor wetland, then redistributing this muck in strips or over the entire surface of the new wetland.

7.3.1.3 Field-harvested plants

In some cases, herbaceous wetland plant stock can be field-harvested and used to plant a new constructed wetland. This is especially true in areas with abundant natural wetlands and high regional water tables, where wetland plants such as cattails and bulrush are common in

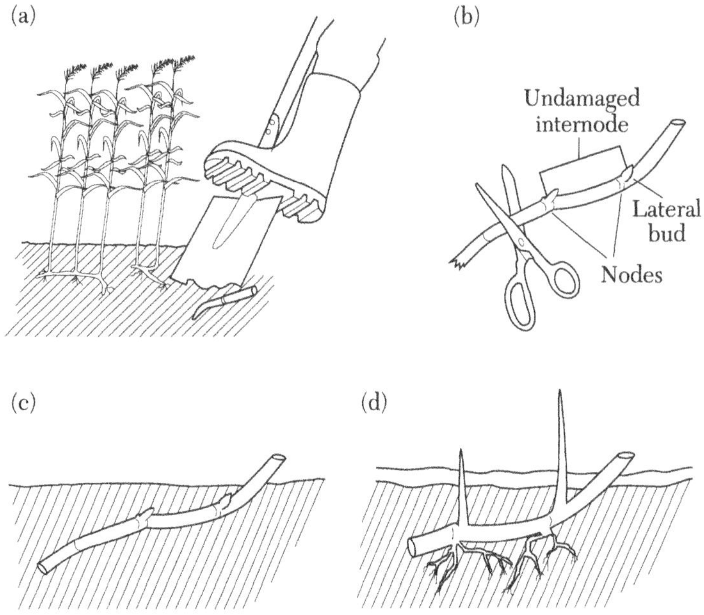

Figure 7.3. Technique for planting rhizome cuttiings (Hawke & José 1996). (a) Dig up rhizomes with a spade or collect from spoil heaps in early spring. (b) Select rhizomes with one undamaged internode and two nodes with lateral buds. Trim off damaged surplus. Rhizomes with a terminal bud can also be used. (c) Plant at an approximately horizontal to 45° angle so that at least one node is buried by approx. 4 cm. Plant in early spring at about four cuttings m^{-2}. (d) Flood to shallow depth (2–5 cm), ensuring that the cut end remains above the surface water. Shoots should appear in early summer.

roadside ditches, in man-made ponds and along canals. A plant collection permit might be required. Field harvesting is done by hand-digging or by using a backhoe or dragline to scoop wetland plants from the ground and spread them on an open, upland area, where they are separated by hand into units of plantable size.

7.3.1.4 Potted seedlings

In some cases, seedlings are planted in containers filled with potting soil to establish older and more robust planting materials. Similarly to field-harvested wetland plants, potted seedlings have a greater advantage for initial growth than seeds or bare-root seedlings. They also have the disadvantage of higher initial cost, making them economically unattractive for most large-scale wetland plantings.

The control of temperature in propagation with seedlings can be very important. A day/night regime is necessary to start germination (see Cooper *et al.* 1996). Various field techniques have evolved for planting and propagating these materials (Figures 7.3 and 7.4).

7.3.2 Propagation

The key requirements for healthy wetland plants to succeed are water, soil, nutrients and light. The first two ingredients must be controlled to some extent by the engineer and the contractor; Nature generally provides the other requirements for plant growth.

7.3.2.1 Climatic factors

Plants have a growing season and are dormant for the rest of the year in temperate and subtropical climates. Annual plants die each autumn or winter and must be re-established from seed the following spring or summer. Fortunately, most plants used in treatment wetlands are perennials, which lose their above-ground tissues and become dormant during the cold season and regrow from storaged reserves in below-ground tissues during the next growing season. In some cases, new growth is timed to take advantage of predictable rainfall and moisture conditions rather than specific day lengths or temperatures.

The best time to establish new plants in a constructed wetland is in the beginning or height of the growing season, which usually coincides with spring or early summer. During this period, available light is increasing daily, and competition from previously established weeds or damage from pests and pathogens is minimal. Most plants are genetically adapted to grow during this period. Also, planting at the beginning of the growing season provides ample time for full plant development and attainment of suitable plant cover before the onset of cold temperatures and declining plant growth rate. Wetland plants can be planted later in the summer or autimn, but their growth will be interrupted by cold weather and decreasing day length, and the below-ground

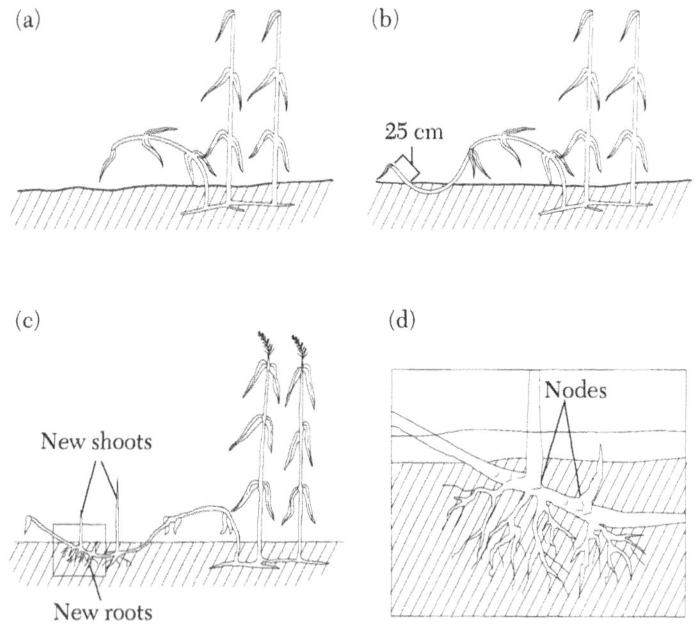

Figure 7.4. Technique of layering (Hawke & José 1996). (a) In May or June a single tall shoot is bent gently over towards the matrix. The matrix should be wet but not flooded. (b) At the point at which the shoot easily contacts the matrix a slot about 10 cm deep is cut, into which the shoot is laid, burying as many nodes as possible. The slot is closed over and gently heeled in. The terminal 25 cm of the shoot should remain above ground. (c) After 3–4 weeks, roots should develop. After another 2–3 weeks, new shoots should develop, as shown. (d) Enlargement of buried stem, showing new shoots and roots arising from the nodes.

plant organs must in turn be protected from killing by frost during the first winter so that they can resume growth the next spring.

A constructed wetland planted with cattails, bulrush or common reed will vegetate quickly if care is taken when selecting planting materials, planting, soil preparation and moisture, and seasonal timing. Plant cover in the range 60–80% is commonly achieved for these species in 3–4 months in most temperate climates.

7.3.2.2 Soil preparation

Emergent wetland plants require suitable soil conditions for rapid initial growth and for long-term propagation and survival. Loamy soils containing a mixture of sand, silt and clays are optimal for the growth of most plants. These soils have adequate texture and organic matter to retain moisture, permit the diffusion of oxygen and carbon dioxide, and retain nutrients for absorption through the plant roots. During the design of a constructed wetland, suitable soil conditions should be incorporated to ensure successful plant growth. Constructed FWS wetland design should incorporate a minimum of 20 cm of topsoil as a rooting medium in all areas that will be planted with emergent macrophytes. Seedlings can be planted in SSF systems without soil additions.

7.3.2.3 Soil moisture

As described above, the incorrect control of soil moisture is the most frequent cause of a failure to establish wetland plants. Inadequate soil moisture results in the desiccation of roots and shoots and causes wetland species to be replaced by weedy upland plant species that might be in the seed bank of the constructed wetland soils. Too much water results in oxygen depletion in the root zone and consequent slow growth or plant death because of insufficient oxygen for root metabolism. The correct amount of moisture can be maintained through adequate planning and attention during the construction period. To maintain suitable soil moisture during plant establishment, there must be a reliable and adequate supply of water for site irrigation.

7.3.2.4 Plant density

The initial density of plant propagules will greatly influence the rate of establishment of plant cover and the cost of planting. When the goal is the establishment of high plant cover (more than 60%) during the first growing season after planting, the minimum density should be *ca.* 10,000 plants ha^{-1} (1 m spacing) for SF systems. SSF systems might require from four to six plants m^{-2}.

7.3.2.5 Detrital development

As wetland plants mature and die, they form

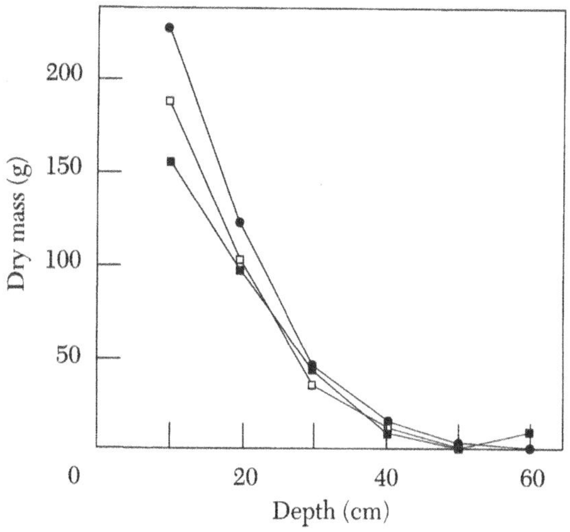

Figure 7.5. Effects of water-level manipulation on the depth distribution of roots and rhizomes of Phragmites in gravel (Daniels & Parr 1990). Hydraulic regime: •, level with matrix surface; ▪, 30 cm below surface in autumn (summer drawdown); ▫, 30 cm below surface in autumn (autumn drawdown).

organic detritus, an essential structural component of a mature FWS constructed wetland treatment system. The standing and fallen dead plants provide a continuing source of organic carbon that is used as substrate by heterotrophic bacteria and fungi. In turn, these microorganisms influence many of the water quality treatment functions important in wetland treatment systems. The organic detritus that is typical of a mature wetland requires from 1 year to more than 5 years to develop, depending largely on the nutritive value of the influent wastewater. For this reason the pollutant transformation functions of wetland treatment systems receiving secondary municipal wastewater will typically mature faster than systems receiving advanced treated, highly renovated and dilute wastewater or runoff from relatively clean watersheds. It is often possible to speed the overall ecosystem development by the placement of imported litter such as straw.

7.3.2.6 Root development

Horizontal rhizomes enable *Phragmites*, *Typha* and *Schoenoplectus* to colonize new areas. Dense mats of underground tissues develop to a depth of *ca.* 30–40 cm (Figure 7.5). In *Phragmites* they can grow downwards for only the first 3 years of establishment. During this period, their deeper penetration is encouraged by moist rather than wet conditions and by a fluctuating water level. These rhizomes live for 3–7 years and sprout thick, deeply penetrating roots. Vertical rhizomes grow upwards from the horizontal rhizomes and can grow to become shoots or branch to form further vertical rhizomes. They are responsible for thickening the growth of reeds in the already colonized areas. They live for about 3 years and sprout shorter and thinner roots that can form dense mats in HF and VF wetlands that have some surface flow.

7.4 Inspection and maintenance

7.4.1 Monitoring

Design specifications for vegetation establishment in constructed wetlands should clearly assign the responsibility for plant maintenance from the time of planting until system start-up. Successful plant establishment requires periodic inspections to document soil moisture conditions, plant survival and plant growth. The frequency of these inspections is project-specific but must be great enough to prevent problems or to detect them relatively quickly after they begin to occur.

Initial plant inspection should examine the viability of the planted propagules, whether they be seeds, seedlings or field-harvested mature plants. Subsequent plant growth is monitored by estimating the percentage cover and average plant height. These non-destructive techniques are used to ascertain the status of plant development before and during the operation of a wetland treatment system. Plant cover is an estimate of the percentage of the total ground area covered by stems and leaves. This can be estimated by walking through or next to a plant stand and visually determining a cover category for the plants.

Cover estimates and observations concerning plant health should be a routine part of operational monitoring in a constructed wetland treatment system. Because plants grow slowly and are important for maintaining the performance of wetland treatment systems used for water quality treatment, problems must be anticipated and prevented before they are serious or have progressed too far. Re-establishing a healthy plant community in a natural or constructed wetland is a slow process when the plants have been irrevocably harmed because of operator neglect.

7.4.2 Troubleshooting

Regardless of the precautions and care with which new wetlands are constructed, problems with plant growth are likely to develop at some time during the project's life. Some of the existing constructed wetland treatment systems have previously had or currently have plant growth problems. Frequently, these problems can be overcome without jeopardizing treatment performance. A few of the major factors in poor plant growth are discussed below.

7.4.2.1 Pathogens or herbivory

Numerous potential stressors on plant growth exist other than low oxygen or the presence of toxins in the root zone. The most visible of these stressors are herbivores such as geese, muskrats, rabbits in the UK, or nutria, which occur in high populations in some constructed and natural wetlands, and insects such as army worms (cattail worms), which are commonly found in monospecific cattail marshes.

In a review of herbivory in wetlands, Lodge (1991) found that invertebrate and vertebrate grazers can remove between 5% and 83% of emergent plant biomass. These organisms generally do not completely eliminate a wetland plant species; instead they contribute an additional stress great enough to eliminate portions of the plant population, resulting in open areas available for weed colonization. This loss of productive biomass adds to the debit side of net production; in concert with other stressors it can result in a chronic loss of wetland plant cover.

7.4.2.2 Water stress (levels too low)

If high water levels cause plants to be planted and root in the upper few inches of soil or gravel in constructed wetlands, the rapid and prolonged decrease in water levels will result in a hostile root environment and plant death under prolonged dry conditions. If water levels are decreased in a gravel bed to promote root penetration, this decrease must not exceed the growth rate of the roots and should generally be less than about 1 cm d^{-1}.

7.4.2.3 Flood stress (levels too high)

As described earlier, all plant species, including all emergent hydrophytes, have some upper tolerance limit for flooding depth. This upper limit is a function of the complex interaction of physical, chemical and biological factors that affect the available oxygen in the plant roots and the effect of the resulting oxygen concentration (or anoxia) on root metabolism and the accumulations of toxic substances.

7.4.2.4 Weather and physical effects

Extremes of hot and cold temperatures, wet and dry conditions and severe winds or hail can all contribute to poor plant survival in wetlands. One side effect of low water levels in gravel-based SSF wetlands during summer conditions is the high temperature of the surface gravel exposed to direct sunlight. These temperatures can physically wilt and damage the wetland plants, exacerbating the desiccation resulting from low water levels. In addition, evapotranspiration is greatly increased during these conditions, further lowering water levels and stressing the plants.

7.5 Weeds

During the early years of establishment of *Phragmites* beds, weeds can make excessive growth, particularly on soil-based beds. The most prevalent species are common weeds of agriculture, including broadleaf species such as *Rumex* spp. and *Polygonum persicaria*, and grasses such as *Holcus lanatus* and *Poa* spp. Provided that the beds are kept sufficiently wet, in the long term *Phragmites* will out-compete all these agricultural weeds. In natural *Phragmites* stands, pure *Phragmites* monoculture occurs only in areas that are flooded for at least two months of the year (Haslam 1972). In drier habitats and in the inner part of littoral *Phragmites* stands, other species are not excluded. Weeds decrease the rate of establishment of *Phragmites* as well as its short-term growth. Studies from the Czech Republic have shown that sparsely planted beds (rhizomes spaced 5–10 m apart in rows 5 m apart) will be covered in 3–4 years, compared with 4–6 years in slightly weed-infested beds (Veber 1978). If water and nutrients are limiting, competitive interference from weeds will be more severe and can lead to an early failure of establishment. The extent to which weed species inhibit the treatment processes is largely unknown. Experience from Danish soil-based systems shows no effects on performance. However, the weeds spoil the appearance of reed beds, and this can be particularly important in pilot schemes in which a good visual impression needs to be made. The most effective method of weed control is flooding. However, *Phragmites* does not tolerate excessive depths of water, particularly during early establishment (Weisner *et al.* 1993). It is therefore essential that beds are flat or nearly flat, so that flooding with 30 cm of water will flood the entire bed. If gravel is used as a medium, weeds are generally not a problem during the time for which the bed is becoming established. However, the seeds used at the banks of the beds might be washed on to the gravel, resulting in considerable growth of grasses.

8 System start-up

The start-up of a new wetland system is a critical time. System start-up comprises the filling and planting of the wetland and a period in which the soil or medium, plants and microbes adjust to the hydrological conditions in the wetland. Like all living systems, wetlands are better able to tolerate change if they have been allowed time to stabilize initially. Some removal processes require only brief periods in which to become fully operative, but others can require months or years to reach stability.

After the initial stabilization period, a gradual increase in wastewater flow to allow the system to adjust to the new water chemistry is often wiser than immediately operating at the ultimate flow. In some parts of Europe, an autimn growing season is often allowed before wastewater is added. Much shorter stabilization periods (several weeks to several months) are typical in the USA. Wastewater should not be added until the plants have shown new growth, indicating that the roots have recovered from being transplanted. Highly concentrated wastewaters, such as some agricultural wastes, require a more gradual introduction than less concentrated waters such as stormwater or pretreated sewage effluent.

8.1 Basin testing

Before accepting the final construction, the wetland should be flooded to design depth to check that water levels and flow distributions meet expectations; all components such as pumps and water control structures should be thoroughly tested to ensure that they are operating properly.

If the wetland basin is lined, a leak test should be performed. One method, known as the Minnesota barrel test (Figure 8.1), is applied to a filled basin with no inflow or outflow. One barrel with perforations near the bottom is used as a stilling well to measure the actual water level in the wetland basin. A second, sealed barrel is used to measure changes in water level due to rain and evaporation. Leakage is determined by difference. A measurement period of 1 month is specified. This leak test is performed after the basin has been sealed but before the rooting soil is placed. Repairs can be made if the test shows excessive leakage.

During the initial operation, any erosion and channelling that develops should be eliminated by raking the substrate and filling by hand. Rills on the dike slopes and spillways should be filled with suitable material and thoroughly compacted. These areas should be reseeded or resodded and fertilized as needed. If there is seepage under or through a dike, an engineer should be consulted to determine the proper corrective measures.

8.2 Antecedent conditions

Small constructed treatment wetlands are often built with imported rooting soils or media with known characteristics. These should be specified to contain no sources of contamination that could jeopardize the performance of the wetland. In this case, the antecedent condition of the wetland soils is known. The type and density of the selected plants are also specified and known, and they compete only with algae for the available space in the system during the grow-in period.

However, larger constructed FWS wetlands can be built without alteration of the original upland soils. Those soils might contain significant amounts of mobile pollutants, depending on the previous land use. For instance, agricultural soils might contain significant amounts of fertilizers. Natural wetlands to be used for treatment contain a whole suite of soils, plants and litter that match the pretreatment conditions of water and nutrient loadings.

These are quite different starting points for the process of ecosystem adaptation to the flows and pollutant loads to be treated. However, in any case, there will be a period of adaptation before stable operation is achieved. Both operators and regulators need to be aware of the consequences of the start-up phenomena.

8.3 Ecological transitions

Constructed wetlands typically require a few months for the establishment vegetation and biofilm, and 1–2 years for development of the litter compartment in FWS. Leaching or sorption of some constituents can also occupy a

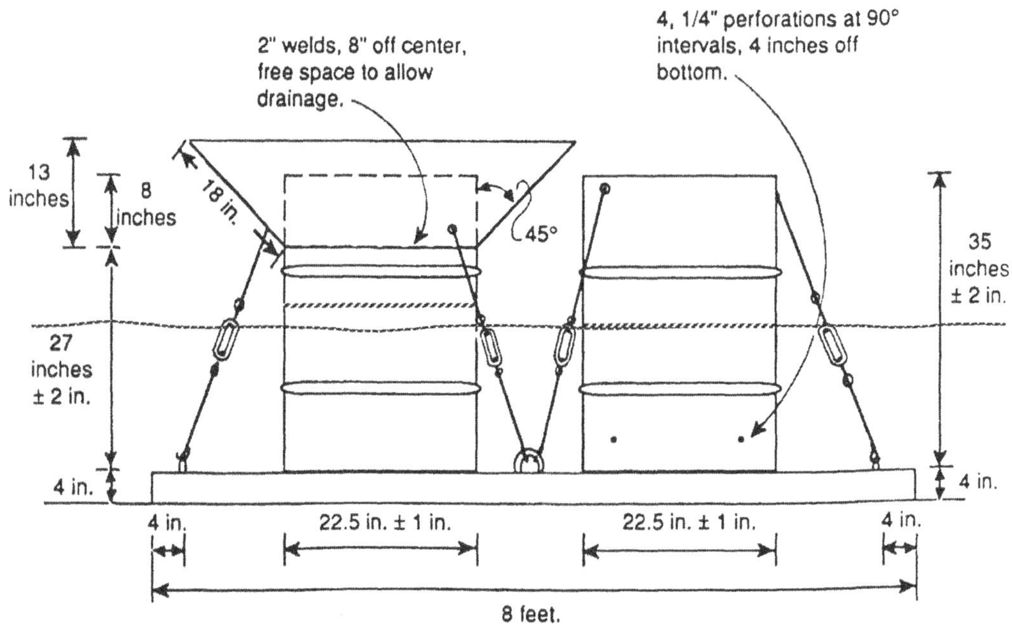

Figure 8.1. The Minnesota barrel test set-up for basin leakage.

period of a year or more if a SSF medium has been selected for sorption capacity.

8.3.1 Soils

8.3.1.1 Previous upland soils

SF wetlands. When an upland mineral soil is inundated, a series of alterations is set in motion. The combination is called gleying and is characterized by a change to a darker colour and low redox potential (Reddy 1994). These processes are fairly rapid and can be complete within a year (Richardson & Tandarich 1992).

The sediments that form in treatment wetlands ultimately form new topsoil layers, and they differ from those that form in natural wetlands for a number of reasons. First, the enhanced activity of various microbes, fungi, algae and soft-bodied invertebrates leads to a greater proportion of fine detritus compared with leaf, root and stem fragments. There is a significant formation of low-density biosolids (sludge). Secondly, there can be a precipitation of metal hydroxides or sulphides, which adds mineral flocs to the sediments.

Table 8.1 provides a description of this layering phenomenon for a 4-year-old FWS wetland built on upland soils (Nolte 1997). Nitrogen levels within the detritus layer (layer A) exceed those in the other two layers by approximately an order of magnitude. NO_3-N concentrations in the detritus layer (layer A) average 70.7 mg kg^{-1} compared with 7.1 mg kg^{-1} in the mat layer (layer B) and 5.6 mg kg^{-1} in the peat layer (layer C). The decrease in NO_3-N concentrations with sediment depth is presumably an indication of the level of denitrification occurring in the anoxic sediment layers. Total Kjeldahl nitrogen concentrations in the detritus layer (layer A) average 9700 mg kg^{-1} compared with 1900 mg kg^{-1} in the mat layer (layer B) and 1000 mg kg^{-1} in the peat layer (layer C).

The sorption capacity of the antecedent soils is re-equilibrated with the new water quality of the incoming water, perhaps along a gradient from inlet to outlet. If there are leachable chemicals, they are depleted and exit from the wetland.

The long hydroperiods of treatment wetlands are conducive to the build-up of organics: first litter and micro-detritus, then the sediments formed from their decomposition, and finally the organic soils generated from those sediments and deposited mineral solids.

In short, the wetland rearranges itself to accommodate the environment created by the designer. The functioning of the wetland after this adaptation is no longer dependent on the previous condition and type of soils, hydrology and biota. It is totally dependent on the new soils, hydrology and biota. It is this new sustainable mode of wetland operation that is the target of most designs.

Available data indicate that the final state of a treatment wetland, and the accompanying suite of water quality functions, are largely independent of the initial condition of the real estate on which it is built. During the interim period of adaptation, antecedent conditions are important because they dictate the short-term performance of the wetland. That period of adaptation seems to extend for up to 2 years for newly constructed wetlands and longer for the alteration of natural wetlands to a treatment function.

SSF wetlands. Some measure of performance control can be exerted by the use of specially tailored bed media for SSF wetlands. In some sense, these are the 'soils' of this type

Table 8.1. Layers in the soil column at the Sacramento wetlands (Nolte 1997)

Layer	Description
A	The A layer consists of a slurry of dark, decomposing, loosely structured detrital material that pours out when the sampler is tipped. The material in the A layer has settled to the bottom but it has not been integrated into the matrix of the basin floor.
B	The B layer is a fairly well consolidated vegetative mat, black in appearance, that holds together and is retained when the A layer is poured off. The B layer is somewhat integrated into the C layer below it.
C	The C layer is less organic than the upper layers and is dominated by greyish-black clay lying beneath the vegetative mat.
D	In some cases, a fourth layer consisting of extremely stiff clay was observed.

of constructed wetland. If sands, soils or gravels are borrowed from natural sources, there will be a period of adaptation as for FWS wetlands; it seems to be of the same general duration as for SSF wetlands. However, a bed material can be chosen that is manufactured to have a very large phosphorus sorption capacity, such as an expanded clay (Jenssen *et al.* 1994). The design philosophy is now quite different from that for most existing treatment wetlands: the intent is to exhaust a short-term capacity, regenerate the wetland and repeat the cycle. This might be a feasible strategy in some cases, provided that the expense of regeneration coupled with its frequency are within acceptable economic bounds.

Solids from the secondary effluents and litter from decaying *Phragmites* will gradually decrease the pore space in tertiary treatment reed beds. Similar processes occur in secondary and storm-treatment reed beds in which sewage-derived solids accumulate together with reed litter. Most of the sewage or effluent-derived solids accumulate at the inlet end of the beds where the pore space can be decreased substantially in 3–4 years. This can cause some surface flow, but this usually only extends for 1 or 2 m across the bed before subsurface flow returns, even when the outlet is kept at 50 mm below the bed surface.

The rate of solids accumulation depends on loading. At Leek Wootton, which was commissioned in June 1990, there has been a build-up of solids of *ca.* 200 mm at the inlet end of the two beds but only a few millimetres of plant litter at the outlet after 6 years of operation. Surface flow is apparent for a distance of *ca.* 1 m during dry weather and 2–3 m when flows are enhanced by rainfall.

8.3.1.2 Previous wetland soils for FWS

The result of high-nutrient waters on existing wetland sediment–soil profiles has often been observed to be a shift to low-density, mushy materials occupying the 'water' column (Kadlec & Bevis 1990; van Oostrom 1994). This might be due to the high ionic strength associated with effluents being treated, reflected in a high content of dissolved salt. The effect of high ionic strength is to alter the structure of the highly hydrated organic materials that compose wetland sediments and soils.

8.3.2 Vegetation

8.3.2.1 Planted constructed treatment wetlands

There are two phases of vegetation adaptation for planted constructed treatment wetlands: a fill-in period and a diversification period. Plants placed in the wetland typically spread rapidly during the growing season. A period from as little as 3–4 months up to 2 years is required to obtain complete plant cover (Figure 8.2). Plants such as *Typha*, *Phalaris* and *Glyceria* might require only one growing season (approximately March to October) to reach 100% coverage. During this span of time, little if any new plant material will die and become standing dead, especially in a northern climate. That die-back and subsequent litterfall occurs in the autumn in northern systems and on a more continual basis in subtropical systems. The turnover time in warm climates is about three to five times per year. Consequently, the erect crop of standing dead and, for FWS, the underwater litter mat take much longer to form, perhaps on the order of a year. Slow decomposition of litter further extends the period to establish a cyclically stable vegetative biomass complex.

Complete root–rhizome development can also span more than one growing season and require 3–5 years.

The submerged litter in a FWS system is a critical component of some treatment processes; hence full start-up can be considerably longer than the time to establish a 'full green' appearance.

Litter in a horizontal SSF wetland is deposited on the surface of the medium and not under water. It is therefore not a critical component in removal processes, except as a minor source of rain-leached chemicals. However, the

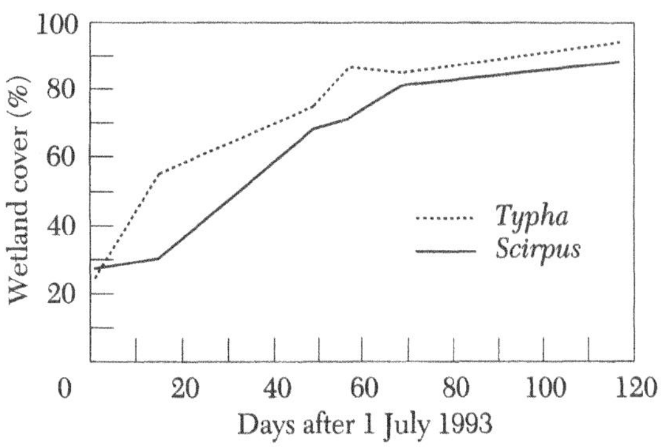

Figure 8.2. Newly constructed wetlands require a start-up period to attain full vegetative cover. Ground-level and aerial reconnaissance were used to follow this process for the project in Tarrant County, Texas, USA (APAI 1995). The litter layer developed subsequently.

slow root–rhizome development can lead to longer start-up transients.

During the fill-in period, other plants can colonize the system if conditions favour their establishment. For instance, in the constructed FWS wetlands in Orange County, Florida, USA, 21 species were planted; 1 year later, 185 species were present (Schwartz *et al.* 1994). There are two schools of thought on the subject of naturally invading species: one holds that a monoculture of a specific plant is a desirable design goal and characterizes the natural invaders as weeds; the other holds that species composition is a secondary influence on treatment and regards diversity as a stabilizing influence.

8.3.2.2 Passive 'volunteer' plant colonization

It is not feasible to plant a very large FWS treatment wetland, such as the Kis-Balaton Hungary system (4000 ha). Such systems are constructed by enclosing a parcel of land within levees and reflooding. The original vegetation might be remnant agricultural crops, such as sugar cane or corn, or an assemblage of terrestrial plants characteristic of colonization during a prolonged dry period after levee construction. The progression to a wetland ecosystem then involves the drowning of terrestrial plants and the recruitment of wetland plants from the existing seed bank within the levees.

This progression can be rapid if an optimal germination environment is maintained. That usually means the maintenance of moist soil conditions during spring. An example is the conversion of agricultural lands at Knight's farm in southern Florida, USA, to the wetlands of the Everglades Nutrient Removal project. Volunteer revegetation to wetland plants was essentially complete 1 year after flooding (Chimney *et al.* 1997).

8.4 Performance transitions

8.4.1 BOD

Wetlands require a period of adaptation to reach a stationary state, from which monotonic time trends are absent. This period includes vegetative areal fill-in, root and rhizome development, litter development and microbial community establishment. The presence of full-sized mature plants in wetlands other than SSF is a necessary but not sufficient condition for the realization of the stationary state for BOD_5 decrease. Operating data are the best indicator of the presence or absence of adaptation trends.

The concept of a wetland as a 'microbial filter' creates the impression that the establishment of the microbial population is the sole determinant of adaptation. Microbial populations are known to adapt rather quickly to their environment, and hence a short adaptation period would be expected. In contrast, the litter decomposition that contributes to the return flux of BOD_5 can require 1 or 2 years to stabilize. Therefore, a newly constructed treatment wetland would be expected to require many months, including at least one full set of seasons, to stabilize.

Data from several locations indicate that this is true in practice. Data from SF wetlands at Listowel, Ontario, indicated a weakly decreasing performance over a period of *ca.* 1 year, with more effect in the wetlands receiving the stronger effluent (Hershkowitz 1986). Other FWS wetlands, such as Iron Bridge, Florida, USA, and West Jackson County, Mississippi, USA, have displayed no adaptation trends over the first few years (NADB).

Vegetated beds in coarse media (n = 16) and in soils (n = 14) were shown to be experiencing adaptation trends into their third year of operation, and performance was improving (Findlater *et al.* 1990). Ten Danish soil-based wetlands displayed a monotonic decrease in performance over 3 years (Schierup *et al.* 1990). The first 3 months of operation of the SSF wetlands in Baxter, Tennessee, USA, were more efficient than the ensuing operations (George *et al.* 1994). Other SSF wetlands seem to have stabilized more quickly. The gravel beds in Richmond, NSW, Australia, experienced little change in performance after decreasing for about 6 months of operation (Bavor *et al.* 1988).

These observations indicate that some weak adaptation effects can be expected for a period of *ca.* 1–2 years, and that performance can decrease during that period. The effect is presumably due to the development of the return flux associated with biomass decomposition.

8.4.2 Nitrogen

Most of the N processes involve microbial mediation and not vegetative interactions. Microbes possess a high potential for population expansion and can quickly colonize the wetlands. Therefore, N processes could start soon in the growth period of the wetland. However, denitrification requires a carbon source to fuel the denitrifying bacteria. In an FWS wetland, decomposing litter provides such a carbon source. As noted above, litter might not be available in a young wetland, and hence denitrification is suppressed. However, the primary effluents usually contain enough carbon to support denitrification.

This transient occurred at the Tres Rios wetlands, Arizona, USA (Wass 1997). An FWS bulrush wetland was established on a mineral substrate in summer 1995. A nitrified secondary effluent was applied, but initially there was no denitrification (Figure 8.3). After 1 year of operation, denitrification suddenly became very efficient (summer) and began a seasonal cycle of efficiency (lower in winter).

Nitrogen requirements for growth are usually satisfied by NH_4-N. Hence, the early vegetation establishment period can show high uptake until the biomass has reached full standing crop. The N removed by plant uptake is negligible in constructed treatment wetlands with emergent vegetation; it therefore usually does not affect the total removal.

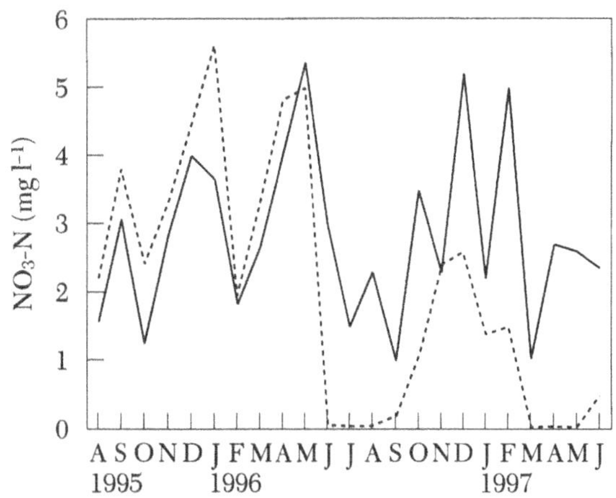

Figure 8.3. Nitrate removal over the history of a Tres Rios treatment wetland. Outlet (dotted line) and inlet (solid line) were nearly the same until early summer 1996, when a seasonal pattern of decrease began.

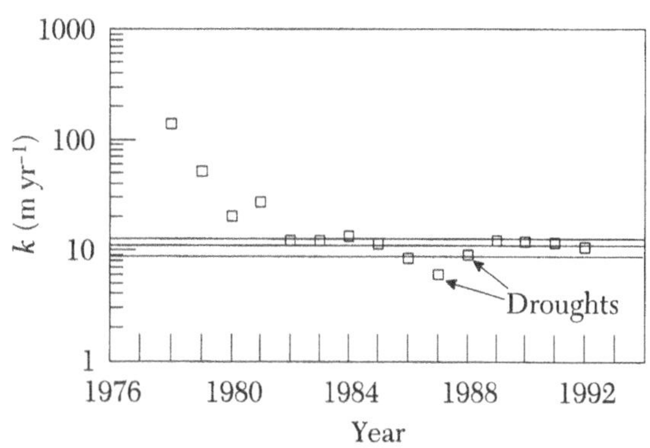

Figure 8.4. Phosphorus removal rate constants (k) over the history of the Houghton Lake Treatment Wetlands. Each point represents the annual average for that year, as determined from transect data and confirmed by input/output data. The middle horizontal line represents k = 11.0 ± 2.2 m yr⁻¹. (Kadlec & Knight (1996).)

8.4.3 Phosphorus

8.4.3.1 Start-up phenomena

The P 'start-up' period for a wetland can extend over various periods, ranging from 1 to 5 years for P removal. During this start-up period, the mass balance model must include the storage of P on sorption sites and in expanded amounts of biomass. For P, the simplified version of the mass balance is

$$\left.\begin{aligned}\left\{\frac{1}{t_\mathrm{m}}\sum_{i=2}^{N-1}\Delta(m_iX_i)+\frac{\Delta(hC)}{t_\mathrm{m}}\right\}+J&=-q\frac{\mathrm{d}C}{\mathrm{d}y}\\&=J_\mathrm{U}\\&=k_\mathrm{U}C.\end{aligned}\right\}\quad(8.1)$$

The terms on the left-hand side of Equation 8.1 cannot be neglected. Of course, it is still possible to execute the calculation of a rate constant, but it will include uptake into, or delivery back from, temporary storages.

If the term in braces on the left is negative, then P is being stripped from static compartments, either from biomass or active soil, leading to decreased uptake. If it is positive, there is extra uptake from water into sorption or expanded biomass. Extra storage in biomass yields a higher value of the uptake rate constant. After the adaptation period passes, the long-term average is attained. An example of this start-up period for the Houghton Lake wetland treatment system is shown in Figure 8.4. Initial k values for this natural wetland were very much higher than the long-term average.

Other patterns of start-up include the rapid vegetation of a bare soil wetland after an initial planting. If that planting is sparse, the start-up period will be characterized by the time for plant fill-in plus the time for litter development. That can be relatively brief, especially in warm climates. At Iron Bridge, Florida, USA, the start-up duration was *ca.* 24 months, during which considerable change in the ecosystem took place. Figure 8.5 shows two of the principal variables: vegetation density and P rate constant. Note the close parallel between

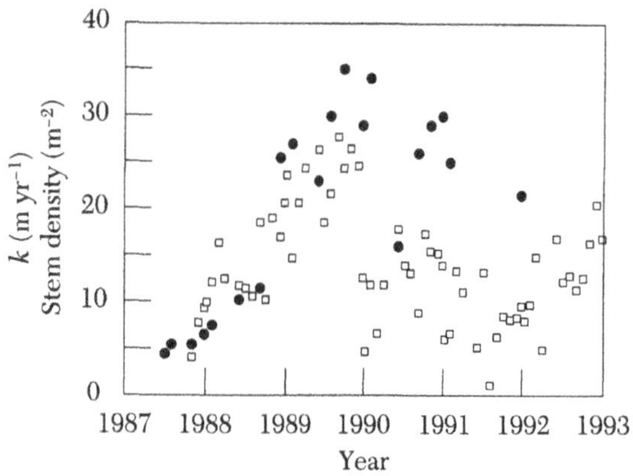

Figure 8.5. Vegetation density for cell 11 (•) and phosphorus rate constant, k, for cells 1–12 (▫) at Iron Bridge, Florida, USA, during start-up. (Kadlec & Knight (1996).)

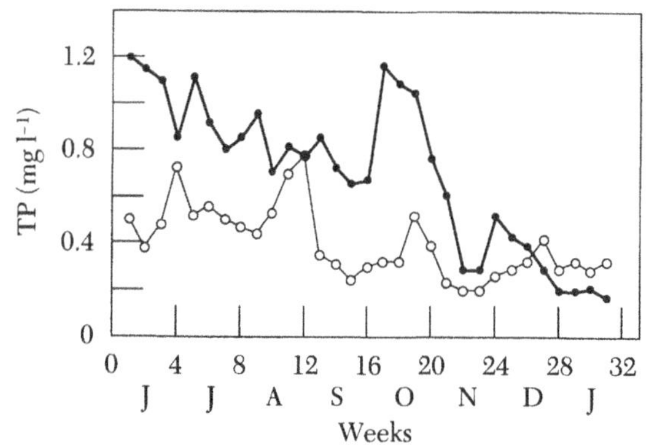

Figure 8.6. The progression of TP concentrations through the start-up period of the natural forested wetland treatment system in system 2 at Walt Disney World, Florida, USA, in 1988/89. The wetland had been damaged by drying and had stored the nutrients from the oxidation of half a metre of peat. Symbols: •, output; ○, input. (Replotted from DeBusk & Merrick (1989).)

the amount of vegetation and the rate constant during the first 2 years. After that period the vegetation density levels off, and the rate constant decreases to its long-term average value of 13.5 m yr^{-1}. The P requirement for building the new standing crop of biomass leads to a peak uptake rate constant that is roughly double the long-term average value.

SSF wetlands are no exception. Wolstenholme & Bayes (1990) showed a large decrease in the k value for P for the reed beds at Valleyfield, Fife, Scotland. The vegetation reached full density by the end of 2 years. The rate constant calculated from their data started at 60 m yr^{-1} and decreased to 13 m yr^{-1} over a 3-year period, with no evidence of levelling out. Because no North American SSF wetland has reported data for more than 3 years, there is a strong chance that all reported US SSF data represents the start-up period. The period of start-up adaptation for an SSF system has also been observed to exceed 2 years in Australia (Mann 1990). However, 10 SSF soil-based wetland systems in Denmark showed no adaptation period for P uptake (Schierup *et al.* 1990).

8.4.3.2 Antecedent phosphorus loads

In some circumstances, a wetland can be constructed or re-established on a site that has a large, mobile P storage already in place. Prime examples are previous peatlands that have undergone drying and peat oxidation. Under such circumstances, the available P can exceed the storage potential under the new water P concentrations. The result is the discharge of P from the antecedent soils into the new overlying water.

A specific case is the reflooding of drained forested peatlands in southern Florida, USA, at Walt Disney World. The site had been isolated from natural surface inflows and outflows for several years. Many tens of centimetres of peat were lost to oxidation during this period, leaving a considerable residue of non-volatile nutrients, including P. Wastewater discharges were begun in 1988. The response of the wetland was to release stored P, creating a higher concentration of TP in the outflow than the inflow (Figure 8.6). Over the course of many weeks, new conditions were established, which displayed P removal.

9 Economics

Constructed wetlands have been developed for the treatment of municipal wastewater and have been widely reported as being low in construction and operating costs (US Environmental Protection Agency 1988; Reed *et al.* 1995). Estimating the initial capital cost of the project is a routine exercise in most respects.

When reviewing costs, the engineer must take into account other factors such as the number of cells. For example, four cells totalling 1 ha will cost more to build than one cell with an area of 1 ha. Equally obvious is that large systems will cost less per unit flow than small systems.

Perhaps the most difficult cost element to assess is the form of pretreatment. If the collection system is a small-diameter system with interceptor tanks, then pretreatment will produce an influent BOD in the range 120–140 mg l^{-1} with anaerobic properties. Influent TSS is averaging 30 mg l^{-1} in some of these small diameter systems. Compare this to a partial-mix aerated lagoon and constructed wetland, in which the influent BOD is typically 30–60 mg l^{-1} and the TSS is 75 mg l^{-1}. The overall energy cost of the second system is considerably higher than the first, but the capital cost of the wetland will be significantly less. Obviously, overall system costs must be considered as well as the treatment goals.

On the basis of the examples above, the first system will require additional treatment steps if nitrogen removal is part of the design goal. Anaerobic pretreatment will produce nitrogen principally in the form of ammonia, and SSF wetlands will generally not nitrify very efficiently, especially in cold weather. The result is ammonia in the wetlands effluent that in the winter is not significantly different than the influent. However, aerobic pretreatment (aerated lagoon) will produce nitrogen principally in the form of nitrates. SSF wetlands do an excellent job of denitrifying.

Wetlands costs can be broken down into the following components:

- excavation
- liner
- plants
- gravel
- distribution and control structures
- fencing
- other.

9.1 Capital costs

Any energy-intensive wastewater treatment technology will always be much more expensive than constructed wetlands to operate. As a general rule, constructed wetlands will be less expensive to build. The basic exchange is energy for land. As the land area of the treatment system increases via the use of wetlands, the energy costs and capital costs decline. As the land area decreases, energy must be added to the wastewater treatment process to accomplish what natural processes accomplish without this assistance.

Because most wetlands have been built in rural areas, where land costs are low, or on land that is not suitable for building, land costs generally have not had much impact on the cost of constructed wetlands. However, if wetlands are considered for urban areas, then the cost–benefit analysis should include land costs as well as the benefits that accrue to the open space, habitat and recreation.

9.1.1 Surface-flow wetlands

Elements of the construction cost for FWS wetlands include excavation, liners, distribution piping, planting and fencing. Although liners might be needed for FWS wetlands, they have rarely been used. More commonly, FWS constructed wetlands are sited where soils are slowly permeable. If a liner is needed, the cost can be significant. The cost of the liner averaged 17% of the total construction cost for the project at Gustine, California, USA (Crites 1997).

The costs for 25 FWS constructed wetlands in the NADB are set out in Figure 9.1. The average cost is US\$58,000 ha^{-1}.

The breakdown of capital costs includes the major categories discussed in preliminary design, but it is generally possible to refine the estimates after final sizing and siting. A more precise economic estimate is possible after final design drawings have been prepared. A sample of a capital-cost estimate based on final

Table 9.1. Estimated capital costs for the SF wetland system at Incline Village, Nevada, USA

Item	Estimated cost (US$)
Site preparation	
Clearing and grubbing	195,000
Fencing	124,000
Dike construction	
Stripping	50,000
Flood embankment°	450,000
Embankment construction°	1,150,000
Erosion control and dike stability requirements†	150,000
Gravel roadway	256,000
Water supply and distribution	
River crossing	50,000
Outfall pipeline	288,000
Distribution piping	318,500
Overflow structures	105,000
Return-flow system	40,000
Miscellaneous	20,000
Site improvements	
Operations Building	95,000
Chain link fence	6,000
Access road and parking lot	10,000
Septic tank/leach field	7,500
Potable water well	7,500
Landscaping	15,000
Wetlands vegetation	50,000
Monitoring	
Monitoring wells	32,500
Initial survey	34,000
Subtotal	**3,454,000**
Contingencies (20%)	691,000
Total	**4,145,000**

This information was developed after the conceptual design was finalized. It does not include engineering costs. (From Culp, Wesner, Culp (1983).)

° Preliminary estimate pending results of more detailed design.

† Allowance based on soils investigation.

conceptual design is shown in Table 9.1. The Incline Village system encompasses 175 ha; the estimated cost was therefore US$23,700 ha^{-1} (US$9600 per acre).

There are two nuances peculiar to FWS treatment wetlands that need consideration: the life expectancy of the items purchased and their value (positive or negative) at the end of service life. In many situations, the wetland alternative is to be compared with other types of process. Traditionally, the life expectancy of a 'conventional' treatment alternative is 20 years, and neither positive nor negative value is assigned to the components after that time.

A treatment wetland has a longer life expectancy than concrete and steel equipment. Although there are no examples of engineered systems with long periods of operation, there are long-lived FWS wetland systems that have retained their effectiveness for up to 80 years, based on *ex post facto* monitoring. Both the Brillion Marsh, Wisconsin, USA (Spangler*et al.* 1976), and Great Meadows Marsh, Massachusetts, USA (Yonika & Lowry 1979), operated for over 70 years and in later years were shown to have retained treatment efficiency. As fully functional ecosystems, treatment wetlands can be expected to sustain their character for as long as appropriate hydrology is maintained. It is common practice to claim no salvage value at the end of project life in a feasibility study for mechanical plants, but this does not make sense in the context of a wetland project. Typically, the entire acquisition price is charged to the project at the outset, and there is no 'salvage' value at the end of 20 years. In contrast with the crumbling concrete and rusted steel left after the mechanical process reaches the end of its useful life, the land associated with the wetland project will probably have a value greater than or equal to that at the time of acquisition. One of the principal components of the wetland project is that it will have appreciated in value. It might be more accurate to remove land cost from the comparison for that reason.

9.1.2 Subsurface-flow wetlands

Gravel beds are more costly on an area basis than FWS wetlands. However, they possess certain advantages in terms of larger rate constants and in terms of nuisance reduction. Therefore, economics must be evaluated in the context of ancillary benefits and values. The cost of the medium is a large fraction of the total cost of gravel-bed wetlands, and this added expense must be weighed against the potential advantages of the SSF system.

The distribution of capital costs for SSF wetlands in the NADB is wide (Figure 9.1). However, the median cost of the SSF systems is US$388,000 ha^{-1}, versus US$58,000 ha^{-1} for FWS wetlands. The reed beds used in the UK average about US$1,000,000 ha^{-1}, which includes pumps, liners, land costs and construction. The land cost is usually not a significant contribution to the total capital cost. Land cost should be excluded from consideration of total capitalized cost because it will appreciate in value at about the rate of inflation. This is current practice for Severn Trent Water (Green & Upton 1994).

Table 9.2 provides an estimate of the relative dollar and percentage costs for a wetland 4600 m^2 in area and 0.6 m deep, using typical unit prices that can be found in many places in the USA. The available information from the

Table 9.2. Example costs of SSF wetlands in the USA

Description	Units	Unit price (US$)	Total cost (US$)	Total cost (%)
Excavation/compaction	m^3	2.30	13,000	10.7
Gravel	m^3	20.00	51,900	42.6
Liner, '30 mil'° PVC	m^2	3.75	19,250	15.8
Plants, 46 cm on centre	Each	0.60	13,330	10.9
Plumbing	–		7,500	6.1
Control structures	–		7,000	5.7
Other	–		10,000	8.2
Total			**121,980**	**100.00**

° 1 mil = 10^{-3} inch.

Czech Republic suggests that prices for gravel and a liner form a similar percentage of the total cost. However, excavation is proportionally more expensive (*ca.* 30%) and plants are cheaper (*ca.* 5%).

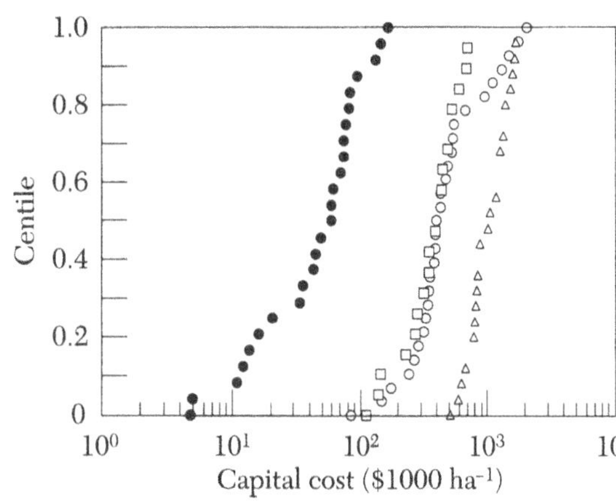

Figure 9.1. Capital costs for treatment wetlands (1995 US dollars). Each point represents one wetland. Sources: NADB; Green & Upton (1992); Vymazal (1994). Median costs are: NADB FWS (•), \$58,000 ha^{-1}; NADB SSF (○), \$388,000 ha^{-1}; Czech SSF (□), \$379,000 ha^{-1}; Severn Trent tertiary SSF (△), \$1,029,000 ha^{-1}.

9.1.2.1 Liners and gravel

Almost without exception, the single most important factor is the cost of gravel, followed by the cost of the liner material. Material costs for both of these items increase as the specifications become more severe, for example: What is an acceptable thickness for a liner? What is an acceptable percentage of fines (i.e. material passing the 200 sieve) in the gravel? How far will the gravel have to be hauled? The cost comparison for these SSF wetlands has not compared specifications, so there will be some obvious differences in costs for these materials.

Gravel, which can be considered to be almost a universal material, will usually cost *ca.* US\$10.50 Mt^{-1} (US\$17.00 m^{-3}) throughout the USA, provided of course that the gravel pit is within 80 km of the project. Hauling costs can add significant amounts to the project, and delivered costs can exceed US\$26 m^{-3}. There are also many areas in the USA where gravel is just not available or is very costly. Some states, for example Florida, are considering the use of recycled concrete rubble.

As a general rule, gravel is 40–50% of the cost of a system for a 4600 m^2 system; the percentage increases as the system gets larger. The reason for the increase is that other costs decrease with increasing system size. For example, the area of the perimeter run-out material in the liner decreases as the percentage of the total area increases. Distribution structure costs are not proportional. Perimeter fencing costs decline for the same reason that liner costs decline.

Liner costs are predicated on the quantity, thickness and type of material that has been specified. A good argument can be made for eliminating liners in certain soils with high clay content, but as regulators focus more attention on groundwater, a reliance on use of *in situ* soils becomes problematic. Even with good soils in place, costs of testing and compaction can exceed the costs of a '30 mil' (0.03 inch, or about 0.8 mm) poly(vinyl chloride) (PVC) liner.

Liner costs have come down significantly in recent years because of demand and because most liners are petroleum-based products. These current low prices could easily skyrocket as they did during the last oil embargo, which would make clay soils and bentonite much more competitive. Generally speaking, it is possible to get a '30 mil' PVC liner installed for approx. US\$3.50 m^{-2} in small systems, and US\$4.00 m^{-2} in systems with 100,000 m^2 or more. This price is generally valid throughout the USA.

Liners generally compose 20–25% of the total costs; this percentage declines as the system gets larger. Soils with angular rocks and rocky terrain might require the use of underlay such as geo-textiles or sand. This will add US\$1.00 m^{-2} to the liner costs. If river-run gravel is not available, sand or geo-textiles should be placed on top of the liner.

9.1.2.2 Excavation and planting

Excavation and/or earthwork is generally the third or fourth largest cost. Obviously this cost is dependent on terrain. Flat sites in Nebraska on sandy loams will be easier to excavate than sites on mountainsides in Colorado. Given this obvious caveat, most SSF wetlands are usually constructed on level sites with good soils. As a result, excavation costs are usually in the range US\$2.00–3.20 m^{-3}.

Plants are generally another minor cost. Because the plants that are used in SSF wetlands (cattails, reeds and bulrushes) are generally available everywhere in the USA, they can be collected and planted in the wetlands. In some cases, planting has been coordinated with county drainage ditch-cleaning operations, and therefore the cost of plants to the project is zero: only labour is required. Planting in the gravel can be accomplished rather easily, with experienced crews planting 600–1000 plants per worker per day.

However, if the project must bear the costs of harvest and separation, cleaning and transport to the job, and splitting them into small plantable units, then the plants are likely to be very expensive. The alternative is to seek wetlands nurseries that are capable of providing the quantity, species and quality of plants that the job requires.

Because of wetlands mitigation work, there are now many nurseries throughout the USA capable of growing and planting the wetlands plants used in constructed wetlands. The advantage of nursery operations is that large quantities of viable plants 30–45 cm high can be grown for ease of harvest and subsequently transplanted by hand or machine with a very high degree of transplant success. Designers can and should expect a minimum of 80% survivability.

Costs of plants will usually run in the range of US\$0.30–1.00 per plant, with most bids at the lower end. The question for the designer is the plant spacing. Plants placed in 90 cm centres will each have to grow to fill in 0.8 m^2, whereas plants on 45 cm centres will have to fill in 0.2 m^2. A 4600 m^2 wetland will require 5555 plants with 90 cm centres or 22,222 plants at 45 cm centres. At 50¢ each, the costs are US\$2777 and US\$11,111, respectively. The problem for the designer is that a 20% loss at 90 cm centres means that there will probably be large unvegetated areas. These will eventually fill in, but can the project wait for the next growing season for these areas to fill in naturally or be replanted? The US\$8000 difference on a project of this size does not seem to be worth the risk, given the importance of viable plants to the overall treatment.

In the past, planting has been a casual affair; success of the planting has relied primarily on the hardiness of the plant species. Cattails and reeds, once started, are very aggressive and are almost impossible to eliminate. Infill of areas that were devoid of vegetation was not particularly important on the large-scale SSF projects, but unvegetated areas on small projects need to be remedied as soon as possible. Replanting is a definite consideration and can in effect be included in the specifications requiring a certain minimum surviving population of plants. Experienced nurserymen are capable of meeting these types of specification and can be called on to replant as part of their contract if necessary. The designer should expect the same type of performance on this part of the contract as from pump suppliers or liner installers.

Other minor costs include piping costs and level control structures, flow distribution structures, flow meters and fencing. In addition, reseeding and erosion control costs should be provided for in any design. Piping materials are generally plastics such as polycarbonate, polyethylene and ABS, commonly available throughout the USA. Plumbing costs are in the range 6–7%. Level control and flow distribution structures can be built out of concrete block, cast-in-place and pre-cast concrete; for smaller systems, reinforced polycarbonate units are commercially available. Depending on the number of cells, these types of structure usually represent *ca.* 5–6% of the total cost.

The life expectancy of SSF systems is limited by the accumulation of mineral solids in the pore space. Blockage by degradable biosolids is also expected, but this is accounted for in hydraulic design. The mineral content of incoming wastewaters is characterized by the non-volatile component of TSS (NVSS). This material will accumulate in pore spaces, preferentially near the bottom of the gravel bed (Kadlec & Watson 1993). This process is very slow when the incoming waters have an NVSS of less than 100 mg l^{-1}. At a loading rate of 30 cm d^{-1} and NVSS = 100, it would take 37 years to fill half the voids in the bed with mineral residues. Depending on whether the location of the material in the pore space, the hydraulic conductivity would be decreased by a factor between 2.0 (all on the bottom) and about 16 (uniform pore blockage).

9.1.3 Cost comparison

Comparisons are made in this section between FWS and SSF wetlands and between constructed wetlands and other technologies. In any comparison, care must be taken to ensure that the elements of a treatment system that are selected to achieve a water quality objective are compared equally. Comparisons between FWS and SSF wetlands can be made on the basis of unit prices. Because of the extra

Table 9.3. Capital cost comparison for FWS and SSF wetlands in the USA

	BOD	TP	NH_4-N	NO_x-N	TN	FC
FWS k Value (m yr^{-1})	34	10	18	50	22	73
SSF k Value (m yr^{-1})	37	10	34	50	27	95
FWS area/SSF area	1.09	1.00	1.89	1.00	1.23	1.30
FWS cost/SSF cost	0.16	0.15	0.28	0.15	0.18	0.19

TSS is presumed to be decreased to backgound by both types. Unit costs: US$388,000 ha^{1} SSF; US$58,000 ha^{-1} FWS. (Modified from Kadlec & Knight (1996).)

cost of gravel, an SF system will be more expensive than an FWS wetland above a certain flow rate.

The costs of FWS and SSF systems depend on the size requirement and the cost per unit area. In turn, the area required is roughly proportional to the inverse of k for a given pollutant. If the median capital costs from the NADB are accepted as norms, the capital cost ratio for a given pollutant can be computed (Table 9.3). The capital cost of the SSF wetland is three to five times that of the FWS wetland to do the same job. On the basis of performance and cost, there is no reason to consider an SSF wetland.

The justification for the added expense of the SSF system lies in the desire to keep the polluted water below the surface of the ground or medium. That desire is usually the result of concerns about mosquito breeding, odours, and pathogen contact for humans and wildlife.

9.2 Operating costs

Wetlands are almost invariably one part of a multiple part treatment system. Determining actual operating costs from the database is therefore difficult because the wetlands labour costs are lumped into the entire overall system costs. However, an estimate of costs can be made by inspection of the design and recognizing that in many respects wetlands are very similar to wastewater stabilization lagoons from a maintenance and operational perspective. There are not many items in wetlands systems that require maintenance or energy.

The operating and maintenance (O&M) costs for an FWS facility include pumping energy, compliance monitoring, dike maintenance, and equipment replacement and repairs. Dike maintenance consists of mowing and the preservation of structural integrity. Equipment replacement and repairs pertain to piping and pipe supports, structures and pumps. It is relatively early in the history of constructed FWS wetland facilities, so there is no track record on frequencies for many of these activities. However, in general terms, pumps and piping can last on the order of 40 years, and repair frequencies are known.

Pumping energy can be quantified accurately, as can the initial level of compliance monitoring, once a permit is issued. Mowing is primarily a matter of aesthetics, with secondary emphasis on the visual detection of snakes and alligators. If public use is encouraged, there might be a need to maintain signage. Nuisance control or removal might be required, most often targeting mosquitoes, burrowing rodents and bottom-stirring fish.

The sum total of these activities is relatively inexpensive. No chemical purchases are involved, and there is no need for highly trained personnel, nor significant time requirements for the necessary semi-skilled employees. Annual costs range from US$5000 to $50,000 yr^{-1} for small systems. However, ancillary research can greatly increase these expenditures. The estimate for the Incline Village system, made at the time of final conceptual design, was US$85,500 per year. Experience is very limited, but all indications are that SSF wetlands need little maintenance. Estimates range from US$2500 to $5000 ha^{-1} yr^{-1}.

9.2.1 System maintenance

Operational costs can be divided into the following general categories:

- operation (testing, level adjustment)
- maintenance (weed control, flow distribution and level adjustment sumps).

The level adjustment function does not usually require any attention. Water levels should be checked periodically (monthly or weekly on small systems and daily on large systems (more than 500 m^3 d^{-1})) to ensure that surfacing has not occurred in the SSF system and that there is indeed some flow through the system. SF systems need the level adjustment to be checked and the level in the wetlands to be inspected visually with a fixed gauge. Water levels can be monitored visually.

Maintenance requirements are similar to those for a waste stabilization lagoon. Weeds should be controlled around the edges, and large weeds should be removed from the gravel bed in the early spring. Plant debris in SF wetlands can be ignored as long as it does not

Table 9.4. Typical minimum monitoring requirements for successful operation of wetland treatment systems (from Kadlec & Knight 1996)

Recommended parameters	Recommended sampling locations	Minimum sampling frequency
Inflow and outflow water quality		
All systems:		
Temperature, dissolved oxygen, pH, conductivity	Inflow(s) and outflow(s)	Weekly
Municipal systems:		
BOD_5, TSS, Cl^-, as an inert tracer	Inflow(s) and outflow(s)	Monthly
Industrial systems:		
COD, TSS	Inflow(s) and outflow(s)	Monthly
Stormwater systems:		
TSS	Inflow(s) and outflow(s)	One storm event per month
Permit parameters as required:		
NO_2 + NO_3-N, NH_4-N, TKN, TP	Inflow(s) and outflow(s)	Monthly
Metals, organics, toxicity	Inflow(s) and outflow(s)	Quarterly
Flow	Inflow(s) and outflow(s)	Daily
Rainfall	Adjacent to wetland	Daily
Water stage	Within wetland	Daily
Plant cover for dominant species	Near inflow, near wetland centre, near outflow	Annually

affect the flow; for example, plant debris after a severe storm might blow downstream and clog the collection piping or level adjustment structures. Regular inspection of the flow distribution devices should be part of the operating requirements for the system. Flow splitters using weirs should be checked and cleaned periodically.

Systems fed from canal conveyance are inclined to collect aquatic weeds from the canal in the inlet structures of the constructed wetland (Brunner 1997; Kosier 1997). This can create the need for regular mechanical harvesting of the floating plant material, to keep the inlet structures operational. This can lead to a significant maintenance cost.

Some systems have incorporated an annual harvest of wetlands plants. In the fall, before the plants become senescent, they are mowed and the litter is removed to a composting operation. The rationale for this operation is that it removes the stored nitrogen that would otherwise be released during the following spring. Although there is a limited amount of information on this type of operation, the value of the harvested nitrogen does not justify the cost.

9.2.2 Monitoring

Monitoring is the most important factor in the successful operation of treatment wetlands. Monitoring information must be collected accurately and consistently and be reviewed frequently by a knowledgeable operator to anticipate the need for operational changes. Incorrect operational control decisions and design errors can cause prolonged periods of poor operational performance of constructed wetlands and significant ecological changes in a natural wetland. The early detection of subtle changes in a treatment wetland's water quality and biological resources requires adequate data collection and frequent data analysis.

All wetland treatment systems should be monitored for at least inflow and outflow water quality, water levels and indicators of biological condition. These parameters are essential for successful system control. Regulatory requirements can also dictate other monitoring requirements. The frequency of operational monitoring for system control is dictated by the size and capacity of the system, the sophistication of the owner's staff and sampling equipment, and site-specific factors related to influent quality variability and climatic factors.

Table 9.4 summarizes a possible monitoring programme for the operation of a wetland treatment system. This list includes measurements of the water quality of all major inflows and outflows associated with the treatment wetland. Inflows include the source(s) of pretreated wastewater entering the wetland as well as natural inflow streams that might have a significant effect on the water quality or the hydrological budget of the natural wetland treatment system. As shown in Table 9.4, the parameter list to be tested at all major inflows and outflows at least monthly includes all regulated pollutants and integrative measures such as BOD_5, TSS, pH, dissolved oxygen, water temperature, conductivity, NO_2 + NO_3-N, NH_4-N, total Kjeldahl nitrogen, P, chloride and sulphate.

Inflow and outflow stations can also be monitored less frequently for selected heavy metals or organics that might be present in the wastewater and for acute and chronic toxicity in the whole effluent. If water quality characteristics are highly variable for any of the inflow or outflow locations, or if there are weekly or monthly permit limits, sampling should be more frequent than monthly or quarterly.

These water quality data as well as any other parameters required by permit should be organized and recorded in computerized spreadsheets for visual analysis of variability and trends. Seasonal and successional changes can be detected by examining trend data regularly. Operational controls are required when trends indicate the potential for future permit violations. Operational modifications should also be made in response to seasonal changes in dissolved oxygen and water temperature.

Flow rate should be measured or estimated daily at the inflow and outflow locations. These data can be collected by installing flow meters at some locations and by collecting stage data and using stage-discharge relationships at non-instrumented locations. Flow estimates are essential for quantifying constituent mass balances in wetland treatment systems.

Rainfall should be monitored at a location next to or near the natural wetland treatment system. Rainfall measurements are used to estimate the wetland water balance and to anticipate elevated flow conditions at the wetland outflow location(s). Evapotranspiration can be estimated with pan evaporation data (corrected by a factor between 0.7 and 0.85) from a regional weather station. Rainfall, evapotranspiration and inflow/outflow measurements can be used to maintain a continuing water balance for the wetland, to detect groundwater exchanges that might be due to leaks in a liner, if one is present.

Water stage (elevation or level) in the wetland should be measured daily near any outflow locations. When combined with a topographic survey of the wetland, stage measurements provide a quantitative tool for assessing the average, maximum and minimum water depths in the wetland and the frequency with which these depths occur. These data are essential for interpreting tracer measurements of hydraulic residence time and for assessing any detrimental hydroperiod effects on biota. Biological monitoring within a wetland treatment system provides the operator with information concerning the structural integrity (health) of the vegetation and fauna. The protection of this biological integrity is important from an environmental habitat perspective and because of the biota's control of wetland operational performance.

The percentage cover of dominant plant species should be recorded in all wetland treatment systems on a quarterly to annual frequency. In addition, quarterly or annual surveys might also be conducted for benthic macroinvertebrate and fish populations at representative stations in wetlands constructed for habitat and in natural treatment wetlands. Quarterly or annual surveys for rare or threatened species might also be conducted when appropriate. This monitoring provides a record of biological changes that occur owing to the altered hydrological regime resulting from the prolonged discharge of pretreated wastewaters.

Testing influent and effluent is generally the single largest cost. This cost will be dependent on the frequency of testing, the number of water quality parameters and the number of samples. For a BOD, TSS, total Kjeldahl nitrogen and NO_3-N and NH_4-N sample, the costs will be *ca.* US$150 per sample.

Actual reported costs for all operations support the notion that wetlands are very low-cost systems. The annual O&M costs for Denham Springs, Louisiana, USA (11 355 $m^3\ d^{-1}$), were US$29,550, which included the costs of operating the aerated lagoon and chlorinator. Mesquite, Nevada, USA (1500 $m^3\ d^{-1}$), has an operating budget of US$10,000. This provides a range of 1–2¢ m^{-3} for an operating budget.

The O&M costs for FWS wetlands are similar in nature to those for SSF wetlands. The components include labour and a generally minor amount of power. Reported O&M costs are few. At Cannon Beach, Oregon, U.S.A., the US$50,000 yr^{-1} budget amounts to 5¢ m^{-3}. For Mt View Sanitary District, Martinez, California, USA, the O&M budget of US$50,000 yr^{-1} amounts to 3¢ m^{-3} for 4920 $m^3\ d^{-1}$.

9.3 Present worth (net present value)

Treatment wetlands often provide very large cost savings because of low O&M costs. The proper evaluation of alternatives therefore requires a consideration of capital and O&M costs. The proper technique for combining the two is a present-worth analysis.

The total cost of a project at the time of inception is the total of capital costs, engineering services and the present worth of O&M costs over the project life. This approach to economic estimating is required when the alternatives under consideration vary greatly in their life expectancies and in their O&M costs. This is the case for wetlands. The overall project evaluation requires the consideration of both capital and O&M costs, and the present-worth technique is the appropriate vehicle for combining the two. The present worth of O&M costs, including equipment repairs and replacements, is the money that needs to be set aside

Table 9.5. Estimated cost comparison (thousands of US dollars) for phosphorus decrease in agricultural runoff

Wetland alternative		Chemical treatment alternative	
Capital costs		*Capital costs*	
Total replacement	129,748	Total replacement	107,770
Land	34,134	Land	2140
Procurement premium	10,330	Procurement premium	375
Pump station capital cost	14,288	Replaceable equipment	
Pump station replacement	522	Pumps piping electrical	18,670
(Present worth, 8% discount rate)		Mixing through thickening	51,390
(25% pump station capital replaced at 25 years)		Equipment replacement (Present worth, 8% discount rate)	
		Pumps piping electrical 25% at 25 years	682
		Mixing through thickening 100% at 20 and 40 years	1948
Land free capital cost	**95,836**	**Land free capital cost**	**108,260**
Operating and maintenance costs		*Operating and maintenance costs*	
Labour	592	Labour	1060
Materials	124	Materials	250
Chemicals	0	Chemicals	560
Energy	228	Energy	228
Monitoring	150	Monitoring	150
Total annual O&M	1094	Total annual O&M	2490
Present worth O&M	**33,443**	**Present worth O&M**	**76,153**
(50 year life span, 8%)		(50 year life span, 8%)	
Total present worth, capital + O&M	**129,279**	**Total present worth, capital + O&M**	**185,637**

Base information from Brown & Caldwell (1993) and Burns & McDonnell 1993.

now, at the prevailing interest rate, to pay for these future costs.

An alternative comparison is illustrated in Table 9.5, evaluating wetland treatment and chemical treatment to remove P from agricultural runoff water in southern Florida, USA. The dollar values in this example are large because the basis is the treatment of a very large flow (*ca.* 200 MGD). The estimates in this table were developed from information available at the time of final conceptual design, and are subject to change during final design and the accompanying modifications. The example is included here to illustrate the unique features of wetland alternatives evaluation.

The chemical treatment alternative is 17% cheaper than the treatment wetland on the basis of the capital expenditures needed to build the project. On the surface, this makes chemical treatment the more attractive alternative. This in-advance comparison presumes a life span short for equipment not to need replacement, nominally 20 years. However, if the life span of the project is taken to be 50 years - which is characteristic of constructed wetlands, and has been demonstrated in the region - the analysis changes. It becomes necessary to consider the salvage value of worn-out components and their replacement costs. In Table 9.5, it is assumed that worn-out equipment has zero value: it is unsellable and there is no charge for disposal. In contrast, land acquired for the project is assumed to maintain its value; no replacement purchases are necessary. It is therefore logical to exclude land costs from the analysis because the land can be sold at the conclusion of the project with no loss in value.

When these factors are taken into consideration, the treatment wetlands are 13% cheaper than chemical treatment. The conclusion of the capital cost analysis is reversed.

Next, the O&M costs are totalled and converted to their present worth. Chemical treatment, as the name implies, requires more energy, materials, labour and supplies than wetland treatment. Monitoring costs would be the same. In this example, and in virtually all cases like it, O&M costs are higher for the equipment-oriented technology. The annual O&M for chemical treatment is twice as expensive. The present worth of O&M is a significant fraction of the capital cost. Consequently, the total present worth of the wetlands project is only 70% of the total present worth of the chemical treatment alternative.

The assumed factors in this example will not apply in all circumstances, but they do serve to indicate that extra care should be taken in economic analysis of a wetland alternative.

10 Case studies

10.1 Free water surface constructed wetlands

10.1.1 Domestic: Vermontville, Michigan, USA

Vermontville is a rural community located 40 km southwest of Lansing, Michigan, USA. The local maple syrup industry is active; each year a festival brings thousands of visitors to this community of 825 residents. The Clean Water Act of the early 1970s dictated that Vermontville upgrade its wastewater treatment capabilities. In common with many other small communities, Vermontville could not afford to own or operate a 'high-tech' physical–chemical wastewater treatment plant. However, it was situated to use the land-intensive natural systems technology, and decided to do so. In 1972, Vermontville opted for facultative lagoons followed by seepage beds. Those seepage beds unexpectedly became wetlands, a system that works remarkably well and is liked by the operators.

The municipal wastewater treatment system at Vermontville consists of two facultative stabilization ponds of 4.4 ha, followed by four diked surface (flood) irrigation fields of 4.6 ha constructed on silty–clayey soils. The system is located on a hill with the ponds uppermost and the fields at descending elevations (Figure 10.1). The irrigation fields are totally overgrown with volunteer emergent aquatic vegetation, mainly cattail. The system was designed for 380 $m^3\ d^{-1}$ and a life of 20 years. It is currently operating successfully in its 25th year.

Pond-stabilized wastewater is released, during the unfrozen season, into each wetland by gravity flow through pipes having several ground-level outlets in each wetland. There is a constant surface overflow from the final wetland, made up of ground-recycled wastewater that vents into the final field.

Weekly monitoring over the period 1989–98 yielded the following outflow water quality:

outflow	65 $m^3\ d^{-1}$
BOD_5	2.1 mg l^{-1}
TSS	5.0 mg l^{-1}
faecal coliforms	122 per 100 ml
TP	0.23 mg l^{-1}
NH_4-N	0.75 mg l^{-1}
pH	7.07
dissolved O_2	6.64 mg l^{-1}.

The Vermontville volunteer wetland system created a marshland habitat suitable for waterfowl production that was otherwise not present in the immediate area. Many other types of bird also nest in the marshes, including red-wing blackbirds, American coot and Amerian goldfinch. Waterfowl (blue-winged teal and mallard), shorebirds (gallinule, killdeer, lesser yellow-legs and sandpiper) and swallows use the wetland pond system for feeding and/or resting during their migration. Great blue heron, green heron, ring-neck pheasant and American bittern have also been seen frequenting the wetlands. These volunteer wetlands are also an important habitat for numerous amphibians and reptiles. These include snapping and painted turtles, garter and milk snakes, green and leopard frogs, bullfrogs and American toads. Muskrats inhabit the wetlands, whereas raccoons, whitetail deer and woodchucks are seen feeding in the wetlands.

Very little wetland maintenance has been required at Vermontville. The berms are mown three or four times per year, for aesthetic reasons only. Water samples are taken on a weekly frequency at the surface outflow. The discharge risers within the wetlands are visited and cleaned periodically during the irrigation season. There is essentially nothing to be vandalized, and no repairs have been required. The dikes are monitored for erosion, which has not been a significant problem. Muskrats build lodges and dig holes in the dikes; woodchucks also dig holes in the berms. A trapper is therefore allowed on the site to remove these animals periodically.

The Vermontville wetlands show a build-up of 0.1–0.2 m of organic residues, largely in the form of cattail straw. There was one attempt to burn the accumulated detritus, which proved to be difficult and of no value in the system's operation or maintenance. The amounts of this material have not compromised the freeboard design of the embankments over the system's 18-year operational period. Tree control has not

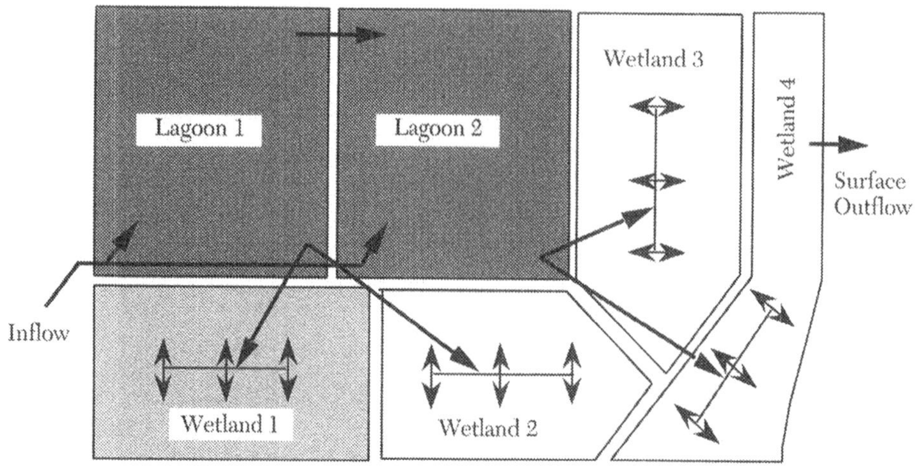

Figure 10.1. Layout of the wastewater treatment system at Vermontville, Michigan, USA. Inflow can be directed to either of the two lagoons. The lagoons discharge to wetlands 1–3. Wetland 4 no longer receives a direct discharge, only seepage from the adjacent uphill cells.

been practised at Vermontville, and the wetlands now contain willow trees up to several metres in height. No hydraulic problems have been experienced owing to these trees, or to any other cause.

The Vermontville ponds and wetlands cost US$395,000 to build in 1972. Much of this expense was incurred in grading, because of the uneven topography of the site. The O&M costs associated with the wetlands portion of the treatment system are quite low. In 1978 these were *ca.* $3500 per year, of which $2150 was labour and field costs, and the balance was for water-quality analytical services. In 1990 these same costs totalled *ca.* $4200, including $3400 for labour and field costs. The vegetation and relatively small surface overflow from the final wetland constitutes an established system providing treatment and wildlife values very economically.

10.1.2 Multiple-benefit

10.1.2.1 Case study: Show Low Wetland, Arizona, USA

Description. The Show Low Wetland is a widely known example of the innovative use of constructed wetland technology. The first wetland in the complex, Pintail Lake, was the first constructed wetland in Arizona to receive municipal wastewater, and began receiving effluent in 1979. The complex has grown to include similar wetlands (Redhead Marsh and Telephone Lake in 1986); in 1994 the constructed wetland complex included 13 cells totalling 75 ha. The Show Low constructed wetlands are located on US Forest Service (USFS) lands under the terms of a cooperative agreement with the Arizona Game and Fish Department (AGF) and the City of Show Low. When a strict discharge limit was imposed on Show Low Creek, the City of Show Low had to look elsewhere to dispose of its treated effluent. The USFS, AGF and the City became partners in this created wetland project as each entity saw opportunities to accomplish shared goals in a cooperative venture. This partnership continues today, and other groups have joined, including the local Audubon Chapter. Table 10.1 summarizes design information for the Show Low wetlands.

The Show Low constructed wetlands were designed with the multiple purposes of effluent management and the improvement of wildlife habitat. The wetlands were designed with large open water with nesting islands for waterfowl, and fringing areas with shallow water levels to favour the growth of emergent vegetation. Planting of a variety of emergent and submergent wetland plant species produced a diverse and productive wildlife habitat. In addition, the constructed wetlands are fenced to exclude domestic livestock grazing.

Operational performance. The Show Low wetlands were designed to improve water quality as water moves through cells arranged in series. Good wildlife habitat depends on good water quality. Water clarity is especially important to allow submergent vegetation to grow in the water column. Wildlife response to the created and improved wetlands is the best indicator of success. Bird surveys conducted during a 16-week period of 1991 found 125 species using the wetlands. So far, there are 14 species of birds that are of special interest because of their rarity in Arizona. Four of these rare bird species nest in the created wetlands.

There is no surface discharge from the Show Low wetland. All water is lost to evaporation and seepage. Table 10.2 summarizes water quality data for the wetlands.

Special features/issues. The Show Low Wetlands were originally designed as zero-

Table 10.1. Case history summary for Show Low, Arizona, USA

Construction start date: Phase 1, Pintail Lake, 1977
Phase 2, Redhead Marsh, 1986
Operation start date: 1979
Construction cost: 1977, US$146,750; 1986, US$300,000
Operation cost: USFS US$9000 yr^{-1} AGF US$3000 yr^{-1} City of Show Low US$12,000 yr^{-1}
Design flow: 5375 m^3 d^{-1} (1,420,000 US gallons d^{-1})
Total constructed wetland area: 75 ha (186 acres)
Wastewater source: Show Low City municipal effluent
Cell design:

	Pintail Lake	Redhead Marsh	Others
Number of cells	3	3	7
Design depth	90 cm (3 ft)	90 cm (3 ft)	0.9–1.8 m (3–6 ft)
Cell areas	23 ha (57 acres)	20 ha (49 acres)	32 ha (80 acres)
Cell aspect	East	West	All
Plant types	Emergent	Emergent	Emergent

Discharge location: no discharge
Time period: year-round discharge
Actual inflow: 2135 m^3 d^{-1} (564,200 US gallons d^{-1})

discharge facilities. Recently, three of the basins have been declared 'Waters of the US'. These wetlands have received a discharge permit from the US Environmental Protection Agency under a special provision that recognizes the net ecological benefits resulting from the effluent discharge. In addition to wildlife, these constructed wetlands attract human visitors. The Pintail Lake Public Use Facility includes a paved trail for handicapped access and an enclosed viewing blind large enough to accommodate 50 students. This facility attracts local, within-state, out-of-state and international visitors, and is a popular outdoor classroom for local students learning about effluent recycling, wetland ecology and wildlife.

Table 10.2. Operational water quality data summary for the constructed wetlands at Show Low, Arizona, USA

BOD_5 in (mg l^{-1})	38
BOD_5 out	None
TSS in (mg l^{-1})	90
TSS out	None
TN in (mg l^{-1})	10.4
TN in wetland (mg l^{-1})	4.0

10.1.3 Petroleum

10.1.3.1 Case study: Chevron Richmond Refinery Wetland, Richmond, California, USA

Description. The Chevron Richmond Refinery Wetland (RRW) originated in 1988 as a pilot study marsh in the Number Two oxidation pond at Chevron's Richmond refinery in Point Richmond, California. The pond was used as a polishing pond for refinery effluent between 1963 and 1985. However, water flow to this pond was decreased during this period; by 1985 the pond no longer provided any positive benefit (Chevron 1996). The pond was dewatered and allowed to dry, serving as a storage basin for stormwater. The mud bottom of the pond became dry and cracked, creating an eyesore. Management at Chevron requested that the visual appearance of the pond be enhanced, so in 1986, the soils were tilled and sampled, and were found to be capable of supporting a variety of vegetation. A two-stage revegetation programme for the pond was implemented with the approval of the California Regional Water Quality Control Board and the help of the California Department of Fish and Game and the National Audubon Society. By 1989, the first stage (12.14 ha) was planted in the pond, and the RRW became operational. The second stage of planting followed with an additional 12.14 ha of wetland plants. The 12.14 ha were kept as a mud flat for shorebird habitat.

The intent of this study was to demonstrate the feasibility of enhancing the effluent water quality by allowing it to pass through a 'created' but 'natural' overland flow wetland (Chevron 1996).

Operational performance. The RRW began operating in 1989 and successfully serves many functions, including water polishing treatment, stormwater storage, habitat for various waterfowl and shorebirds, and design and water quality performance data for the RRW. From 1989 to 1992 vegetation and sediments were

Table 10.3. Summary of design and performance of Richmond Refinery Wetland, Richmond, California, USA (Chevron 1996)

Operation start date: 1988
Constructed wetland area
Total: 36.42 ha
Pass 1: ~12.14 ha
Pass 2: ~12.14 ha
Pass 3: ~12.14 ha
Typical flow: 9500 $m^3 d^{-1}$
Wastewater source: refinery effluent

Parameter	Influent quality (average)	Effluent quality (average)
BOD (mg l^{-1})	12.2	7.1
TSS (mg l^{-1})	35.9	34.1
TDS (mg l^{-1})	2.6	2.9
TP (mg l^{-1})	89.8	73.3
TN (mg l^{-1})	5.5	1.9

TDS, total dissolved solids.

sampled annually for accumulation of heavy metals. Bird use and reproduction have been conducted at the RRW since its inception. A study of the aquatic invertebrate population living in the RRW was conducted in 1991, and a detailed study of shorebird use of the RRW was conducted in 1994 and 1995.

Operation of the RRW from 1988 to 1991 resulted in a decrease in several water quality parameters as summarized in Table 10.3.

A total of 8 orders and 53 families of invertebrates contributing to the food chain at the RRW were identified during the invertebrate survey (Chevron 1996). The wetland has demonstrated the ability to improve water quality while providing significant habitat for numerous waterfowl.

Special features/issues. The single most important design factor contributing to the physical success of the RRW is the ability to control water flow rates and levels. Proper water management is the key to optimizing plant propagation, water quality and habitat use.

A complete census of the wildlife species using the wetland was taken and logged during 1990 and 1991 by Chevron wetland staff and members of the National Audubon Society. The estimated total number of birds using the wetland during 1991 was over 2 million individuals, based on a daily average of *ca.* 5600 individuals. The heaviest use was during the spring and autumn migrations, when huge numbers, sometimes 25,000 per day, of transient shorebirds were on the wetland. Up to 85 different species of birds were sighted. These birds included those that have special status afforded them by either state or federal agencies, such as the California clapper rail (*Rallus longirostris*), common yellowthroat (*Geothlypis trichas*) and osprey (*Pandion haliaetus*). Among the 85 species, ground-nesting resident birds include mallard (*Anas platyrhynchos*), gadwall (*Anas strepera*), northern pintail (*Anas acuta*), Canada goose (*Branta canadensis*), black-necked stilt, American avocet (*Recurvirostra americana*) and killdeer (*Charadrius vociferus*). These birds have been recorded as having successfully raised broods in successive years and, for the most part, can be seen all year round (Chevron 1996).

10.1.4 Urban stormwater

10.1.4.1 Case study: Hidden River Corporate Office Park, Tampa, Florida, USA

Background. Urban stormwater has been identified as a major source of pollutant loadings to surface waters in Florida (Livingston 1989) and in the USA (US Environmental Protection Agency 1989). Both constructed and natural wetlands are being used for stormwater quality management in Florida (Rushton *et al.* 1997; Kehoe *et al.* 1994). Constructed wetlands and wet detention systems are being widely used for stormwater management in the USA (Schueler 1992; Strecker *et al.* 1992). A detailed study of the performance of one of these stormwater treatment wetlands was conducted from May 1991 to October 1993 in Tampa, Florida (Carr & Rushton 1995).

Project description. The Hidden River stormwater treatment wetland in Tampa consists of two constructed inlet basins with a combined surface area of *ca.* 750 m^2 and a natural herbaceous marsh of *ca.* 1.21 ha (Figure 10.2). This wetland system receives runoff from a 6.2 ha drainage basin that includes multi-story offices and parking lots. The watershed:wetland ratio for this system is *ca.* 4.8:1.

Forty wetland plant species were documented in the Hidden River wetland during this study period. Dominant plant species were maidencane (*Panicum hemitomon*), pickerelweed (*Pontederia cordata*) and fragrant waterlily (*Nymphaea odorata*).

Results. Data were collected for a total of 81 storm events. The mean rainfall for all events was 1.5 cm. Mean rainfall for the 81 sampled events was 2.9 cm. The mean antecedent period between these monitored events was 9.9 days and the mean event duration was 3.04 h. The average inflow volume to the wetland during the 29-month monitoring period was 136 $m^3 d^{-1}$ and the outflow volume was 52.4 $m^3 d^{-1}$. The average inlet HLR to this wetland system was 1.06 cm d^{-1}.

Water quality performance results are summarized for the Hidden River stormwater treatment wetlands in Table 10.4. Inlet concentrations for most pollutants were relatively low, and mass removal efficiencies were high for NH_4-N, NO_2-N + NO_3-N, P, TSS and several trace metals (zinc, copper, cadmium and lead). A significant amount of the observed pollutant mass decreases were due to groundwater losses from the wetland. The wetland treatment system was ineffective at decreasing the loads of sodium, manganese, magnesium, iron and chloride.

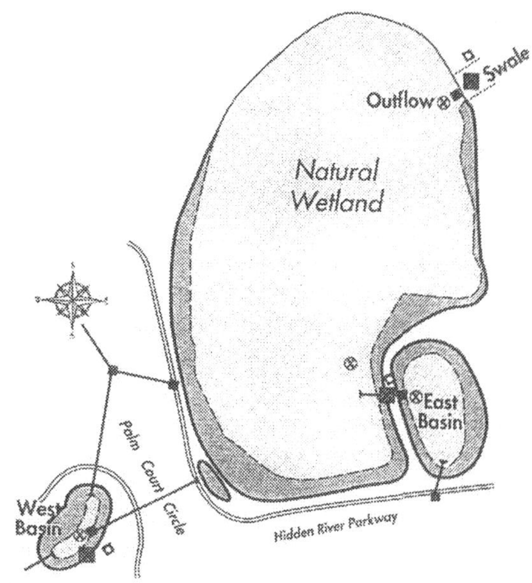

Figure 10.2. Hidden River Corporate Office Park stormwater treatment wetland, Tampa, Florida, USA. (From Carr & Rushton (1995).)

10.1.5 Animal wastewater

10.1.5.1 Case study: Oregon State University Dairy Farm Treatment Wetlands, Corvallis, Oregon, USA

Background. Oregon State University designed and built six wetland demonstration/research systems. The project was designed to determine the treatment efficiencies of the constructed wetlands by season at various hydraulic and nutrient, solids and organics loading rates and for the final polishing of pretreated wastewater before discharge. The design was based on the 1988 US Environmental Protection Agency publication *Design manual on constructed wetlands and aquatic plant systems for municipal wastewater treatment* (EPA/625/1-88/022). They were constructed at a site south of the university dairy barns in spring 1992 and started up in autumn 1993 (Gamroth *et al.* 1993). The site has an average of 60 cm soil depth of Amity silty clay and Bashaw clay loam. A poorly drained mottled clay layer is below the soil layer. The cell bottoms are compacted Bashaw clay and the cell berms are compacted Amity clay.

The funding for this project was obtained from USDA (60%), the US Environmental Protection Agency (20%), State Experiment Station (15%) and the Oregon Dairy Farmers Association (5%) (percentages of budget are shown in parentheses).

Design. Each system contains a single cell that measures 26.7 m × 5.5 m (147 m^2) with a length:width ratio of *ca.* 5:1. Liquid depths range from 30 to 45 cm and the average slope of cells from upstream to downstream is 0.5%. The constructed wetlands treat stored animal wastewater from a flush dairy operation. The dilution ratio during the 1994 and 1995 operating periods was 94.5% recycle water to 5.5% pretreated wastewater. The recycle water is treated effluent from the wetland cells and the wastewater is pretreated in a solids separator. The site was designed to deliver a fixed volume of 5790 l d^{-1} of wastewater to each system for a total wastewater flow of 34,750 l d^{-1}. The flow to the systems is a fixed proportion of the total flow generated by the livestock operation, allowing the researchers to maintain the hydraulic and nutrient loading rates at the design levels throughout the year. Dilution of the wastewater ensured that maximum loading rates for biochemical oxygen demand of 74 kg $ha^{-1} d^{-1}$ and concentrations of 100 mg NH_3 l^{-1} and total solids of 1500 mg l^{-1} would not be exceeded. The wastewater that is delivered to each system makes a single pass through the cell and is collected and returned to the main dairy wastewater storage system.

Operations and maintenance. Two of the systems were planted with cattails (*Typha latifolia*) and four, including the two systems with deep zone areas, with hardstem bulrush (*Scirpus acutus*) to determine the effects of different types of vegetation and the use of deep zone areas on removal rates. After the vegetation had become established, nutria (*Myocastor coypus*), a rodent that is native to South America, created problems for this

Table 10.4. Summary of operational data from the Hidden River Stormwater treatment wetland in Tampa, Florida, USA, from May 1991 to October 1993 (from Carr & Rushton 1995)

	Concentration (mg l^{-1})			Pollutant mass (kg)		
Constituent	East in	West in	Out	Total in	Total out	Efficiency (%)
NH_4-N	0.049	0.066	0.041	4.82	0.14	79
TKN	0.986	0.523	1.200	47.24	3.11	34
NO_3 + NO_2-N	0.063	0.157	0.025	10.42	0.07	94
TN	1.130	0.680	1.235	58.62	31.81	46
Orthophosphate	0.042	0.047	0.014	2.24	0.07	67
TP	0.145	0.071	0.045	6.2	0.19	70
TSS	13.8	4.6	3.0	528	74.9	86
TOC	6.8	4.1	16.4	286	261	9
SO_4	5.4	5.14	3.88	96.4	45.4	53
Calcium	18	10.3	8.4	192	118	39
Chloride	1.15	0.90	2.63	11.63	33.55	-189
Potassium	0.167	0.107	0.106	2.17	2.08	4
Sodium	0.393	0.173	0.828	8.40	10.29	-23
Cadmium	0.002	0.002	0.001	0.128	0.016	88
Copper	0.005	0.004	0.003	0.278	0.059	79
Iron	0.384	0.109	0.337	14.18	13.50	5
Lead	0.002	0.001	0.001	0.116	0.02	83
Manganese	0.014	0.012	0.016	0.697	0.685	2
Zinc	0.017	0.096	0.018	2.95	0.465	84

wetland site in the early stages of operation by destroying most of the plants and burrowing into the berms. A welded wire fence with 5 cm × 7.5 cm holes that extended 5 cm below the ground surface was erected around the site and has been successful in keeping the nutria out. An electric fence wire runs the perimeter of the fence *ca.* 15 cm above the ground. No noticeable damage has occurred from nutria since the installation of the fence.

No major operating problems were experienced since the start-up of the system. However, minor intermittent problems were encountered with uneven distribution of the wastewater across the cells since switching the operation of two cells to a 2-day detention time and four cells to a 7-day detention time. In spite of pretreating the wastewater by allowing the wastewater to flow through a screen, solids have entered the piping system and have restricted the flow by partly blocking the flow control valves. This problem was remedied in 1996 by routing all of the wastewater flow for the systems to one central distribution box containing six adjustable steep-angle V-notch weirs to permit frequent visual inspection and easier servicing.

Study results. Performance data from the Oregon State University wetland systems showed an increase in removal efficiencies from the first year of operation to the second year for faecal coliforms (80–90% compared with 89–95%), BOD (40–50% compared with 59–72%), total Kjeldahl nitrogen (50–55% compared with 54–69%), COD (40–50% compared with 53–59%) and TP (40–50% compared with 54–69%). Total solids (40–50%) removal efficiencies were relatively unchanged in the second year of operation. There was no noticeable improvement in treatment efficiency in the two wetland systems that had the deep centre sections or between cells with differing mixes of plant populations (Moore *et al.* 1995). Average concentrations and percentage change during warm and cold-weather conditions are presented in Table 10.5. Maximum and minimum concentrations of selected parameters are shown in Table 10.6.

High BOD_5 and NH_4-N concentrations of *ca.* 700 and 130 mg l^{-1} respectively have led to oxygen depletion in the wetland cells with the consequent decreased nitrogen removal rates owing to the inhibition of nitrification. In the late summer of 1995, a further dilution of the wastewater to decrease the BOD_5 concentration to *ca.* 100 mg l^{-1} and an increase in the hydraulic retention time to 7 days in four of the cells maintained sufficient oxygen supply in the wetland to improve the nitrification efficiency of these systems. This loading rate represents 10% of the 1994 loading rate.

A study was conducted that found that high ammonia concentrations of up to 71,000 mg l^{-1} did not inhibit wetland plant seed germination. It is speculated that plant mortality might have been due to volatile acids in the wastewater.

Table 10.5. Average concentrations or readings and percentage changes for the Oregon State University treatment wetland systems, Corvallis, Oregon, USA, 1993–1995 (Moore et al. 1995)

Parameter	Season	Average concentration or reading		
		Influent	Effluent	Change (%)
BOD_5 (mg l^{-1})	Warm	981	290	70
	Cold	471	208	56
COD (mg l^{-1})	Warm	2812	1245	56
	Cold	1686	896	47
NH_3 + NH_4-N (mg l^{-1})	Warm	166	82	51
	Cold	88	52	41
Org-N (mg l^{-1})	Warm	225	109	52
	Cold	117	68	42
TP (mg l^{-1})	Warm	44.9	22.7	50
	Cold	20.6	12.4	40
PO_4-P (mg l^{-1})	Warm	–	–	–
	Cold	4.9	1.9	61°
TSS (mg l^{-1})	Warm	748	144	81
	Cold	336	140	58
Dissoved oxygen (mg l^{-1})	Warm	2.72	0.15	94
	Cold	5.14	0.28	95
Faecal coliforms (per 100 ml)	Warm	907,000	78,000	91
	Cold	1,520,000	211,000	86
pH	Warm	7.43	7.14	4
	Cold	7.50	7.10	5
Water temperature (°C)	Warm	12.9	12.1	–
	Cold	7.6	7.3	–
Total solids (mg l^{-1})	Warm	3329	1736	48
	Cold	1586	958	35

° Only eight samples.

Table 10.6. Maximum and minimum wetland outlet concentrations for selected parameters at the pilot dairy waste treatment wetlands at Oregon State University, Corvallis, Oregon, USA, 1993–1995 (Moore et al. 1995)

Parameter	Maximum concentration (mg l^{-1})	Minimum concentration (mg l^{-1})
NH_3+NH_4-N	301	12
TSS)	1705	75
TP	115	3.5
PO_4-P	12.0	1.2

A vegetation competition cell was established in April 1994 in the fifth cell to determine which species of vegetation would survive and flourish in the various areas of the cell with changes in contaminant concentration from point of influent to point of effluent. The cell was divided into a 4 × 19 grid totalling 76 sections, each 1.5 m × 1.5 m in size. Each section contained a single plant species of *Typha latifolia*, *Alisimo plantago-aquatica*, *Scirpus microcarpus*, *Elocharis palustris* or *Hydracotyl ranunculoides*. Representative plants of each species were planted along the length of the cell in multiple locations to ensure that all plant species were exposed to the decreasing wastewater strength as the water passed through the wetland system. The percentage vegetation cover by species as of 9 October 1994 in each of the treatment wetland cells is summarized in Table 10.7.

Conclusion. The shape of the wetland cell bottom with a deep centre section does not have any impact on the treatment efficiency of common water quality parameters. Although the treatment provided by plants differs with species, the difference between mixed plant populations in wetland cells seems to be similar. Wetland cells providing 7 days of detention time and treating flush water from a dairy can remove between 45% and 70% of the major water quality parameters and up to 90% of the faecal coliforms. The removal of P needs to be

Table 10.7. Percentage coverage by plant species at the treatment wetland site at Oregon State University, Corvallis, Oregon, USA, in October 1994 (Moore et al. 1995)

Cell no.	*Typha latifolia*	*Scripus acutus*	Grass*	*Hydrocotyle*	*Lemna* sp.
1	30	5	20	0	45
2	30	10	60	0	0
3	30	20	50	0	0
4	10	5	80	0	5
5	35	20	5	30	10
6	0	20	5	0	75

* *Glyceria occidentalis* and *Alopecurus geniculatus*.

studied for longer to confirm long-term removals (Moore *et al.* 1995).

10.1.6 Agricultural runoff

10.1.6.1 Case study: Everglades Nutrient Removal project, Florida, USA

A 1543 ha freshwater wetland has been constructed in south Florida, USA, to remove nutrients from runoff from 280,000 ha of former peatlands now in agriculture. The receiving ecosystem is the Everglades, an ultra-oligotrophic wetland complex. The Everglades Nutrient Removal (ENR) project is a prototype for, and comprises 10% of, several other buffer wetlands now in design and operation. These stormwater treatment areas (STAs) are the first phase of the Everglades protection project. The design goals are the removal of 75% of the P load and to decrease TP concentrations to less than 50 $\mu g\ l^{-1}$. During the first 50 months of flow-through operation, after 12 months of accommodation at total recycle, removals have been 82% and effluent concentrations have averaged 20 $\mu g\ l^{-1}$. Four-year results confirm the design basis for the entire complex of STAs.

The Everglades is a vast freshwater wetland located in south Florida that, before 1900, encompassed more than 10,000 km^2 and extended from the south shore of Lake Okeechobee to the mangrove estuaries of Florida Bay. Agricultural and urban development during the past 80 years has decreased the present size of the Everglades by almost 50%, of which 3400 km^2 has been impounded within shallow, diked reservoirs known as Water Conservation Areas (WCAs). However, the remaining wetland still contains a variety of habitats (such as tree islands, wet prairies and aquatic sloughs) that support unique biotic communities and is widely recognized as an ecosystem of immense regional and international importance.

Historically, the Everglades is thought to have been an ultra-oligotrophic system, with particularly low surface-water concentrations of P and other micronutrients. Contemporaneous levels of TP at remote sites in the interior of the Everglades typically range from 10 to 30 $\mu g\ l^{-1}$, whereas levels of soluble reactive P are often 4 $\mu g\ l^{-1}$ at most. Eutrophication of the system in recent decades has been attributed to excessive P loading in runoff from the 2800 km^2 Everglades Agricultural Area (EAA), which is located to the north and west of the remaining natural wetland. Present-day TP concentrations at inflows to the WCAs range from 100 to 250 $\mu g\ l^{-1}$. Species shifts in microbial, plant and macroinvertebrate communities have been linked with this degradation of surface water quality.

Legislation requires the South Florida Water Management District (SFWMD) to mitigate environmental problems in the Everglades associated with eutrophication and hydroperiod disruption. A key component of SFWMD's proposed US\$800M Everglades Restoration Plan is the construction of enormous wetlands (*ca.* 16,000 ha in total) to serve as STAs, which will temporarily hold runoff from the EAA and decrease P concentrations to acceptable levels before the water is released southwards into the WCAs. The long-term nutrient removal mechanism that will operate in the STAs involves the initial incorporation of P into macrophyte tissues and the subsequent burial, with minimal decomposition, of this biomass in the bottom sediments as peat. The ENR project was built as a technology demonstration project and was designed to operate as an STA to decrease TP levels in agricultural runoff.

The wetlands were fully inundated and flowing in a total recycle mode for one year (September 1993 to August 1994) before the initiation of through-flow. During this period, leachable nutrients, associated with antecedent land uses, were redissolved. Initial concentrations of 100–400 $\mu g\ l^{-1}$ decreased to 20–30 $\mu g\ l^{-1}$ after a few months, presumably owing to utilization by the regrowing wetland vegetation.

The ENR is operated as a once-through treatment system and has the capacity to process about one third of the annual runoff that would otherwise be pumped directly into the Refuge. Water is first pumped from the inflow pump station (six electric pump units with a

Table 10.8. Water and phosphorus loading rates for Everglades Nutrient Removal wetland, Florida, USA, 18 August 1994 to 31 July 1998

Loading source	Water (cm d^{-1})	Water (%)	Phosphorus (g m^{-2} yr^{-1})	Phosphorus (%)
Inflow pumps	3.08	71.5	1.26	91.3
Seepage return pumps	0.67	15.5	0.05	3.6
Rainfall	0.44	10.2	0.03	2.0
Dry deposition	–	–	0.03	2.2
Incoming seepage	0.12	2.8	0.01	0.9
Total	**4.62**	**100.0**	**1.38**	**100.0**

total capacity of 16.99 m^3 s^{-1}) into the buffer cell (53 ha) and then distributed by gravity flow to two independent, parallel treatment trains separated by a transverse levee (treatment cells 1 and 3, and 2 and 4). Treatment cells 1 and 2 are intended to remove the bulk of the nutrient load that enters the ENR, whereas treatment cells 3 and 4 accomplish the final polishing of the water to lower nutrient concentrations. Water is discharged from the ENR at the outflow pump station (six electric pump units with a total capacity of 12.74 m^3 s^{-1}) over the L-7 levee into WCA-1. A canal located on the outside of the western and northern section of the perimeter levee collects groundwater seepage from the ENR and returns it to the seepage return pump station (three electric pump units with a total capacity of 5.66 m^3 s^{-1}), where it is pumped back into the headworks of the project.

Treatment cells 1 and 2 have been allowed to revegetate naturally; the dominant emergent macrophyte is cattail (*Typha domingensis* and *T. latifolia*). Treatment cell 3 is a mixture of naturally recruited cattail and areas that were planted with wetland species common to south Florida, i.e. arrowhead (*Sagittaria latifolia* and *S. lancifolia*), spikerush (*Eleocharis interstincta*), maidencane (*Panicum hemitomon*), pickerelweed (*Pontederia cordata*) and sawgrass (*Cladium jamaicense*). Treatment cell 4 has been actively maintained through the selective use of herbicides as an open-water periphyton/submerged macrophyte community that is dominated by coontail (*Ceratophylum demersum*) and southern naiad (*Najas quadalupensis*).

A water budget was generated for the ENR on the basis of: (1) daily flow measurements at the inflow and outflow pump stations, (2) total daily rainfall collected at a network of automated tipping-bucket gauges located throughout the project, and (3) daily estimates of surficial seepage entering the ENR from WCA 1 along the L-7 levee. These water budget data were the basis for nutrient mass balance budgets.

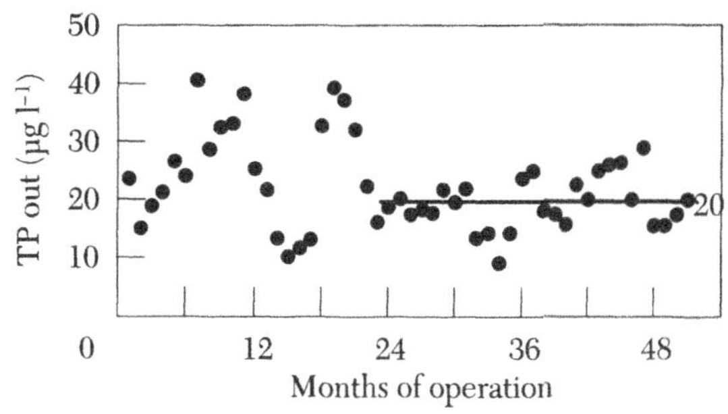

Figure 10.3. TP into and out of the Everglades Nutrient Removal agricultural stormwater treatment wetlands.

Figure 10.3 shows that TP concentrations were decreased from *ca.* 120 to 20 µg l^{-1}. The HLR during this period was *ca.* 3 cm d^{-1}, but flows were episodic, as necessitated by patterns of rainfall and runoff. Table 10.8 shows the incoming loads and their allocations. About 82% of the phosphorus load was retained within the wetlands (*ca.* 1.13 g m^{-2} yr^{-1}). Phosphorus decreases have exceeded expectations for the long-term sustainable removal rate. The temporary processes of biomass increase and sorptive saturation are presumably complete, because the wetlands have had over five years to develop. In 1999 the ENR project was augmented by the addition of a fifth cell, and new inlet and outlet facilities.

10.1.7 Leachate

10.1.7.1 Case study: Isanti–Chisago Leachate Treatment, Cambridge, Minnesota, USA

The Isanti–Chisago Sanitary Landfill, an unlined municipal solid waste facility located near Cambridge, Minnesota, USA, was closed in 1992. Leaching of soluble wastes had contaminated the surficial and increasingly deeper aquifers with toxic organic compounds and heavy metals. The Minnesota Pollution Control Agency requested an innovative treatment system with O&M costs far below a conventional system. The selected approach was a natural

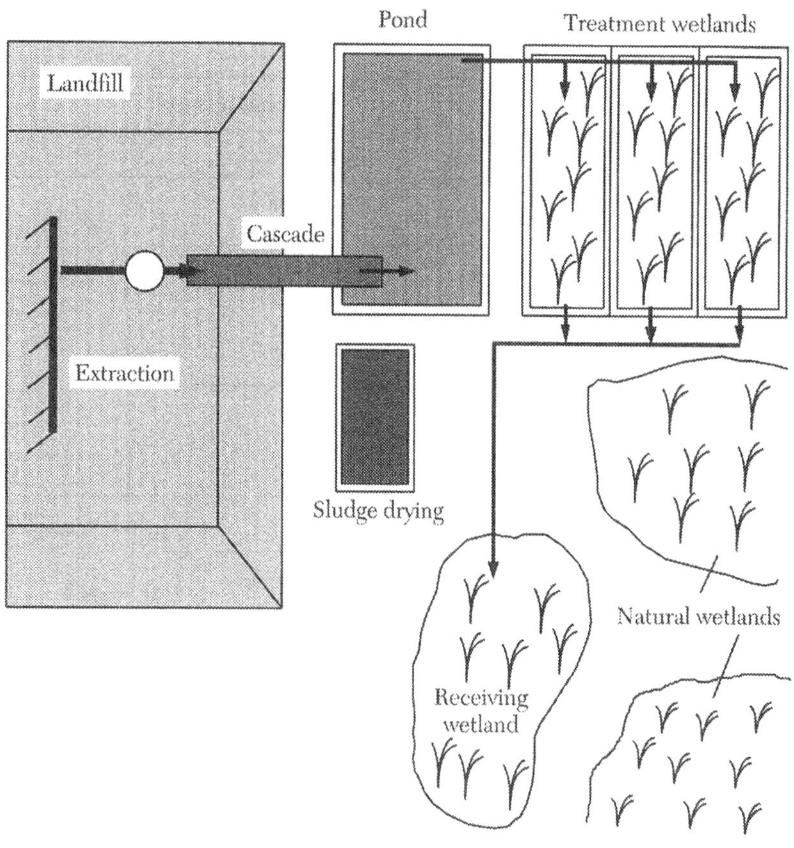

Figure 10.4. Isanti–Chisago Leachate Treatment System, Cambridge, Minnesota, USA.

systems engineering design that relies on existing topography for gravity flow (with the exception of ground water pumping), solar and wind energy inputs rather than electrical, and natural biological, chemical, and physical interactions rather than petrochemical inputs.

The system layout includes several features (Figure 10.4). Volatile organic compounds (VOCs) are removed by cascading the water down the side of the landfill in a polypropylene 'step aerator', which is designed also to increase dissolved oxygen concentrations in the water to oxidize ferrous to ferric iron, thereby precipitating the hydroxide and other solids. Co-precipitation of heavy metals also occurs in this stage.

A sedimentation basin was selected as the second component of the treatment train, to continue aeration via natural surface agitation, and oxidation/degradation via UV mechanisms; and to allow the settling of insoluble metals and other inorganic and organic solids after aeration with a cascade. The basin was designed for a 6-day residence time at a pumping rate of 600 m^3 d^{-1} (110 US gallons min^{-1}) and a liquid depth of 1.2 m (4 ft) (including settled sludge). Basin size and residence time were selected on the basis of settling rate studies, expected sludge volume generation calculations and land availability. The sedimentation basin is constructed of earthen materials and a soil-covered polypropylene liner to minimize infiltration through the base.

The next component of the treatment train is a 0.6 ha (1.5 acre) FWS constructed wetland. Three parallel-flow cells were seeded with cattails in autumn 1995 and developed into a dense stand during summer 1996. The wetland provides 3 days of residence time at a pumping rate of 600 m^3 d^{-1} and an average free water depth of 30 cm (1 ft). Continued treatment occurs via aeration, sorption, biological storage and transformation, and trapping of solids. The wetland was constructed of earthen materials and polypropylene liner. Water from the sedimentation basin enters the constructed wetland by means of gated inlet pipes, which promote evenly distributed flow. A mid-cell, deep-water channel recreates sheet flow to the second half in case channelling occurs through the vegetation. The water level is controlled by adjustable stoplogs at the cell outlets.

Discharge from the constructed wetland is to a borrow pit wetland (pond), modified to infiltrate treated water into the surficial aquifer from which the contaminated groundwater was initially removed. At the completion of the first season of treatment, results indicate that the system efficiency ranges from 85% to 100% for VOCs and from 95% to 99% for iron, with

Table 10.9. First-year pollutant removals at the Isanti–Chisago Sanitary Landfill treatment wetland, Cambridge, Minnesota, USA

Parameter	Average system removal efficiency (%)	Average system removal rate (kg ha^{-1} per season)
Volatile organic compounds	97	0.081
Iron	97	3.8
Zinc	93	0.044
Manganese	91	0.36
Arsenic	89	0.0064
Lead	80	0.00021
Mercury	75	0.000037
Chromium	67	0.001
Cadmium	65	0.00056
Nickel	19	0.013
Copper	–	0.00056

varying decreases in heavy metals (Table 10.9). O&M requirements of this simple, natural system are minimal, as was desired.

10.2 Soil-based reed beds

10.2.1 Domestic secondary treatment

10.2.1.1 Case study: Uggerhalne, Denmark

Description. The site was one of the first reed beds to be constructed in Denmark after the RZM was introduced in the early 1980s (Brix 1994). The design used was therefore based mainly on the German ideas (Kickuth 1980). It was believed that the root system of the reeds would increase the hydraulic conductivity of the soil to accommodate the hydraulic loading over a period of 3 years. Furthermore, it was prescribed that the soil should contain at least 20% clay to secure a good removal of phosphorus. Kickuth's representative in Denmark designed the system. The catchment area of the reed bed includes the village Uggerhalne, which is a small residential area north of Aalborg, Denmark. There are only small industries, such as petroleum tanks connected to the sewerage system. The sewerage system in the village is a combined system receiving rainwater as well as the domestic sewage. The system is dimensioned for secondary treatment of the sewage from 400 PE.

Constructed. August–November 1985.

Operational. November 1985 to the present.

Costs. Approximately 1 million DKr (1985) (*ca.* US$150,000).

Process description. The sewage is pretreated in an existing sedimentation tank before the inlet to the reed bed. The effluent from the sedimentation tank is pumped into the middle of the 80 m long inlet trench with open water. After passage through the reed bed, the effluent is collected in a gravel-filled effluent trench through a drainage pipe positioned in the bottom of the effluent trench; from there the effluent is led to the recipient.

Dimensions. The system consists of a single bed 33 m long and 80 m wide (surface area 2640 m^2). The depth of the bed is 0.60–0.65 m. The slope of the bed is 1.2%.

Medium. It was prescribed by the designer that the medium in the bed should consist of an imported soil containing *ca.* 20% clay and organic soil mixed in the proportion 2:1. However, a grain size analysis of the actual soil in the bed shows that the composition is 25% silt and 75% of sand (Schierup *et al.* 1990). The organic content of the soil is 5.9%, and the contents (on a dry mass basis) are: TN, 1.71 mg g^{-1}; TP, 0.34 mg g^{-1}; iron, 8.6 mg g^{-1}; calcium 2.9 mg g^{-1}; aluminium 9.4 mg g^{-1}.

Plants. Phragmites australis imported from Germany, planted in November 1985.

Liner. High-density polyethylene, 2 mm.

Inlet distribution. Open trench with gravel in the bottom.

Outlet collection. Gravel-filled trench with a 145 mm diameter poly(vinyl chloride) (PVC) drainage pipe.

Effluent standards. Effluent standards for the system were less stringent during the initial 3 years of operation, i.e. during 1986–88, because of the time needed for the vegetation to develop (Table 10.10).

Performance. The performance of the system is controlled six to twelve times a year by taking 24 h samples proportional to volume at the inlet and the outlet of the reed bed. The inlet sampling is done after the settler, i.e. the performance data presented in Table 10.11 include only the actual reed bed. The standards listed in Table 10.10 have been fulfilled throughout the whole period of operation. However, the removal of N and P is poor (*ca.* 30%) and the system does not produce a nitrified effluent.

Table 10.10. Effluent standards for the soil-based constructed wetland at Uggerhalne, Denmark

Parameter	Units	Initial 3 years	After 3 years
Amount of effluent			
During dry weather	$m^3 d^{-1}$	<150	<150
	$m^3 h^{-1}$	<15.5	<15.5
During rain	$l s^{-1}$	<10	<10
Temperature	°C	<30	<30
pH		6.5–8.5	6.5–8.5
BOD_5 (modified)	$mg l^{-1}$	40	10
Settleable sludge (2 h)	$ml l^{-1}$	0.5	0.5
TSS	$mg l^{-1}$	30	15

10.3 Horizontal subsurface flow

10.3.1 Domestic wastewater tertiary treatment

10.3.1.1. Case study: Leek Wootton, Warwickshire, UK

Description. This site was the first of a new generation of tertiary treatment reed beds built by Severn Trent Water (Green & Upton 1995; Green *et al.* 1995; Cooper *et al.* 1996). Until that time there had been very few tertiary systems in the UK, so this was very much a site on which the tertiary treatment design was being tested. The catchment area of the sewage treatment works includes the villages of Leek Wootton and Hill Wootton, where there is a resident population of 1007 with two village inns, a golf club and a training college.

Constructed. 1990.

Operational. June 1990 to the present.

Process description. The existing works consisted of a biological filter together with primary and secondary settlement tanks. These were refurbished and two new tertiary reed beds were built. There are two beds, each 15 m (length) × 28 m (width). The total area is 825 m^2 because the end of one of the beds was shaved off to fit it into the existing site. The two beds are built back to back in the house roof style with a common central inlet distributor to permit parallel inlets 28 m wide. The guidelines at the time that the system was designed in 1989 were for 1 m^2 per PE, but this was not possible on the site and it was decided to take a calculated risk with the 0.8 m^2 per PE available. As a result of the success of this design, the Severn Trent Water design is now set at 0.7 m^2 per PE at all their new tertiary beds.

Dimensions. Two beds, each 15 m (length) × 28 m (width).

Medium. Gravel, 5–10 mm.

Plants. Phragmites australis.

Liner. Monarflex low-density polyethylene (LDPE).

Inlet distribution. Riser pipes.

Outlet collection. Large stones and agricultural drainage pipe.

Standard. As 95th centiles (which allows only one failure in twenty samples):

BOD_5	20 $mg l^{-1}$
NH_4-N	10 $mg l^{-1}$
TSS	30 $mg l^{-1}$
DWF	180 $m^3 d^{-1}$.

Performance. The long-term performance data for the system are shown in Table 10.12. They clearly demonstrate the ability of the tertiary beds to remove peaks in BOD and TSS that might come from the biological filter.

The beds also managed to achieve some nitrification and some denitrification. Table 10.13 shows that the systems also manage to remove between 1.5 and 2 orders of magnitude for *Escherichia coli* and total coliforms.

10.3.2 Domestic secondary treatment

10.3.2.1 Case study: Little Stretton, Leicestershire, UK

Description. Little Stretton is a small village near Leicester. Before 1987 it was served by a communal septic tank, the overflow from which went into a drainage ditch. There were a number of similar situations in the region, and this design served as the model for others. The system was designed by WRc and was built by Severn Trent Water.

Constructed. Summer 1987.

Operational. July 1987 to the present.

Process description. The population in the village at the start was 40, but in addition 20 PE were allowed for the run-off from drainage known to come from the nearby dairy farm. (At times in the first 2 years of operation, the system had to treat up to 200 PE in BOD_5 terms.)

The system is built down the side of a hill, which permits the use of gravity flow. The system comprises eight beds, each 12.5 m × 2.0 m × 0.6 m at the bed inlet. It was the second system to be built with the use of gravel and the first to be planted with pot-grown seedlings.

The system was preceded by the existing

Table 10.11. Annual average performance data for the soil-based constructed wetland at Uggerhalne, Denmark

Year	n	q (mm d^{-1})	TSS (mg l^{-1})		COD (mg l^{-1})		BOD$_5$ (mg l^{-1})	
			Inlet	Outlet	Inlet	Outlet	Inlet	Outlet
1986	13	35	110	38.4	207	78	89	33.8
1987	11	42	113	12.9	245	110	99	14.2
1988	10	53	89	13.1	244	100	99	16.2
1989	12	34	127	7.4	314	70	164	10.1
1990	10	46	103	8.8	215	46	120	5.9
1991	8	33	179	7.1	140	30	224	5.0
1992	9	50	219	6.0	–	–	159	3.3
1993	7	27	165	5.9	450	24	225	4.8
1994	7	90	232	5.1	–	–	193	7.0
1995	8	39	125	6.1	403	77	176	3.9
1996	6	52	148	6.8	408	93	150	9.5
1997	10	39	180	5.3	377	65	184	4.5
1998*	4	39	158	6.4	330	63	115	6.0

Year	n	q (mm d^{-1})	TN (mg l^{-1})		NH$_4$-N (mg l^{-1})		TP (mg l^{-1})	
			Inlet	Outlet	Inlet	Outlet	Inlet	Outlet
1986	13	35	27.9	23.2	–	–	7.3	6.2
1987	11	42	28.3	20.3	–	–	9.1	6.5
1988	10	53	26.8	20.8	–	–	8.8	7.1
1989	12	34	37.2	20.3	–	–	12.1	7.8
1990	10	46	29.1	18.6	27.0	15.0	6.7	4.0
1991	8	33	24.0	14.0	12.0	18.2	3.7	2.1
1992	9	50	–	–	33.2	12.6	–	–
1993	7	27	94.0	31.0	28.6	14.2	9.0	7.0
1994	7	90	–	–	13.5	13.6	–	–
1995	8	39	–	–	20.9	11.6	–	–
1996	6	52	35.6	23.0	24.9	15.6	8.3	7.1
1997	10	39	38.7	20.2	28.1	13.6	9.8	6.6
1998*	4	39	22.5	16.8	17.3	12.5	4.8	4.8

n, number of samples; q, hydraulic loading rate.

Table 10.12. Annual average performance data for Leek Wootton tertiary treatment HF RBTS, Warwickshire, UK (Cooper et al. 1996)

	BOD (mg l^{-1})		COD (mg l^{-1})		TSS (mg l^{-1})		NH$_4$-N (mg l^{-1})		TON (mg l^{-1})	
	In	Out	In	Out	In	Out	In	Out	In	Out
1990/91	11.6	4.8	75.7	32.1	27.6	6.1	7.6	5.8	32.8	23.4
1991/92	11.9	2.0	76.7	34.0	19.1	3.7	5.4	1.9	29.7	20.8
1992/93	15.4	2.7	109.0	55.5	24.2	5.3	7.0	2.8	20.4	8.7
1993/94	9.1	1.5	93.8	48.3	16.3	4.4	7.2	3.0	25.6	16.8
1994/95	9.1	1.0	82.1	46.6	18.4	4.5	6.6	1.9	25.7	18.4

septic tank. Although the major problem with drainage from the dairy farm was sorted out in 1989/90, there is still some dilution provided by run-off from the fields in which the beds are placed.

The system was designed at 3.0 m^2 per PE because this fitted the area most easily available in the existing drainage ditch that took the flow from the septic tank. The original intention had been to use this ditch but it proved impossible to widen it because of tree roots, so a parallel ditch was cut and the removed soil was put in the old ditch.

Dimensions. Eight beds, each 12.5 m (length) × 2.0 m (width) × 0.6 m (depth) at inlet.

Medium. Washed gravel, 5–10 mm.

Liner. LDPE.

Plants. Phragmites australis.

Performance. A calculated risk was taken in designing at 3 m^2 per PE (in contrast with the usual UK HF design of 5 m^2 per PE) in the knowledge that, because this was a multi-stage

Table 10.13. Dry weather survey for the reed beds at Leek Wootton, Warwickshire, UK, 1–5 May 1995: daily average total coliform and Escherichia coli (colony-forming units per 100 ml) and removal efficiencies (Cooper et al. 1996)

Day	Influent		Effluent		Removal (%)		Removal (log)	
	Total coliforms	*E. coli*	Total coliforms	*E. coli*	Total coliforms	*E. coli*	Total coliforms	*E. coli*
1	3.30×10^5	7.67×10^4	8.05×10^3	2.14×10^3	97.6	97.2	1.61	1.55
2	1.20×10^5	3.90×10^4	5.03×10^3	1.23×10^3	95.8	96.8	1.38	1.50
3	1.53×10^5	3.64×10^4	3.29×10^3	6.88×10^2	97.8	98.1	1.67	1.72
4	1.92×10^5	5.34×10^4	5.63×10^3	1.45×10^3	97.1	97.3	1.53	1.57
5	1.22×10^5	3.10×10^4	4.39×10^3	6.50×10^2	96.4	97.9	1.44	1.68
Overall mean	1.87×10^5	4.81×10^4	5.16×10^3	1.26×10^3	97.1	97.3	1.55°	1.61°

° Geometric mean.

Table 10.14. Annual average performance data from the HF RBTS at Little Stretton, Leicestershire, UK, since July 1987 (Cooper et al. 1996)

	BOD_5 (mg l^{-1})		TSS (mg l^{-1})		NH_4-N (mg l^{-1})		TON (mg l^{-1})	
Year	Inlet	Outlet	Inlet	Outlet	Inlet	Outlet	Inlet	Outlet
1987°	147	29	132	19	10.0	10.0	15.0	1.0
1988	112	33	85	24	12.2	13.8	12.2	3.4
1989	162	34	127	43	14.9	11.3	9.1	3.0
1990	112	3.9	93	28	24.8	12.1	2.2	6.2
1991	55	4.1	70	28	14.7	5.9	9.0	6.2
1992	26	1.7	41	22	8.0	0.4	22.2	16.6
1993	35	1.7	30	8	8.4	0.2	16.5	11.4
1994	58	2.4	62	16	15.8	0.8	4.9	8.8
1995	78	7.3	65	16	19.5	3.6	6.1	5.1

° July to December.

system, there was the likelihood of better aeration resulting from the change of stages. This has proved to be accurate: Table 10.14 shows that after 1989/90 when the major flows from the dairy farm had been diverted the plant not only produces an effluent with less than 10 mg BOD_5 l^{-1} but it also achieves almost full nitrification (and considerable denitrification).

10.3.2.2 Case study: Middleton, Shropshire, UK

Description. Middleton is a small village in Shropshire, England. There is no major infiltration in the village, so the sewage is strong (more than 300 mg BOD_5 l^{-1}) after the settlement tank.

Constructed. 1991, by Severn Trent Water.

Commissioned. September 1991; parallel operation changed in 1993 to series operation.

Process description. There are two small HF reed beds. The system was designed at 5.6 m^2 per PE.

Dimensions. Two beds, each 10.5 m (length) × 8.0 m (width) × 0.6 m (depth) at inlet. Total area 168 m^2.

Medium. Washed gravel, 5–10 mm.

Liner. LDPE.

Plants. Phragmites australis.

Inlet distribution. Riser pipe in stone gabion.

Outlet collection. Agricultural drainage pipe in large stones.

Standard. 50 mg BOD_5 l^{-1}, 90 mg TSS l^{-1}, 21 m^3 d^{-1}.

Performance. Initially the beds were set to operate in parallel. Although a significant removal was achieved, it was not good enough. In 1993 the arrangement was changed to allow series operation. This resulted in much better treatment, possibly as a result of more aerobic conditions in the second bed.

Table 10.15 shows the performance results from parallel operation in 1992 and the series operation in 1993. The series operation produces superior removals with respect to BOD, TSS and also NH_4-N. A small degree of nitrification was achieved.

10.3.2.3 Case study: Kolodeje, Prague, Czech Republic

Description. Kolodeje is a small village that was administratively connected to Prague 10 years ago. In 1992, local authorities decided not to

Table 10.15. Annual average performance for the HF RBTS at Middleton, Shropshire, UK, in 1992 (parallel operation) and 1993 (series operation) (Cooper et al. 1996)

	BOD_5 (mg l^{-1})			TSS (mg l^{-1})			COD (mg l^{-1})		NH_4-N (mg l^{-1})			Kjeldahl N (mg l^{-1})		TON (mg l^{-1})		
	Feed	Effl.		Feed	Effl.		Feed	Effl.	Feed	Effl.		Feed	Effl.	Feed	Effl.	
1992	306	46		105	25		774	154	61.1	43.3		81.4	47.5	0.5	0.11	
		Bed			Bed					Bed					Bed	
	Feed	1	2	Feed	1	2			Feed	1	2			Feed	1	2
1993	333	109	22	109	43	15			64.9	58.9	39.7			0.4	0.13	3.4

connect the newly built sewerage to the Central Prague wastewater treatment plant because the connector to the central sewerage system was too expensive. It was decided that a local treatment plant should be built, and a constructed wetland was selected.

The population was set at 900. In the beginning *ca.* 500 people were connected. In 1998, almost the full design capacity was achieved.

Constructed. 1993.

Operational. March 1994 to the present.

Pretreatment. Screens and Imhoff tank.

Dimensions. Four beds, each 27 m (length) × 41.6 m (width) with a total area of 4500 m^2. The layout was set as two parallel cells in series.

Depth of the beds. Inlet, 0.65 m; outlet, 0.92 m.

Slope. Bottom, 1%; surface, none.

Medium. Beds, sand 1–4 mm; inlet and outlet zones, stones 50–150 mm.

Liner. PVC-covered geotextile.

Plants. Phragmites australis from seedlings grown in the nursery; density, four seedlings per m^2.

Inlet distribution. The influent distribution pipework formerly comprised PVC pipe 100 mm in diameter with T pieces at 2 m intervals and was laid on the surface of the distribution zone. However, this layout did not work satisfactorily and was replaced with perforated PVC pipes 150 mm in diameter after 2 years of operation. The pipes are laid on the surface and covered with stones.

Outlet collection. Collection pipework laid in the bottom of the outflow zone comprised drainage pipes 100 mm in diameter, which were connected to the polypropylene oulet chambers with a weir. The weirs worked poorly, and the material did not prove to be suitable. After 3 years the polypropylene chambers were replaced with concrete outlet sumps; flexible hoses connected to drainage pipes replaced weirs.

Standard

BOD_5	15 mg l^{-1} (av.), 20 mg l^{-1} (max.)
TSS	15 mg l^{-1} (av.), 20 mg l^{-1} (max.)
COD	60 mg l^{-1} (av.), 80 mg l^{-1} (max.).

Performance. The design flow was set at 190 m^3 d^{-1}, resulting in an HLR of 4.2 cm d^{-1}. The actual average flow varied during 1994–96 between 151 and 192 m^3 d^{-1} and the HLR between 3.4 and 4.3 cm d^{-1}. Table 10.16 shows the performance results from the operation during 1994–97. The efficiency was very high and the quality of the effluent was steady despite the wide fluctuation in the influent quality. The discharge limits were met easily in all 4 years of operation.

10.3.3 Combined sewer overflow

10.3.3.1 Case study: Lighthorne Heath, Warwickshire, UK

Description. The village of Lighthorne Heath comprises an older area with combined sewerage, a newer development with separate sewerage having soakaways for roof and hard surface run-off and a visitor centre for heritage cars, which attracts a large number of tourists. The natural drainage from Lighthorne Heath discharges to a small watercourse, the Tach Brook, which has a high River Quality Objective, 5 mg BOD_5 l^{-1} and 0.7 mg NH_4-N l^{-1} as 95th centile (i.e. only one failure allowed in twenty samples). This led the Environment Agency to improve an effluent consent (standard) with 95th centiles of 10 mg BOD_5 l^{-1}, 20 mg TSS l^{-1} and 5 mg NH_4-N l^{-1}. The storm overflow consent provided for storm overflow when the flow to the works reached 9.7 l s^{-1}, with an absolute quality of 40 mg BOD_5 l^{-1}, 60 mg TSS l^{-1} and 15 mg NH_4-N l^{-1}. It was decided to provide a reed bed to ensure complete compliance. The population was 1154 but a design value of 1400 PE was used.

Constructed. 1992, by Severn Trent Water.

Operational. November 1992 to the present.

Process description. There are three tertiary treatment reed beds at Lighthorne Heath and two additional storm overflow reed beds. The two storm overflow beds have a total area of 642 m^2 providing 0.5 m^2 per PE for the population of 1154 PE. The beds are operated in parallel.

The storm overflow is set at 9.7 l s^{-1}.

Table 10.16. Annual average performance data from the Kolodeje constructed wetland in the Czech Republic

Year	BOD_5 ($mg\ l^{-1}$)		COD ($mg\ l^{-1}$)		TSS ($mg\ l^{-1}$)		TN ($mg\ l^{-1}$)		TP ($mg\ l^{-1}$)	
	In	Out	In	Out	In	Out	In	Out	In	Out
1994	85	8.2	226	55	129	5.1	42	23	10.6	2.9
1995	61	8.5	155	35	66	7.0	37	26	5.1	2.4
1996	106	12.7	266	43	159	12.3	54	20	7.0	3.3
1997	129	9.0	292	36	162	12.7	62	31	8.2	4.1

Inflow denotes pretreated wastewater, i.e. inflow to the vegetated beds.

Table 10.17. Summary of total flow loads and percentage removal of BOD_5, TSS, NH_4-N and TON for three storm surveys at Lighthorne Heath, Warwickshire, UK (Cooper et al. 1996)

Date	Flow (m^3)	BOD_5 ($mg\ l^{-1}$) kg In	kg Out	Rem. (%)	TSS ($mg\ l^{-1}$) kg In	kg Out	Rem. (%)	NH_4-N ($mg\ l^{-1}$) kg In	kg Out	Rem. (%)	TON ($mg\ l^{-1}$) kg In	kg Out	Rem. (%)
11–13/6/93	237	8.2	2.0	76	20.2	3.7	82	0.97	0.57	42	1.56	0.91	41
13–15/11/93	417	18.7	4.3	77	45.3	7.2	84	1.76	0.73	58	2.13	1.33	38
5–6/1/94	301	15.2	3.1	80	38.2	7.8	80	1.64	0.86	48	1.42	0.92	35

Dimensions. Two beds, each 12.5 m (length) × 25.7 m (width) × 0.6 m (depth) at inlet. Total area 642.5 m^2.

Medium. Washed gravel, 5–10 mm.

Plants. Phragmites australis.

Liner. LDPE, 0.75 mm.

Inlet distribution. Riser pipes encased in the stones of the inlet gabion.

Outlet collection. Agricultural drainage pipes in stone gabion.

Performance. The performance of the system during three storm surveys in June and November 1993 and June 1994 is shown in summary in Table 10.17. More details of the hydrographs and the BOD_5, TSS, NH_4-N and TON profiles are available in Cooper *et al.* (1996).

10.4 Vertical subsurface flow

10.4.1 Tertiary nitrification

10.4.1.1 Case study: WRc, Medmenham, Buckinghamshire, UK

Description. The system at the Water Research Centre's Medmenham site was used for tertiary nitrification after a conventional biological filter (Job *et al.* 1996; Cooper *et al.* 1996, 1997).

Constructed. April 1993.

Operational. May 1993 to December 1996.

Process description. Two VF stages operated in series. Each stage comprises four beds, each 4 m × 4 m × 0.7 m (depth). One bed was used in each of the two stages each day, with the other three beds in each stage being rested. The beds in operation on Fridays were left to operate over the weekend. The site had 290 people working there in April 1993, but by 1994 this had decreased to 207. The site is not residential. It contains research laboratories, a large marine (salt water) laboratory, a kitchen and a restaurant. The flows from the toilets and showers go to the sewage treatment together with laboratory flows. The design population was 150 PE to allow for the fact that the staff only spend part of the day on site. Staff are on site between 08:00 and 18:00 Monday to Friday, with only a few at weekends (hence the possibility of feed flowing to one bed in each stage from Friday to Monday morning).

In 1992 it was realized that the biological filter was close to failing its discharge consent (standard) for NH_4-N. This would have been very embarrassing for WRc because the national regulatory body, the Environment Agency, is one of the main funders of work at Medmenham. A reed bed system was put into place as a quick and cheap way of maintaining the consent and to buy time while the biological filter and settlement tanks were refurbished.

The reed beds ensured that the effluent was always within standard. They were shut off in December 1996 when the Medmenham laboratory and the local village were connected to the sewerage network for the first time. At the time of design in 1992 the following feed conditions prevailed:

- average NH_4-N concentration in sewage effluent, 11 $mg\ l^{-1}$
- peak NH_4-N concentration in sewage effluent, 25 $mg\ l^{-1}$

Table 10.18. Performance of the WRc system at Medmenham, UK, in May 1993 to June 1995 (Cooper et al. 1996)

	BOD (mg l^{-1})	TSS (mg l^{-1})	NH_4-N (mg l^{-1})
14 May 1993 to 29 October 1993			
Biofilter effluent (11 samples)	16.6	16.6	6.6
Reed bed 1 effluent (23 samples)	5.6	5.3	1.7
Reed bed 2 effluents (23 samples)	5.9	6.0	1.8
15 April 1994 to 22 July 1994			
Biofilter effluent (10 samples)	27.1	29.3	7.4
Reed bed 1 effluent (10 samples)	11.7	11.0	3.9
Reed bed 2 effluents (10 samples)	4.7	9.1	1.5
28 July 1994 to 24 May 1996			
Biofilter effluent (18 samples)	10.6	7.3	5.1
Reed bed 1 effluent (18 samples)	2.6	3.4	1.0
Reed bed 2 effluents (18 samples)	2.3	2.7	0.87

average flow (for 8 h/d) (sewage effluent plus marine waste flow), 4.2 m^3 h^{-1}
peak flow, 10.8 m^3 h^{-1}.

In practice, during tests in 1995, the average flow rate was 5 m^3 h^{-1} for a period of 8 h d^{-1}.

Dimensions. Two stages, four beds in each stage, each 4 m × 4 m × 0.7 m (depth).

Medium. Sharp sand (top), 5 cm; 5–10 mm gravel, 35 cm; 30–60 mm rounded stones (bottom), 30 cm.

Plants. Phragmites australis.

Liner. Monarflex LDPE.

Inlet distribution. Intermittent flooding to set depth via pipe network with pumps working on level controls.

Outlet collection. Large stones, 30–60 mm, plus agricultural drain pipes.

Standard. Maximum values that must not be exceeded (100th centiles) set by the Environment Agency:

BOD	20 mg l^{-1}
TSS	30 mg l^{-1}
NH_4-N	10 mg l^{-1}
pH	6.5–8.5.

Performance. During the period April to July 1994, the old biological filter plus the primary and secondary settlement tanks were being refurbished; at times the reed beds were treating sewage rather than effluent, hence the higher feed concentrations.

Tests done in July and August 1995, in which two concentration levels of ammonium nitrate (5 and 8 mg NH_4-N l^{-1}) in the feed were imposed on the system, allowed WRc to estimate the nitrification rate in these tertiary systems (Cooper *et al.* 1997). At temperatures of 20–22 °C the removal rate was 30–55 g NH_4-N d^{-1} m^{-3} of bed, but because three out of four beds were being rested each day the effective rate was equivalent to one-quarter of this, i.e. 7.5–14 g NH_4-N d^{-1} m^{-3} of bed. The daily performance is recorded on the WRc/Severn Trent Water Database (Job *et al.* 1996). Table 10.18 summarizes the performance for the period from May 1993 to June 1995.

10.4.1.2 Case study: Strengberg, Lower Austria

Description. Strengberg is a village in western Lower Austria that has a conventional activated sludge process wastewater treatment plant designed for 1500 PE. Because the receiving watercourse is very small, a final nutrient removal stage was required by the authorities. A VF constructed wetland was chosen, designed and investigated by a team of IWGA (Institute for Water Provision) of the University of Agricultural Sciences in Vienna and the Austrian Research Centre, Seibersdorf.

Constructed. Autumn 1994.

Operational. Spring 1995 to the present.

Process description. The conventional treatment plant effluent runs into a storage tank (6 m^3) where a pump is used to load the four parallel VF beds intermittently. Each bed can be operated independently. The flushes have a programmable volume and therefore occur at variable intervals. Two beds have a 120 cm main layer of sand and gravel (0–8 mm); the other two beds have an 80 cm main layer of the same material, to check whether the height of the bed influences the treatment performance. The outlet pipes are adjustable at different heights in the outlet manholes.

Dimensions. Four beds, two being 23 m (length) × 6.5 m (width) × 1.2 m (depth) and two being 23 m (length) × 6.5 m (width) × 0.8 m (depth).

Medium. Sand and gravel, 0–8 mm.

Liner. Polyethylene plastic liner, 2 mm.

Plants. Phragmites australis.

Table 10.19. Performance of the tertiary treatment constructed wetland at Strengberg, Lower Austria, from 1995 to 1997 at increasing hydraulic loading rates

Hydraulic load (mm d^{-1})	COD In (mg l^{-1})	COD Out (mg l^{-1})	COD Elim. (%)	NH_4-N In (mg l^{-1})	NH_4-N Out (mg l^{-1})	NH_4-N Elim. (%)	PO_4-P In (mg l^{-1})	PO_4-P Out (mg l^{-1})	PO_4-P Elim. (%)
100	39	33	15	0.9	0.14	84	0.2	0.2	0
118	40	28	30	0.5	0.04	92	1.1	0.6	45
154	35	28	20	0.8	0.05	93	1.1	0.9	18
200	70	30	57	7.3	0.05	99	4.6	0.9	80
250	41	30	28	0.2	0.05	79	0.2	0.2	0
333	41	35	15	6.3	0.11	98	1.5	0.7	53
500	–	–	–	1.8	0.27	85	0.2	0.2	0

Inlet distribution. Above-ground steel pipes with 8 mm holes at 3 m intervals; plates are situated below the holes to prevent erosion.

Outlet collection. Perforated drainage pipes (150 mm in diameter) in a 26 cm gravel layer (16–32 mm grain size).

Performance. Table 10.19 shows the performance results at increasing HLRs from 1995 to 1997.

The nitrification capacity was very high, even at loading rates of 500 mm d^{-1}. The NH_4-N outlet concentration was below 0.3 mg l^{-1} without dependence on the inlet concentration within the investigated range of up to 73 mg l^{-1}. Two tests with ammonia shock loadings (14 g NH_4-N m^{-2} d^{-1}) showed that the system could decrease the peak concentrations from 47 mg l^{-1} in the inlet to 9 mg l^{-1} in the outlet. The phosphate elimination was clearly dependent on the inlet concentration, with relatively high elimination rates only at high inlet concentrations.

10.4.2 BOD removal and partial nitrification

10.4.2.1 Case study: Camphill Village Trust, Oaklands Park, Gloucestershire, UK

Description. This is a community-based project that was designed and built by the staff of a charity that cares for handicapped people in a beautiful rural location. The charity is keen to encourage 'green' methods in both agriculture and wastewater treatment. The reed bed system is preceded by a standard septic tank. There is also a sludge-drying reed bed on the site that takes sludge from the septic tank (Cooper *et al.* 1996).

Constructed. July 1989.

Operational. July 1989 to the present.

Process description. The Oaklands Park system consists of five stages: two VF stages in series followed by two HF stages, again in series, and finally a pond. All the reed beds are constructed with gravel.

Settlement in a septic tank
stage 1 six parallel beds VF beds
stage 2 three parallel beds VF beds
stage 3 one HF bed
stage 4 one HF bed
stage 5 pond.

In stages 1 and 2 there is one bed in each stage in operation on rotation, with the other beds being rested.

The daily flow rate is *ca.* 9.8 m^3 d^{-1}; the population served is 65 PE.

Dimensions
Stage 1 VF six beds, each 8 m^2
stage 2 VF three beds, each 5 m^2
stage 3 HF one bed, 8 m^2
stage 4 HF one bed, 20 m^2.

Medium. All the beds use gravel. The VF beds are built up as follows:

8 cm of sand
15 cm of 6 mm washed gravel
10 cm of 12 mm round washed gravel
15 cm of 30–60 mm round washed gravel.

Plants
Stage 1 *Phragmites australis*
stage 2 *Iris*, *Schoenoplectus*, *Phragmites*
stage 3 *Iris*
stage 4 *Accorus*, *Carex*, *Schoenoplectus*, *Sparganium*.

Liner. Believed to be LDPE.

Inlet distribution. In the VF beds by overflow channels.

Outlet collection. With agricultural drainage pipes.

Performance. The 2-year study (August 1989 to September 1991) was constructed by WRc on behalf of the water companies. Table 10.20 summarizes the average performance over the 2-year period.

The first two VF stages show good removal of BOD and partial nitrification. The first stage had 0.74 m^2 per PE and the second stage 0.23 m^2 per PE, making a total of just under

Table 10.20. Performance data (averages) from the secondary treatment RBTS at Oaklands Park, Gloucestershire, UK, August 1989 to September 1991 (Cooper et al. 1996)

	Concentration (mg l^{-1})					
		Effluents				
	Influent	Stage 1	Stage 2	Stage 3	Stage 4	Stage 5
BOD_5	285	57	14	15	7	11
TSS	169	53	17	11	9	21
NH_4-N	50.5	29.2	14.0	15.4	11.1	8.1
TON	1.7	10.2	22.5	10.0	7.2	2.3
PO_4-P	22.7	18.3	16.9	14.5	11.9	11.2

Table 10.21. Performance of the VF constructed wetland for the Wolfern/farmhouse Schillhuber, Upper Austria, from 1992 to 1997

	Hydraulic load (mm d^{-1})	BOD_5			COD			NH_4-N			TP		
		In (mg l^{-1})	Out	Elim. (%)	In (mg l^{-1})	Out	Elim. (%)	In (mg l^{-1})	Out	Elim. (%)	In (mg l^{-1})	Out	Elim. (%)
1992	24	143	11	92	378	54	86	48	8.6	82	10.8	3.0	72
1993	32	186	9	95	533	47	91	88	16.4	81	12.5	3.6	71
1994	27	139	3	98	366	36	90	71	1.3	98	11.5	6.1	47
1995	30	120	3	98	383	30	92	63	4.9	92	14.0	7.0	51
1996	37	157	<3	99	436	30	93	49	1.5	97	9.4	6.2	34
1997	30	278	<3	99.6	549	25	95	59	5.5	91	9.6	4.9	49
Average	30	171	5	97	441	37	91	63	6.4	90	11.3	5.1	54

1 m^2 per PE. This area per PE is large enough to bring the average BOD below 20 mg l^{-1} but it left *ca.* 14 mg NH_4-N l^{-1} in the effluent from the two VF stages. Significantly the two HF stages achieve significant denitrification, decreasing the TON from 22 to 7.2 mg l^{-1}.

The Oaklands Park system achieved a nitrification rate of *ca.* 8 g NH_4-N m^{-3} d^{-1}.

10.4.2.2 Case study: Wolfern/farmhouse Schillhuber, Upper Austria

Description. The farmhouse Schillhuber is situated in the hilly region of Upper Austria. It is too far from the village Wolfern to be connected to the sewer line, so it was selected within a pilot project to be an example for the many other farms in the area. It was designed by the IWGA (Institute for Water Provision) of the University of Agricultural Sciences in Vienna.

Constructed. Spring/summer 1991.

Operational. September 1991 to the present.

Process description. Settlement tank (4.4 m^3) as pretreatment, followed by a feeding tank (2.7 m^3) and a 40 m^2 VF bed; the system was designed at 5 m^2 per PE. Intermittent feeding is accomplished by an automatic valve that opens four times a day. Since 1998 the wastewater has been fed intermittently by a mechanical device without electric power. The outlet pipe is adjustable to different heights in the outlet manhole.

Dimensions. One bed, 6.5 m × 6.5 m × 0.8 m (depth).

Medium. Sand and gravel, 0–8 mm.

Liner. Polyethylene plastic liner, 2 mm.

Plants. Phragmites australis.

Inlet distribution. Above-ground PVC pipes with 8 mm holes; plates are situated below the holes to prevent erosion.

Outlet collection. Perforated drainage pipes (100 mm in diameter) in a 20 cm gravel layer (16–32 mm grain size).

Standard. 25 mg BOD_5 l^{-1}, 90 mg COD l^{-1}, 30 mg TSS l^{-1}, 10 mg NH_4-N l^{-1} (at wastewater temperatures above 12 °C in the outlet).

Performance. Table 10.21 shows the performance results from 1992 to 1997; the Austrian effluent standards are being met easily. The elimination of P decreased from 72% to *ca.* 40–50% within the operation time owing to the limited adsorption potential of the substrate. During 1995 an experiment to increase the total elimination of N was undertaken: a recirculation pump was installed in the effluent, which pumped the nitrified effluent into the settlement tank of the pretreatment. An 80% recirculation rate increased the TN elimination to 72% (from the original 40% without recirculation).

10.4.2.3 Case study: Hörbach, Upper Austria

Description. Hörbach is a small village in a rural area of Upper Austria. Because there was

Table 10.22. Performance of the VF constructed wetland at Hörbach, Upper Austria, from 1995 to 1998

	Hydraulic load (mm d^{-1})	BOD_5 In (mg l^{-1})	BOD_5 Out (mg l^{-1})	BOD_5 Elim. (%)	COD In (mg l^{-1})	COD Out (mg l^{-1})	COD Elim. (%)	NH_4-N In (mg l^{-1})	NH_4-N Out (mg l^{-1})	NH_4-N Elim. (%)	TP In (mg l^{-1})	TP Out (mg l^{-1})	TP Elim. (%)
1995	14	556	5	99	1167	28	98	63	10.8	83	10.3	1.6	85
1996	22	176	8	96	426	56	87	48	17.9	63	9.1	2.5	72
1997	21	207	3	99	437	30	93	45	5.3	88	9.2	3.3	64
1998	21	442	<3	99	497	24	95	42	1.4	94	7.9	4.2	61
Average	20	345	5	98	632	35	93	50	8.9	82	9.1	2.9	71

no biological treatment plant it was selected within a pilot project to be an example for other small communities that are too far away to be connected to a central wastewater treatment plant. The constructed wetland was designed by the IWGA (Institute for Water Provision) of the University of Agricultural Sciences, Vienna.

Constructed. Spring 1995.

Operational. June 1995 to the present.

Process description. A screen and a three-chambered settlement tank (40 m^3) have been installed as pretreatment for the constructed wetland designed for 230 PE at 7 m^2 per PE. The intermittent feeding is done by an automatic valve, which opens in response to the water level in the settlement tank. The four beds of the treatment plant can be operated in parallel or as two pairs in series. The outlet pipes are adjustable to different heights in the outlet manholes of the single beds.

Dimensions. Four beds, each 20 m × 20 m; two beds are 1.2 m deep and two are 0.6 m to test the influence of depth on treatment performance.

Medium. Sand and gravel, 0–8 mm.

Liner. Polyethylene plastic liner, 2 mm.

Plants. Phragmites australis.

Inlet distribution. Above-ground PVC pipes with 8 mm holes; plates are situated below the holes to prevent erosion.

Outlet collection. Perforated drainage pipes (100 mm in diameter) in a 20 cm gravel layer (16–32 mm grain size).

Standard. 25 mg BOD_5 l^{-1}, 90 mg COD l^{-1}, 30 mg TSS l^{-1}, 10 mg NH_4-N l^{-1} (at wastewater temperatures above 12 °C in the outlet).

Performance. Table 10.22 shows the performance results from 1995 to 1998: the Austrian effluent standards were easily met for organic pollutants. However, sufficient nitrification occurred only from 1997 onwards. The reasons for the high outlet concentrations of ammonia during the first 2 years of operation were operational problems with the mechanical pretreatment (sludge was pumped on the fields, which resulted in partial soil clogging and therefore a decreased supply of oxygen to the substrate) and freezing problems during winter (outlet holes in the distribution system that were too small). After some changes to the settlement tank and the distribution system, the constructed wetland now works at high treatment levels. The elimination of P is still high after 4 years of operation.

10.4.2.4 Case study: Dhulikhel hospital, Nepal

Description. The Dhulikhel hospital is located in the Kathmandu valley in the small town Dhulikhel. The climate is subtropical, with a monsoon season from June to September. The construction of the hospital started in 1995. It includes 60 beds with nursing quarters and a washing hall. The constructed wetland is the first in Nepal and was designed by a team of local engineers of the Environment and Public Health Organisation and the IWGA (Institute for Water Provision) of the University of Agricultural Sciences, Vienna.

Constructed. April–July 1997.

Operational. July 1997 to the present.

Process description. A three-chambered settlement tank (16.7 m^3) is installed for pretreatment, followed by an HF bed (140 m^2) as the first stage and a 120 m^2 VF bed as the second stage. The average wastewater quantity is 11 m^3 d^{-1}, which results in an HLR of 79 mm d^{-1} for the horizontal bed and 92 mm d^{-1} for the vertical bed during serial operation. Both beds are fed intermittently by mechanical devices without electric power. The outlet pipes of both beds are adjustable to different heights in the outlet manholes. As well as serial operation, parallel operation is also possible; however, it turned out to be not as efficient as serial operation. The sludge of the settlement tank is dried in a sludge drying bed.

Dimensions. One HF bed, 7 m (length) × 20 m (width) × 0.6 m (depth); one vertical bed, 11 m × 11 m × 0.85 m.

Medium. Broken gravel (5–20 mm) in the horizontal bed; washed sand (0–2 mm) in the vertical bed.

Table 10.23. Performance of the constructed wetland system for Dhulikhel hospital, Nepal, during serial operation

		Settlement tank			
		In	Out	HF bed out	VF bed out
BOD_5	(mg l^{-1})	118	67	25	2
	Elim. (%)		43	79	98
COD	(mg l^{-1})	261	162	45	10
	Elim. (%)		38	83	96
TSS	(mg l^{-1})	159	57	19	1.5
	Elim. (%)		64	88	99
NH_4-N	(mg l^{-1})	32	32	27	0.08
	Elim. (%)		0	16	99.8
NO_3-N	(mg l^{-1})	0.2	0.2	0.4	27
TP	(mg l^{-1})	4.6	4.4	2.6	1.4
	Elim. (%)		4	43	70
E. coli	(per 100 ml)	1.64×10^7	1.71×10^6	4353	20
	Elim. (%)		90	99.97	99.9999

Liner. Plastic.

Plants. Phragmites karka.

Inlet distribution. Above-ground PVC pipes with 8 mm holes; plates are situated below the holes to prevent erosion.

Outlet collection. Perforated drainage pipes (100 mm in diameter) in a 15 cm gravel layer.

Standard. Not yet established in Nepal.

Performance. Table 10.23 shows the performance results from August 1997 to July 1998 during serial operation (first stage HF bed, second stage VF bed).

The HF bed is used as a kind of pretreatment for the VF bed. Within the HF bed most of the organic compounds are eliminated, as are the TSS. The VF bed is therefore used as nitrification stage and final treatment, especially for bacterial pollution. Even at the relatively high HLRs (92 mm d^{-1}), no clogging problems have been observed in the VF bed, which is also due to the enhanced pretreatment.

10.5 Integrated natural treatment system

10.5.1 Case study: potato processing water treatment, Connell, Washington, USA

Potato processing wastewater contains high concentrations of COD, TSS and total Kjeldahl nitrogen. A combination of natural, land-intensive technologies including SF wetlands, intermittent VF wetlands, ponds and land application have been used for treatment. This engineered system balances irrigation requirements, nitrogen supply and seasonal growth patterns to provide effective year-round operation. A first pilot wetland was operated to determine operability, effectiveness and plant survival at high COD and nitrogen concentrations. A second pilot system of four wetlands in series was operated to obtain design and operating information. Two SF wetlands provided TSS and COD decreases and ammonified the Org-N. Subsequently, nitrification occurred in the VF wetlands, followed by denitrification in an SF wetland. The design target was a balanced nitrogen and irrigation supply for application to crops. Winter storage was used to match the crop application period to the growing season. Both pilot projects met design objectives, and a full-scale system has begun operation.

The integrated system involves SF wetlands W1 and W2, planted with *Scirpus* spp. and *Typha* spp., subsurface vertical downflow wetland W3, unplanted, and SF wetland W4, planted with *Scirpus* and *Typha*. (Figure 10.5). The medium for W3, 0.8 m deep, was a coarse sand (D_{10} (number mean diameter) = 0.8 mm), available on site. Wastewater was pumped from the clarifier to W1, and proceeded by gravity to W2. Effluent from W2 was sprayed on the surface of W3 subcells, intermittently in rotation, and collected in under drains. Application and draining periods were variable, approximating a 50% duty cycle. Flow to W4 was by gravity, and the final effluent was stored for land application in summer. Operation started in autumn 1995 and continues to the present.

Winter storage of wastewater is one of the main goals of the overall project, because land application is fully effective only during the growing season. Design criteria involved only volumetric considerations, and consequently this portion of the full-scale project was built during the operation of pilot 2. This process element is ultimately to be used at the end of the sequence of units, and will then contain treated water, low in COD and TSS and moderately low in NH_4-N. In the interim, it

Table 10.24. Project results for the pilot system at Connell, Washungton, USA

		Summer (June to October)						Winter (November to March)					
Parameter	Units	Influent	W1	W2	W3	W4	%	Influent	W1	W2	W3	W4	%
HLR	cm d^{-1}	–	7.7	6	8.9	6.8	–	–	5.1	2.9	6	3.9	–
T	°C	27	16	16	17	15	–	24	5	4	5	4	–
COD	mg l^{-1}	2986	1056	601	209	161	95	3309	1400	958	385	287	91
TSS	mg l^{-1}	607	85	72	48	37	94	531	104	62	42	93	82
Org-N	mg l^{-1}	91	10	3	13	12	87	79	18	2	2	13	84
NH_4-N	mg l^{-1}	73	129	116	26	29	60	92	106	111	56	42	54
TKN	mg l^{-1}	164	139	119	39	41	75	171	124	113	58	55	68
NO_3-N	mg l^{-1}	1	1	1	43	13	–	1	1	1	27	1	–
TN	mg l^{-1}	164	139	119	39	41	75	171	124	113	58	55	68
DO	mg l^{-1}	1	0	1	4	1	–	2	1	3	5	1	–
EC	mS	2.5	2.9	2.9	2.6	2.8	–	2.5	2.1	2.5	2.3	2.4	–
pH		5.7	6.7	7	6.8	7.1	–	5.5	6.6	6.9	7	7.1	–

DO, dissolved oxygen; EC, electrical conductivity; TKN, total Kjeldahl nitrogen.

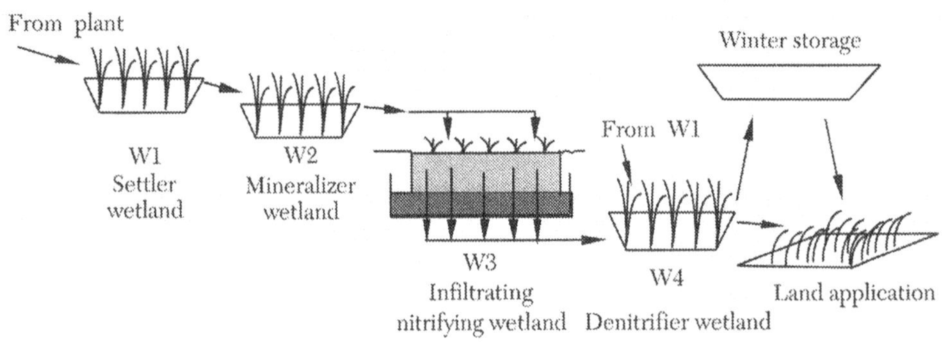

Figure 10.5. Layout of the integrated natural system at Connell, Washington, USA.

has served to store partly treated water for summer irrigation. The potential for odour problems in the storage of untreated effluent led to the early construction of the W3 units, which were used on an interim basis for the pretreatment of clarifier effluent before storage. Solids accretion was expected to occur in the W3 wetlands during this interim operational mode, because the W1 and W2 prefilters were not yet built. Some surface caking did occur, leading to ponding in some portions of the W3 units. These W3 units were cleaned and rebuilt in summer 1996, before the start of operations. Wetlands W1 and W2 were constructed and planted in late 1995. Spring and summer 1996 were used to patch-in lost autumn plantings and to 'ramp up' the system to wastewater flows. Operation was better than design during 1997.

Table 10.24 indicates some seasonal effects on percentage decreases in nitrogen species. There were smaller decreases in NH_4-N and Org-N, presumably because of lower temperatures (4 °C compared with 15 °C). The apparently greater decrease in NO_3-N was due to the initiation of carbon supply via a feedforward from the process inlet, which greatly augmented denitrification owing to autochthonous carbon in the W4 wetland. There was a slight decrease in COD removal during winter. TSS removals were the same in the first three units in winter and summer, but unit W4 exported some TSS during the winter season.

This integrated natural treatment system provides a low-capital-cost alternative for managing potato processing wastewater. There are no chemical costs, and some of the energy requirement is met by gravity and solar sources. Pilot and full-scale performance has been acceptable, but optimal operating conditions are not yet completely specified. For instance, the duty cycle of the W3 nitrifiers is subject to optimization, and the nitrogen loss in the storage pond has not been quantified. A single treatment process does not possess the versatility of this integrated system.

10.6 Floating aquatic plant system

10.6.1 Case study: sludge lagoon supernatant, Hornsby Bend Facility, Austin, Texas

Description. The City of Austin used water hyacinths (*Eichhornia crassipes*) seasonally to

Table 10.25. Operational data from the Hornsby Bend FAP system, Austin, Texas, USA, for 1987–88 (from US Environmental Protection Agency 1988)

			Concentration ($mg\ l^{-1}$)							
	pH		BOD_5		TSS		VSS		NH_3-N	
Date	In	Out	In	Out	In	Out	In	Out	In	Out
September 1987	8.4	7.1	97	30	140	31	90	28	22.9	38.6
October 1987	8.3	7.8	39	11	120	19	169	22	26.5	43.0
November 1987	8.3	7.8	153	9	245	21	240	17	26.1	39.3
December 1987	8.2	7.7	106	14	142	24	111	14	41.9	39.1
January 1988	8.1	7.6	79	18	127	17	96	16	121.1	31.0
February 1988	8.1	7.7	84	45	84	36	71	12	95.6	36.4
March 1988	8.1	7.6	–	–	155	41	91	37	77.6	42.0
April 1988	7.9	7.6	357	139	182	47	180	49	76.8	42.5
May 1988	7.9	7.4	143	34	121	26	68	8	43.5	21.9
June 1988	8.0	7.7	156	30	117	30	79	23	47.0	33.9
July 1988	8.1	7.7	99	28	132	19	104	12	24.7	37.4
Average	8.1	7.6	131	36	142	28	118	22	54.9	36.8

VSS, volatile suspended solids.

upgrade lagoon effluent from 1977 to 1990. In February 1986, the City's Hornsby Bend facility expanded the floating aquatic plant (FAP) technology to include three water-hyacinth ponds that were entirely enclosed in a 2 ha glass greenhouse. The water-hyacinth cells had a total surface area of 1.6 ha and a length of 265 m, and ranged in size from 0.48 to 0.64 ha. Basin depths ranged between 90 cm at the upstream end to 150 cm at the downstream end. The design flow rate was 7570 $m^3\ d^{-1}$ for an average HLR of 47 $cm\ d^{-1}$. This system provided additional polishing of sludge lagoon supernatant to meet discharge standards of 30 $mg\ l^{-1}$ for BOD_5 and 90 $mg\ l^{-1}$ for TSS on a year-round basis. In 1990 the use of water hyacinths was stopped owing to plant maintenance and harvesting difficulties; duckweed (*Lemna* sp.) was planted to provide an alternative FAP cover.

The Austin FAP system was designed for natural mosquito control through the use of predator species such as mosquito fish (*Gambusia affinis*), grass shrimp (*Palemonetes kadiakensis*), and several species of frog. Eight open-water exclosures were located in each of the FAP cells to maintain an available oxygenated habitat for the fish and shrimp. A 3.4 m cascade provided passive aeration of the effluent before final discharge.

Operational performance. Performance data for a one-year period from 1987 to 1988 have been published for the Hornsby Bend water hyacinth facility (Table 10.25). Effluent pH was lower than the influent pH, with monthly effluent pH averages between 7.1 and 7.8. Influent BOD_5 averaging 131 $mg\ l^{-1}$ was decreased to an average outflow concentration of 36 $mg\ l^{-1}$. Average monthly TSS concentrations were reduced from 142 to 28 $mg\ l^{-1}$. Approximately 77% of this effluent TSS was organic as measured by the volatile suspended solids test. Influent and effluent NH_4-N concentrations for the water-hyacinth facility were high, with monthly average effluent concentrations exceeding inflow concentrations during some months, apparently because of mineralization of organic nitrogen.

Estimated costs. The estimated capital cost of the water-hyacinth system at Hornsby Bend, Texas, was US$1.2M, at US$750,000 ha^{-1}.

11 Bibliography

11.1 Major literature sources on wetlands to improve water quality (focus on pollution control)

European constructed treatment wetlands (ed. J. Vymazal *et al.*) Backhuys Publishers, Leiden, 1998.

Wetlands for water quality management and habitat enhancement: policy and permitting issues. US EPA Environmental Technology Initiative Program, Draft, July 1997.

Wetland systems in water pollution control 1996 (ed. R. Haberl, R. Perfler, J. Laber and P. Cooper). *Wat. Sci. Technol.* **35** (5) (1997) (347 pages.).

Habitat assessment of treatment wetlands. US EPA Environmental Technology Initiative Program. Draft, October 1996. (80 pages plus diskette.)

Proceedings of the Fifth International Conference on Wetland Systems for Water Pollution Control, Vienna, Austria. Universität för Bodenkultur Wein, Vienna, 1996.

Constructed Wetlands for Livestock Wastewater Management CH2M Hill & Payne Engineering, December 1996. (100 pages plus diskette.)

Reed beds and constructed wetlands for wastewater treatment, by P.F. Cooper, J.D. Job, M.B. Green and R.B.E. Shutes. WRc, Swindon, UK, 1996. (Book, bibliography and UK database diskette.)

Nutrient cycling and retention in wetlands and their use for wastewater treatment (ed. J. Vymazal). Institute of Botany of the Czech Republic and Ecology and Use of Wetlands, Praha, 1996. (246 pages.)

Constructed wetlands in cold climates; Proceedings of Niagara-on-the-Lake Conference. Friends of Fort George, June 1996.

Surface flow constructed wetlands for wastewater treatment: a technology assessment. US EPA, Draft Report, 1996. (154 pages.)

Arizona guidance manual for constructed wetlands for water quality improvement. Arizona DEQ, 1995. (200 pages.)

Treatment wetlands, by R.H. Kadlec and R.L. Knight. CRC Press, Boca Raton, FL, USA, 1996. (893 pages.)

A handbook of constructed wetlands: a guide to creating wetlands for agricultural wastewater, domestic wastewater, coal mine drainage, and stormwater, by L. Davis. USDA-NRCS and US EPA, 1995. (143 pages.)

Wastewater and stormwater applications of wetlands in Canada, by J.H. Pries. North American Wetlands Conservation Council, 1995. (66 pages.)

Wetland systems in water pollution control 1994 (ed. R.H. Kadlec and H. Brix). *Wat. Sci. Technol.* **32** (3). (1995) (376 pages.)

Versatility of wetlands in the agricultural landscape (ed. K.L. Campbell). St Joseph, MI, USA: American Society of Agricultural Engineers, 1995. (755 pages.)

Constructed wetlands for wastewater treatment and wildlife habitat: 17 case studies. US EPA, EPA 832-R-93-005, 1995.

Constructed wetlands for animal waste management (ed. P.J. DuBowy and R.P. Reaves). Purdue University, Lafayette, IN, USA, 1994. (188 pages.)

Wetland systems in water pollution control (ed. H.J. Bavor and D.S. Mitchell). *Wat. Sci. Technol.* **29** (4) (1994). (336 pages.)

Proceedings of the Fourth International Conference on Wetland Systems for Water Pollution Control, Guangzhou, China (ed. C. Jiang), 1994. (798 pages.) ICWS '94 Secretariat, Guangzhou, China.

Global wetlands: Old World and New (ed. W.J. Mitsch). Elsevier, Amsterdam, 1994. (967 pages.)

Pflanzenkläranlagen: Abwasserreinigung mit Bepflantzen Bodenkörpern (ed. B. Saurer). Wasserwirtschaft Land Steiermark, Graz, Austria, 1994. (172 pages.) [In German.]

Naturliga system för avloppsrening och resursutnyttjande i tempererat klimat (ed. H.B. Wittgren and K. Hasselgren). VA-VORSK, Stockholm, Sweden, 1993. (68 pages.) [In Swedish.]

Constructed wetlands for water quality improvement (ed. G.A. Moshiri) (Proceedings of an International Symposium). Lewis Publishers, Chelsea, MI, USA, 1993. (632 pages.)

Database: North American wetlands for water quality treatment, phase II report, by R.L. Knight, R.W. Ruble, R.H. Kadlec and S.C. Reed. Prepared for US EPA, September 1993. (824 pages.) (Diskette available from D.S. Brown, US EPA, Risk Reduction Engineering Laboratory, Cincinnati, OH, USA 45268, USA.)

General design, construction, and operation guidelines: constructed wetlands wastewater treatment systems for small users including individual residences. Technical Report TVA/WM-93/10, Tennessee Valley Authority, Chattanooga, TN, USA, May 1993. (42 pages.)

Subsurface flow constructed wetlands for wastewater treatment: a technology assessment. Report EPA 832-R-93-001, US EPA Office of Water, July 1993. (72 pages.)

Wetland systems in water pollution control (Proceedings of an International Symposium, Sydney, Australia, 1992) (ed. Organizing Committee of Wetland Systems in Water Pollution Control Conference, Conference Secretariat). Artarman, NSW, Australia, 1992.

Design of stormwater wetland systems by T.R. Schueler. Metropolitan Washington Council of Governments, Washington, DC, USA, 1992. (133 pages.)

The use of wetlands for controlling stormwater pollution by E.W. Strecker, J.M. Kersnar, E.D. Driscoll and R.R. Horner. EPA/600, Terrene Institute, Washington, DC, USA, 1992. (66 + 230 pages.)

Ecological engineering for wastewater treatment (ed. C. Etnier and B. Guterstam). Bokskogen, Gothenburg, Sweden, 1991. (365 pages.)

Spildevandsrensning i rodzoneanlæg by H.-H. Schierup, H. Brix and B. Lorenzen. Botanical Institute, Aarhus University, Denmark, 1990. (87 pages.)

An inventory of constructed wetlands used for wastewater treatment in the United States by S.C. Reed. Prepared for US EPA, March 1990. (163 pages.)

Moerassen voor de zuivering van water [*Wetlands for the purification of water*], Post Academic Course, published as *The Utrecht Plant Ecology News Report,* no. 11, October 1990. (269 pages.) [About 50% in English.] (University of Utrecht, Lange Nieuwstraat 106, 3512 PN Utrecht, The Netherlands)

Behandlung von häslichen Abwasser in Pflanzenbeeten ATV, Gesellschaft zur Föderung der Abwassertechnik e.V., Postfach 1160, Markt 71, D-5205 St Augustin 1, Germany. (English translation of this design guideline document, 1989.) (12 pages.)

European design and operations guidelines for reed bed treatment systems (ed. P.F. Cooper). Prepared by EC/EWPCA Emergent Hydrophyte Treatment Systems Expert Contact Group, Water Research Centre, Swindon, UK, 1990. (33 pages.)

Constructed wetlands in water pollution control (ed. P.F.

Cooper and B.C. Findlater). Pergamon Press, London, UK, 1990. (605 pages.)

Manual of practice: natural systems, wetlands chapter. MOP FD-16 WPCF, Water Pollution Control Federation, February 1990. (270 pages.)

Water: laws and management (ed. F.E. Davis). AWRA, Bethesda, MD, USA, 1989.

Wetlands: concerns and successes (ed. D.W. Fisk). AWRA, Bethesda, MD, USA, 1989. (565 pages.)

Constructed wetlands for wastewater treatment (conference proceedings) (ed. D.A. Hammer). Lewis Publishers, Chelsea, MI, USA, 1989. (831 pages.)

Design manual: constructed wetlands and aquatic plant systems for municipal wastewater treatment. US EPA 625/1-88/022, September 1988. (83 pages.)

Natural systems for waste management and treatment by S.C. Reed, E.J. Middlebrooks and R.W. Crites. McGraw-Hill, NY, USA, 1995. (431 pages.)

The ecology and management of wetlands (ed. D.D. Hook *et al.*). Timber Press, Portland, OR, USA, 1988. (992 pages.)

Report on the use of wetlands for municipal wastewater treatment and disposal. US EPA 430/09-88-005, October 1987.

Aquatic plants for water treatment and resource recovery (ed. K.R. Reddy and W.H. Smith). Magnolia, Orlando, FL, USA, 1987. (1032 pages.)

Ecological considerations in wetland treatment of municipal wastewaters by E.R. Kaynor, P.J. Godfrey and J. Benforado. Van Nostrand, 1985. (473 pages.)

Freshwater wetlands for wastewater management handbook US EPA 904/9-85-135, September 1985. (507 pages.)

The ecological impacts of wastewater on wetlands: an annotated bibliography, US EPA 905/3-84-002, February 1984. (257 pages.)

Technology assessment of wetlands for municipal wastewater treatment by H.C. Hyde, R.S. Ross and F. Demgen. US EPA 600/2-84-154, NTIS no. PB 85-106896, November 1984. (98 pages.)

Design principles for wetland treatment systems by D.E. Hammer and R.H. Kadlec. US EPA 600/2-83-026, 1983. (243 pages.)

Phase I report: freshwater wetlands for wastewater management. US EPA 904/9-83-107, 1983.

The effects of wastewater treatment facilities on wetlands in the Midwest. US EPA 905/3-83-002, 1983. (377 pages.)

An overview of municipal wastewater aquaculture by D.M. Eisenberg and J.R. Benneman. US EPA, Draft Final Report, contract no. DM41USC252(C), March 1982. (84 pages.)

The use of wetlands for water pollution control by E. Chan, T.A. Bursztynsky, N. Hantsche and Y.J. Litwin. US EPA 600/2-82-036, September 1982. (261 pages.)

Proceedings of the Seminar on Aquaculture Systems for Wastewater Treatment (ed. R.K. Bastian and S.C. Reed). US EPA publication no. MCD-67, September 1979. (485 pages.)

Wetland functions and values: the state of our understanding (ed. P.E. Greeson, J.A. Clark and J.E. Clark). AWRA, Minneapolis, MN, USA, 1979. (674 pages.)

Freshwater wetlands and sewage effluent disposal (Proceedings of NSF/RANN Conference) (ed. D.L. Tilton, R.H. Kadlec and C.J. Richardson). NTIS PB259305, University of Michigan, Ann Arbor, MI, USA, 1976. (343 pages.)

11.2 Wetlands: some general references

Algae and element cycing in wetlands by J. Vymazal. Lewis Publishers, Boca Raton, FL, USA, 1995. (689 pages.)

Everglades: the ecosystem and its restoration (ed. S.M. Davis and J.C. Ogden). St Lucie Press, Delray Beach, FL, USA, 1994. (826 pages.)

Wetlands, industry and wildlife (ed. A. Merritt). The Wildfowl and Wetlands Trust, Slimbridge, Gloucester, UK, 1994. (182 pages.)

Global wetlands: Old World and new (ed. W.J. Mitsch). Elsevier, Amsterdam, The Netherlands, 1994. (967 pp.)

Applied wetlands science and technology (ed. D.M. Kent). Lewis Publishers, Boca Raton, FL, USA, 1994. (436 pages.)

Wetlands by W.J. Mitsch and J.G. Gosselink. Van Nostrand Reinhold Co., New York, 1993. (722 pages.)

Freshwater wetlands and wildlife (ed. R.R. Sharitz and J.W. Gibbons). US Department of Energy, NTIS no. DE90005384, 1989. (1265 pages.)

Wetlands ecology, productivity and values (ed. S.K. Majumdar, F.J. Brenner, R.P. Brooks and R.W. Tiner). Pennsylvania Academy of Science, Easton, PA, USA, 1988. (395 pages.)

The ecology and management of wetlands (ed. D.D. Hook). Croom Helm, Beckenham, UK, 1987. (2 volumes; 988 pages.)

Freshwater marshes by M.W. Weller. University of Minnesota Press, Minneapolis, MN, USA, 1981. (146 pages.)

Wetland functions and values: the state of our understanding American Water Resources Association, Minneapolis, MN, USA, 1979. (674 pages.)

Freshwater wetlands (ed. R.E. Good, D.F. Whigham and R.L. Simpson). Academic Press, New York, 1978. (378 pages.)

11.3 References

Allender, B.M. (1984) Water quality improvement of pulp and paper mill effluents by aquatic plants. *Appita* **37**, 303–306.

Anderson, P.A. (1996) Advanced treatment of sugar beet process water using constructed wetlands. In *Proceedings of the 5th International Conference on Wetland Systems for Pollution Control, Vienna* (ed. R. Haberl and J. Laber), session 10, pp. 2-1–2-8. Institute for Water Provision, Water Ecology and Waste Management, Universität für Bodenkultur Wien, Vienna, Austria.

Andrews, T.L. and Cockle, C. (1996) Surface flow wastewater wetlands: how the Ouray, CO Project compares to the US EPA Technology Assessment. Paper prepared for RMSAWWA/RMWEA 1996 Conference, Steamboat Springs, CO, USA. AWWA, Washington, DC, USA.

APAI (Alan Plummer Associates, Inc.) (1995) *Use of constructed wetlands for protection of water quality in water supply reservoirs.* Prepared for Tarrant County Water Control and Improvement District Number One in conjunction with American Water Works Association Research Foundation. Washington, DC, USA.

API (American Petroleum Institute) (1998) *The use of treatment wetlands for petroleum industry effluents.* Prepared by CH2M HILL for the API Biomonitoring Task Force, Health and Environmental Sciences Department. Washington, DC, USA: API.

Armstrong, J. and Armstrong, W. (1988) *Phragmites australis* – preliminary study of soil-oxidising sites and internal gas transport parthays. *New Phytol.* **108**, 373–382.

Armstrong, J. and Armstrong, W. (1990) Pathways and mechanisms of oxygen transport in *Phragmites australis.* In *Constructed wetlands in water pollution control* (ed. P.F. Cooper and B.C. Findlater), pp. 529–534. Oxford, UK: Pergamon Press.

Armstrong, W. (1967) The use of polarography in the assay of oxygen diffusion from roots in anaerobic media. *Physiol. Plant.* **20**, 540–553.

Armstrong, W. (1978) Root aeration in the wetland environment. In *Plant life in anaerobic environments* (ed. D.D. Hook and R.M.M. Crawford), chapter 9, pp. 269–297. Ann Arbor: Ann Arbor Science.

Armstrong, W. (1979) Aeration in higher plants. *Adv. Bot. Res.* **7**, 225–331.

Armstrong, W., Armstrong, J. and Beckett, P.M. (1990) Measurement and modeling of oxygen release from roots of *Phragmites australis.* In *Constructed wetlands in water pollution control* (ed. P.F. Cooper and B.C. Findlater), pp. 41–52. Oxford, UK: Pergamon Press.

ATV (1989) *Behandlung von häuslichen Abwasser in Pflanzenbeeten.* ATV, Gesellschaft zur Föderung der Abwassertechnik e.V., Postfach 1160, Markt 71, D-5205 St Augustin 1, Germany. (English translation of this design guideline document, 1989.) (12 pages.)

Bahlo, K.E. and Wach, F.G. (1990) Purification of domestic sewage with and without faeces by vertical intermittent filtration in reed and rush beds. In *Constructed wetlands in water pollution control* (ed. P.F. Cooper and B.C. Findlater), pp. 215–221. Oxford, UK: Pergamon Press.

Barber, D.A. and Martin, J.K. (1976) The release of organic substances by cereal roots into soil. *New Phytol.* **76**, 69–80.

Barko, J.W., Gunnison, D. and Carpenter, S.R. (1991) Sediment interactions with submersed macrophyte growth and community dynamics. *Aquat. Bot.* **41**, 41-65.

Batchelor, A., Scott, W.E. and Wood, A. (1990) Constructed wetland research programme in South Africa. In *Constructed wetlands in water pollution control* (ed. P.F. Cooper and B.C. Findlater), pp. 373-382. Oxford, UK: Pergamon Press.

Bavor, H.J., Roser, D.J., Fisher, P.J. and Smalls, I.C. (1989) Performance of solid-matrix wetland systems, viewed as fixed-film bioreactors. In *Constructed wetlands for wastewater treatment: Municipal, industrial, and agricultural* (ed. D.A. Hammer), Chapter 39k, pp. 646-656. Chelsea, MI, USA: Lewis Publishers.

Bavor, H.J., Roser, D.J., McKersie, S.A. and Breen, P. (1988) *Treatment of secondary effluent.* Report to Sydney Water Board, Sydney, NSW, Australia. Water Research Laboratory, Hawkesbury Agricultural College, Richmind, NSW, Australia.

Bedford, B.L., Bouldin, D.R. and Beliveau, B.D. (1991) Net oxygen and carbon-dioxide balances in solutions bathing roots of wetland plants. *J. Ecol.* **79**, 943-959.

Beven, K. and Germann, F. (1982) Macropores and water flow in soils. *Wat. Resour. Res.* **18**, 1311-1325.

Birkbeck, A.E., Reil, D. and Hunter, R. (1990) Application of natural and engineered wetlands for treatment of low-strength leachate. In *Proceedings of the International Conference on the Use of Constructed Wetlands in Water Pollution Control, Cambridge, UK, 24-28 September 1990* (ed. P.F. Cooper and B.C. Findlater), pp. 411-418. Oxford, UK: Pergamon Press.

Bishay, F.S., Bayley, S.E. and Nix, P.G. (1995) Ammonia removal in constructed wetlands receiving oil sands wastewater, Fort McMurray, AB. Paper presented at SETAC Conference, Vancouver, BC, Canada.

Bock, E., Koops, H.P. and Harms, A. (1986) Cell biology of nitrifying bacteria. In *Nitrification* (Society for General Microbiology publication) (ed. J.J. Prosser), pp. 17-38. Oxford, UK: IRL Press.

Boller, M., Schwager, A., Eugster, J. and Mottier, V. (1993) Dynamic behaviour of intermittent buried filters. *Wat. Sci. Technol.* 28 (10), 99-107.

Boon, A.G. (1985) *Report of a visit by members and staff of WRc to Germany to investigate the Root Zone Method for treatment of wastwaters.* WRc Report 376-S/1, Stevenage, UK.

Börner, T., von Felde, K., Gschlössl, E., Gschlössl, T., Kunst, S. and Wissing, F.W. (1998) Germany. In *Constructed wetlands for wastewater treatment in Europe* (ed. J. Vymazal, H. Brix, P.F. Cooper, M.B. Green and R. Haberl), pp. 169-190. Leiden, The Netherlands: Backhuys Publishers.

Boyd, J., McDonald, O., Hatcher, D., Portier, R.J. and Conway, R.B. (1993) Interfacing constructed wetlands with traditional wastewater biotreatment systems. In *Constructed wetlands for water quality improvement* (International Conference on Constructed Wetlands, Pensacola, Florida, October 1991) (ed. G.A. Moshiri), pp. 453-459. Boca Raton, FL, USA: Lewis Publishers.

Brix, H. and Schierup, H. (1989) Sewage treatment in constructed reed beds-Danish experiences. *Wat. Sci. Technol.* **21**, 1665-1668.

Brix, H. and Schierup, H. (1990) Soil oxygenation in constructed reed beds: the role of macrophyte and soil-atmosphere interface oxygen transport. In *Proceedings of the International Conference on the Use of Constructed Wetlands in Water Pollution Control, Cambridge, UK* (ed. P.F. Cooper and B.C. Findlater), pp. 53-66. Oxford, UK: Pergamon Press.

Brix, H. (1987) Treatment of wastewater in the rhizosphere of wetland plants-the Root-Zone Method. *Wat. Sci. Technol.* **19**, 107-118.

Brix, H. (1988) Light-dependent variations in the composition of the internal atmosphere of *Phragmites australis* (Cav.) Trin. ex Steudel. *Aquat. Bot.* **30**, 319-329.

Brix, H. (1990) Gas exchange through the soil-atmosphere interphase and through dead culms of *Phragmites australis* in a constructed reed bed receiving domestic sewage. *Wat. Res.* **24** (2), 259-266.

Brix, H. (1994*a*) Functions of macrophytes in constructed wetlands. *Wat. Sci. Technol.* **29**, 71-78.

Brix, H. (1994*b*) Use of constructed wetlands in water pollution control: Historical development, present status, and future perspectives. *Wat. Sci. Technol.* **30**, 209-223.

Brix, H. (1994*c*) Constructed wetlands for municipal wastewater treatment in Europe. In *Global wetlands: Old World and new* (ed. W.J. Mitsch), pp. 325-333. Amsterdam: Elsevier.

Brix, H. (1997) Do macrophytes play a role in constructed treatment wetlands? *Wat. Sci. Technol.* **35** (5), 11-17.

Brix, H. (1998) Denmark. In *Constructed wetlands for wastewater treatment in Europe* (ed. J. Vymazal, H. Brix, P.F. Cooper, M.B. Green and R. Haberl), pp. 123-152. Leiden, The Netherlands: Backhuys Publishers.

Brix, H., Sorrell, B.K. and Orr, P.T. (1992) Internal pressurization and convective gas flow in some emergent freshwater macrophytes. *Limnol. Oceanogr.* **37**, 1420-1433.

Brodie, G.A., Britt, C.R., Tomaszewski, T.M. and Taylor, H.N. (1993) Anoxic limestone drains to enhance performance of aerobic acid drainage treatment wetlands. In *Constructed wetlands for water quality improvement* (ed. G.A. Moshiri), pp. 129-138. Boca Raton, FL, USA: Lewis Publishers.

Broome, S.W., Seneca, E.D. and Woodhouse, W.W., Jr. (1988) Tidal salt marsh restoration. *Aquat. Bot.* **32**, 1-22.

Burka, U. and Lawrence, P.C. (1990) A new community approach to wastewater treatment with higher plants. In *Constructed wetlands in water pollution control* (ed. P.F. Cooper and B.C. Findlater), pp. 359-371. Oxford, UK: Pergamon Press.

Caffrey, J.M. and Kemp, W.M. (1991) Seasonal and spatial patterns of oxygen production, respiration and root-rhizome release in *Potamogeton perfoliatus* L. and *Zostera marina* L. *Aquat. Bot.* **40**, 109-128.

Carr, D.W. and Rushton, B.T. (1995) *Integrating a native herbaceous wetland into stormwater management.* Stormwater Research Program, Southwest Florida Water Management District, Brooksville, Florida, USA. (131 pages + appendices.)

Caswell, P.C., Emerick, J.C. and Gelb, D. (1992) *Engineered wetlands-based biological treatment systems for produced water.* Research Project sponsored by Wyoming DEQ and Marathon Oil Co. and conducted by the Dept of Environmental Science and Engineering, Colorado School of Mines, Golden, CO, USA.

CH2M HILL & Payne Engineering (1997) *Constructed wetlands for livestock wastewater management. Literature review, database and research synthesis.* Gulf of Mexico Program, Stennis Space Center, MI, USA.

CH2M HILL. (1992) *Carolina Bay natural land treatment program.* Final report prepared for Grand Strand Water and Sewer Authority, Myrtle Beach, SC, USA.

CH2M HILL. (1994) *Experience with the use of constructed wetland effluent treatment systems in the pulp and paper industry.* Draft Report for the National Council of the Paper Industry for Air and System Improvement, Inc., Gainesville, FL, USA.

Chappell, K.R. and Goulder, R. (1994) Seasonal variation of epiphytic extracellular enzyme activity on two freshwater plants, *Phragmites australis* and *Elodea canadensis. Arch. Hydrobiol.* **132**, 237-253.

Chimney, M.J. and Moustafa, M.Z. (1999) Effectiveness and Optimization of Stormwater Treatment Areas for Phosphorus Removal. In *Everglades interim report* (ed. G. Redfield), chapter 6. South Florida Water Management District, West Palm Beach, FL, USA.

Conley, L.M., Dick, R.I. and Lion, L.W. (1991) An assessment of the Root Zone Method of wastewater treatment. *Wat. Pollut. Control. Fed. Res. J.* **63** (3), 239-247.

Conn, W.M. and Langworthy, A.C. (1984) *Practical operation of a small scale aquaculture.* (Proceedings from Water Reuse III), vol. 2. American Water Works Association.

Coombes, C. (1990) Reed bed treatment systems in Anglian Water. In *Constructed wetlands in water pollution control* (ed. P.F. Cooper and B.C. Findlater), pp. 223-234. Oxford, UK: Pergamon Press.

Cooper, P. (1998) A review of the design and performance of vertical-flow and hybrid reed bed treatment systems. In *Proceedings of 6th International Conference on Wetland Systems for Water Pollution Control, Aguas de Sao Pedro, Brazil, 27 Sept. to 2 Oct. 1998* (ed. S.M. Tauk-Tornisielo), pp. 228–242. Universidade Estadual Paulista, São Paulo State, Brazil.

Cooper, P.F. and Findlater, B.C. (eds) (1990) Constructed wetlands in water pollution control. In *Proceedings of the international conference on the use of constructed wetlands in water pollution control, Cambridge, U.K., 24–28 September*. Oxford, UK: Pergamon Press.

Cooper, P.F. and Green, M.B. (1995) Reed bed treatment systems for sewage treatment in the United Kingdom: the first 10 years experience. *Wat. Sci. Technol.* **32** (3), 317–327.

Cooper, P.F. (Ed.) (1990) *European design and operations guidelines for reed bed treatment systems.* Prepared by EC/EWPCA Emergent Hydrophyte Treatment Systems Expert Contact Group, Water Research Centre, Swindon, UK.

Cooper, P.F., Job, G.D., Green, M.B. and Shutes, R.B.E. (1996) *Reed beds and constructed wetlands for wastewater treatment.* Medmenham, Marlow, UK: WRc Publications.

Cooper, P.F., Smith, M.W. and Maynard, H. (1995) *The performance of a full-scale nitrifying vertical-flow reed bed treatment system.* Paper presented at the Workshop Nutrient Cycling and Retention in Wetlands and their use for Wastewater Treatment, at Trebon, Czech Republic.

Cooper, P., Griffin, P., Humphries, S. and Pound, A. (1998) Design of a hybrid reed bed system to achieve complete nitrification and denitrification of domestic sewage. In *Proceedings of 6th International Conference on Wetland Systems for Water Pollution Control, Aguas de Sao Pedro, Brazil, 27 Sept. to 2 Oct. 1998.*

Craft, C.B. and Richardson, C.J. (1993) Peat accretion and phosphorus accumulation along a eutrophication gradient in the Northern Everglades. *Biogeochemistry* **22**, 133–156.

Crites, R.W., Watson, R.C. and Williams, C.R. (1995) *Removal of metals in constructed wetlands* (Proceedings of WEFTEC '95, Miami, Florida). WEF, Alexandria, VA, USA.

Culp, Wesner, Culp (1983) *Draft design memorandum, Incline Village General Improvement District Wetlands Enhancement Project.* CWC, 3461 Robin Lane, Cameron Park, CA 95682, USA.

Daniels, R.E. and Parr, T.W. (1990) *1989 survey of reed growth in UK reed bed treatment systems.* WRc/NERC Contract Report T02058F1, The Institute of Terrestrial Ecology, Wareham, Dorset, UK.

Davidsson, T. (1997) *Nitrogen transformations in wetlands: effects of water flow patterns.* Ph.D. Dissertation, Department of Ecology and Limnology, University of Lund, Lund, Sweden.

DeBusk, T. and Merrick, P. (1989) *Summary of operational and research data for the Reedy Creek improvement district wetland-based wastewater treatment system* (WTS2). Report of March 1989.

DeBusk, T.A., Burgoon, P.S. and Reddy, K.R. (1989) Secondary treatment of domestic wastewater using floating and emergent macrophytes. In *Constructed wetlands for wastewater treatment: municipal, industrial, and agricultural* (ed. D.A. Hammer), chapter 38d, pp. 525–529. Chelsea, MI, USA: Lewis Publishers.

Delgado, M., Biggeriego, M. and Guardiola, E. (1993) Uptake of Zn, Cr, and Cd by water hyacinths. *Wat. Res.* **27** (2), 269–272.

Dickson, R. (1995) Bluther Burn drainage improvement scheme wetland reed bed sewage treatment works. Paper presented at the Reed Bed Seminar of the Association of Municipal Engineers, Dundalk, Ireland. (Unpublished.)

Dong, K. and Lin, C. (1994*a*) Treatment of petrochemical wastewater by the system of wetlands and oxidation ponds and engineering design of the system. In *Proceedings of the Fourth International Conference on Wetland Systems for Water Pollution Control, Guangzhou, China*, pp. 628–633. ICWS '94 Secretariat, Guangzhou, China.

Dong, K. and Lin, C. (1994*b*) The purification mechanism of the system of wetlands and oxidation ponds. In *Proceedings of the Fourth International Conference on Wetland Systems for Water Pollution Control, Guangzhou, China*, pp. 230–236. ICWS '94 Secretariat, Guangzhou, China.

DuBowy, P. and Reaves, R. (Eds) (1994) *Constructed wetlands for animal waste management.* Proceedings of workshop sponsored by the Conservation Technology Information Center, the U.S. Department of Agriculture Soil Conservation Service, U.S. Environmental Protection Agency Region V, and Purdue University Agricultural Research Program, 4–6 April 1994, Purdue University, Lafayette, IN, USA.

DuBowy, P.J. (ed.) (1997) *Proceedings of the Second National Workshop on Constructed Wetlands for Animal Waste Management.* Texas AandM University, College Station, TX, USA.

Duda, P.J., II. (1992) *Chevron's Richmond refinery water enhancement wetland.* Report to The Regional Water Quality Control Board, Oakland, CA, USA.

Edwards, M.E. (1993) Unpublished data from Tennessee Valley Authority.

Eger, P., Melchert, G., Antonson, D. and Wagner, J. (1993) The use of wetland treatment to remove trace metals from mine drainage. In *Constructed wetlands for water quality improvement* (ed. G.A. Moshiri), pp. 283–292. Boca Raton, FL, USA: Lewis Publishers.

Ewel, K.C. and Odum, H.T. (eds.) (1984) *Cypress swamps.* Gainesville, FL, USA: University of Florida Press.

Farmer, J.M. *et al.* (*ca.* 1996) Non-US oil terminal. Unpublished draft of British Petroleum internal report on constructed wetlands.

Finlayson, C.M., Von Oertzen, I. and Chick, A.J. (1990) Treating poultry abattoir and piggery effluents in gravel trenches. In *Constructed wetlands in water pollution control* (ed. P.F. Cooper and B.C. Findlater), pp. 559–562. Oxford, UK: Pergamon Press.

Fisher, P.J. (1990) Hydraulic characteristics of constructed wetlands at Richmond, NSW, Australia. In *Constructed wetlands in water pollution control* (ed. P.F. Cooper and B.C. Findlater), pp. 21–32. Oxford, UK: Pergamon Press.

Gambrell, R.P., Khalid, R.A. and Patrick, W.H., Jr. (1987) Capacity of a swamp forest to assimilate the TOC loading from a sugar refinery wastewater stream. *J. Wat. Pollut. Control. Fed.* **39** (10), 897–904.

Gearheart, R.A. (1992) Use of constructed wetlands to treat domestic wastewater, City of Arcata, California. *Wat. Sci. Technol.* **26**, 1625–1637.

Geller, G. (1996) 'Horizontal flow systems' for waste-water treatment: long term scientific and practical experiences; recommendations. In *Proc. 5th Int. Conf. Wetland Systems for Water Pollution Control*, chapter III/2. Universität für Bodenkultur Wien, Austria.

George, D.B., Kemp, M.C., Caldwell, A.S., Winfree, S.K. and Tsai, P.J. (1994) Design considerations for subsurface flow wetlands treating municipal wastewater. In *Proceedings of the 67th annual conference of the water environment federation*, vol. 8, pp. 13–24. Alexandria, VA, USA: Water Environment Federation.

Gersberg, R.M., Elkins, B.V. and Goldman, C.R. (1984) Use of artificial wetlands to remove nitrogen from wastewater. *J. Wat. Pollut. Control Fed.* **56**, 152–156.

Gersberg, R.M., Elkins, B.V., Lyon, S.R. and Goldman, C.R. (1986) Role of aquatic plants in wastewater treatment by artificial wetlands. *Wat. Res.* **20**, 363–368.

Gersberg, R.M., Gearhart, R.A. and Ives, M. (1989*b*) Pathogen removal in constructed wetlands. In *Constructed wetlands for wastewater treatment: municipal, industrial and agricultural* (ed. D.A. Hammer), pp. 431–446. Chelsea, MI, USA: Lewis Publishers.

Gersberg, R.M., Lyon, S.R., Brenner, R. and Elkins, B.V. (1989*a*) Integrated wastewater treatment using artificial wetlands: a gravel marsh case study. In *Constructed wetlands for wastewater treatment: municipal, industrial and agricultural* (ed. D.A. Hammer), pp. 145–152. Chelsea, MI, USA: Lewis Publishers.

Gosselink, J.G. and Turner, R.E. (1978) The role of hydrology in freshwater wetland ecosystems. In *Freshwater wetlands: ecological processes and management potential* (ed. R.E. Good, D.F. Whigham and R.L. Simpson), pp. 63–78. New York, USA: Academic Press.

Grant, W.D. and Long, P.E. (1981) *Environmental microbiology.* Glasgow, UK: Blackie & Son.

Grant, W.D. and Long, P.E. (1985) Environmental microbiology. In *The handbook of environmental chemistry* (ed.

O. Hutzinger), vol. 1, part D (*The natural environment and biogeochemical cycles*), pp. 125–237. Berlin, Germany: Springer-Verlag.

Gray, K.R. and Biddlestone, A.J. (1995) Engineered reed-bed systems for wastewater treatment. *Trends Biotechnol.* **13** (7), 248–252.

Green, M.B. and Martin, J.R. (1994) Constructed reed beds clean up stormwater overflows on small works. In *Proceedings of the Water Environment Federation 67th Annual Conference and Exposition*, vol. 8, pp. 113–124. WEF, Alexandria, VA, USA.

Green, M.B. and Upton, J. (1995) Constructed reed bed: appropriate technology for small communities. *Wat. Sci. Technol.* **32** (3), 339–348.

Green, M., Friedler, E., Ruskol, Y. and Safrai, I. (1997) Investigation of alternative method for nitrification in constructed wetlands. *Wat. Sci. Technol.* **35** (5), 63–70.

Green, M.B. and J. Upton. (1994) Constructed reed beds: A cost-effective way to polish wastewater effluents for small communities. *Wat. Envir. Res.* **66**, 188–192.

Gries, C.L., Kappen, L. and Lösch, R. (1990) Mechanism of flood tolerance in reed *Phragmites australis* (Cav.) Trin. ex Streudel. *New Phytol.* **114**, 589–593.

Gully, J.R. and Nix, P.G. (1993) *Wetland treatment of oil sands operation waste waters.* Preprint of paper presented at Conference on Biogeochemical Cycling in Wetlands, February, 1993, Baton Rouge, LA, USA.

Gumbricht, T. (1993*a*) Nutrient removal processes in freshwater submersed macrophyte systems. *Ecol. Engng* **2** (1), 1–30.

Gumbricht, T. (1993*b*) Nutrient removal capacity in submersed macrophyte systems. *Ecol. Engng* **2** (1) 49–61.

Haberl, R. (1997) Humedales construidos en Europa con enfasis en Austria [Constructed wetlands in Europe with emphasis on Austria]. In *Proceedings, 3rd Int. Seminar of Experts on the Treatment of Industrial Effluents and Residues, 3–9 April 1997, Mexico City* (ed. C.D. Duran and I.R. Burgos), pp. 219–245. Universidad Nacional Autonoma de Mexico, Faculdad de Chimica.

Haberl, R., Perfler, R. and Laber, J. (1998*a*) *Abwasserreinigung mit Pflanzen.* Annual Report 1997. Amt der OÖ Landesregierung, Abteilung Siedlungswasserbau, A 4020 Linz, Kärntnerstrasse 12, Austria.

Haberl, R., Perfler, R., Laber, J. and Grabher, D.(1998*b*) Constructed wetlands for wastewater treatment in Austria. In *Constructed wetlands for wastewater treatment in Europe* (ed. J. Vymazal, H. Brix, P.F. Cooper, M.B. Green and R. Haberl), pp 67–93. Leiden, The Netherlands: Backhuys Publishers.

Haberl, R., Perfler, R. and Mayer, H. (1995) Constructed wetlands in Europe. *Wat. Sci. Technol.* **32** (3), 305–316.

Hall, B.R. and Freeman, G.E. (1994) Study of hydraulic roughness in wetland vegetation takes new look at Manning's n. *Wetlands Res. Prog. Bull.* **4** (1), 1–4. USACOE, Waterways Experiment Station, Vicksburg, MS, USA.

Hammer, D.A., Pullin, B.P., McMurry, D.K. and Lee, J.W. (1993) Testing color removal from pulp mill wastewaters with constructed wetlands. In *Constructed wetlands for water quality improvement* (ed. G.A. Moshiri), pp. 449–452. Boca Raton, FL, USA: Lewis Publishers.

Hammer, D.E. and Kadlec. R.H. (1986) A model for wetland surface water dynamics. *Wat. Resour. Res.* **22** (13), 1951–1958.

Haslam, S.M. (1971*a*) Community regulation in *Phragmites communis* Trin. I. Monodominant stands. *J. Ecol.* **59**, 65–73.

Haslam, S.M. (1971*b*) The development and establishment of young plants of *Phragmites communis. Trin. Ann. Bot.* **35**, 1059–1072.

Haslam, S.M. (1972) Biological flora of the British Isles. *J. Ecol.* **60**, 585–610.

Hauck, R.D. (1984) Atmospheric nitrogen chemistry, nitrification, denitrification, and their relationships. In *The handbook of environmental chemistry* (ed. O. Hutzinger), vol. 1, part C (*The natural environment and biogeochemical cycles*), pp. 105–127. Berlin, Germany: Springer-Verlag.

Hawke, C.J. and José, P.V. (1996) *Reedbed management for commercial and wildlife interests.* Ref. 24/571/95. UK: Royal Society for the Protection of Birds.

Hendrey, G.R., Clinton, J., Blumer, K. and Lewin, K. (1979) *Lowland recharge project operations, physical, chemical, and biological changes 1975–1978.* Final report to the town of Brookhaven. Brookhaven National Laboratory, Brookhaven, NY, USA.

Hershkowitz, J. (1986) *Listowel artificial marsh project report.* Ontario Ministry of the Environment, Water Resources Branch, Toronto, ON, Canada. (251 pages.)

Hinchman, R.R., McNally, R.G. and Negri, M.C. (1993) *Biotreatment of produced waters.* Research project sponsored by the Gas Research Institute, Chicago, IL, USA; conducted at Argonne National Laboratory, Argonne, IL, USA.

Hofmann, K. (1986) Wachstumverhalten von Schilf (*Phragmites australis* [Cav.] Trin. ex Steudel) in klärschlammbeschickten Filterbeeten. *Arch. Hydrobiol.* **107**, 385–409.

Hokosawa, Y. and Horie, T. (1992) Flow and particulate nutrient removal by wetland with emergent macrophyte. *Sci. Tot. Environ.* (Suppl. 1992), 1271–1282.

Honig, R. (1988) *Tenneco's use of a rock-reed filter at a natural gas pipeline compressor station.* Poster paper at International Conference on Constructed Wetlands for Wastewater Treatment, Chattanooga, TN, USA.

Hotchkiss, N. (1972) *Common marsh, underwater, and floating-leaved plants of the United States and Canada* New York: Dover Publications.

Howard, E.A., Emerick, J.C. and Wilderman, T.R. (1988) Design and construction of a research site for passive mine drainage treatment in Idoho Springs, Colorada. In *Constructed wetlands for wastewater treatment: municipal, industrial and agricultural* (ed. D.A. Hammer), pp. 761–764. Chelsea, MI, USA: Lewis Publishers.

Kadlec, R.H. and Bevis, F.B. (1990) Wetlands and wastewater: Kinross, Michigan. *Wetlands* **10** (1), 77–92.

Kadlec, R.H. and Knight, R.L. (1996) *Treatment wetlands.* Boca Raton, FL, USA: Lewis-CRC Press.

Kadlec, R.H. and Tilton, D.L. (1979) The use of freshwater wetlands as a tertiary wastewater treatment alternative. *CRC Crit. Rev. Envir. Control* **9** (2), 185–212.

Kadlec, R.H. and Watson, J.T. (1993) Hydraulics and solids accumulation in a gravel bed treatment wetland. In *Constructed wetlands for water quality improvement* (ed. G.A. Moshiri), pp. 227–236. Boca Raton, FL, USA: Lewis Publishers.

Kadlec, R.H. (1990*a*) Modelling nutrient behavior in wetlands used for wastewater treatment. *Utrecht Plant Ecology News Report* **11**, 104–129.

Kadlec, R.H. (1990*b*) Overland flow in wetlands: vegetation resistance. *ASCE J. Hydraul. Engng* **116** (5), 691–706.

Kadlec, R.H. (1994*a*) Detention and mixing in free water wetlands. *Ecol. Engng* **3** (4), 1–36.

Kadlec, R.H. (1994*b*) Unpublished data from wetlands near Holly, MI, USA.

Kadlec, R.H. (1996) An autobiotic wetland phosphorus model. *Ecol. Engng* **8** (2), 145–172.

Kalin, M. (1996) Paper presented at WEAO conference, Toronto, ON, Canada.

Kehoe, M.J., Dye, C.W. and Rushton, B.T. (1994) *A survey of the water-quality of wetlands-treatment stormwater ponds* Final report, Southwest Florida Water Management District, Brooksville, FL, USA. (42 pages + appendices.)

Kemp, W.M. and Murray, L. (1986) Oxygen release from roots of the submerged macrophyte *Potamogeton perfoliatus* L.: regulating factors and ecological implications. *Aquat. Bot.* **26**, 271–283.

Kickuth, R. (1977) Degradation and incorporation of nutrients from rural wastewaters by plant rhizosphere under limnic conditions. In *Utilization of manure by land spreading*, pp. 335–343. EUR 5672e. London, UK: Commission of the European Communities.

Kleinman, R.L.K.P. and Girts, M.A. (1987) Acid mine water treatment wetlands: an overview of an emergent technology. In *Aquatic plants for water treatment and resource recovery* (ed. K.R. Reddy and W.H. Smith), pp. 255–262. Orlando, FL, USA: Magnolia Publishing.

Knight, R.L. and Ferda, K.A. (1989) Performance of the Boggy Gut Wetland Treatment System, Hilton Head, South Carolina. In *Proceedings of the symposium on wetlands: concerns and successes* (ed. D. Fisk), pp. 439–450. Bethesda, MD, USA: American Water Resources Association.

Knight, R.L. and Iverson, M.E. (1990) Design of the Fort Deposit, Alabama constructed wetlands treatment system. In *Constructed wetlands in water pollution control* (ed. P.F. Cooper and B.C. Findlater), pp. 521-524. Oxford, UK: Pergamon Press.

Knight, R.L. (1993) Operating experience with constructed wetlands for wastewater treatment. *Tappi J.* **76** (1), 109-112.

Knight, R.L. (1997) Wildlife habitat and public use benefits of treatment wetlands. *Wat. Sci. Technol.* **35** (5), 35-43.

Knight, R.L., Hilleke, J. and Grayson, S. (1994) Design and performance of the Champion pilot-constructed wetland treatment system. *Tappi J.* **77**, 240-245.

Knight, R.L., Ruble, R.W., Kadlec, R.H. and Reed, S. (1993) Wetlands for wastewater treatment performance database. In *Constructed wetlands for water quality improvement* (ed. G.A. Moshiri), pp. 35-58. Boca Raton, FL, USA: Lewis Publishers.

Knight, R.L., Ruble, R.W., Kadlec, R.H. and Reed, S.C. (1993) *Database: North American wetlands for water quality treatment.* Phase II report. Prepared for US EPA, September, 1993. (824 pages.)

Laan, P., Smolders, A., Blom, C.W.P.M. and Armstrong, W. (1989) The relative roles of internal aeration, radial oxygen loss, iron exclusion and nutrient balance in flood tolerance of *Rumex* species. *Acta. Bot. Neer.* **38**, 131-145.

Laanbroek, H.J. (1990) Bacterial cycling of minerals that affect plant growth in waterlogged soils: a review *Aquat. Bot.* **38**, 109-125.

Laber, J., Haberl, R. and Perfler, R. (in press) Advanced nitrogen elimination in vertical flow constructed wetlands in cold climates. In *Proceedings, VII International Congress of Ecology (Intecol), 19-25 July 1998, Florence, Italy* (ed. Ü. Mander and P.D. Jenssen). Wessex Institute of Technology, Constructed Wetlands for Wastewater Treatment in Cold Climates.

Laber, J., Haberl, R. and Shresta, R. (1999) Two stage constructed wetland for treating hospital wastewater in Nepal. *Wat. Sci. Technol.* (In the press.)

Laber, J., Perfler, R. and Haberl, R. (1997) Two strategies for advanced nitrogen removal in vertical flow constructed wetlands. *Wat. Sci. Technol.* **35** (5), 71-77.

Lakatos, G. (1998) Hungary. In *Constructed wetlands for wastewater treatment in Europe* (ed. J. Vymazal, H. Brix, P.F. Cooper, M.B. Green and R. Haberl), pp. 191-206. Leiden, The Netherlands: Backhuys Publishers.

Lawson, G.J. (1985) *Cultivating reeds* (Phragmites australis) *for root zone treatment of sewage.* Contract Report to the Water Research Centre, IRE project 965, Grange-over-Sands, Cumbria, UK.

Lienard, A., Boutin, C. and Esser, D. (1990) Domestic wastewater treatment with emergent hydrophyte beds in France. In *Constructed wetlands in water pollution control* (ed. P.F. Cooper and B.C. Findlater), pp. 183-192. Oxford, UK: Pergamon Press.

Lienard, Esser, D., Deguin, A. and Virloget, F. (1990) Sludge dewatering and drying in reed beds: an interesting solution. In *Constructed wetlands in water pollution control* (ed. P.F. Cooper and B.C. Findlater), pp. 257-267. Oxford, UK: Pergamon Press.

Litchfield, D.K. (1989) Constructed wetlands for wastewater treatment at Amoco Oil Company's Mandan, North Dakota refinery. In *Constructed wetlands for wastewater treatment: municipal, industrial and agricultural* (ed. D.A. Hammer), pp. 233-237. Chelsea, MI, USA: Lewis Publishers.

Litchfield, D.K. (1993) Constructed wetlands for wastewater treatment at Amoco Oil Company's Mandan, North Dakota, refinery. In *Constructed wetlands for water quality improvement* (ed. G.A. Moshiri), pp. 485-488. Chelsea, MI, USA: Lewis Publishers.

Livingston, E.H. (1989) Use of wetlands for urban stormwater management. In *Constructed wetlands for wastewater treatment: municipal, industrial and agricultural* (ed. D.A. Hammer), pp. 253-262. Chelsea, MI, USA: Lewis Publishers.

Lodge, D.M. (1991) Herbivory on freshwater macrophytes. *Aquat. Bot.* **41**, 195-224.

Maehlum, T., Jenssen, P.D. and Warner, W.S. (1995) Cold-climate constructed wetlands. *Wat. Sci. Technol.* **32** (2), 95-102.

Mann, R.A. (1990) Phosphorus removal by constructed wetlands: substratum absorption. In *Constructed wetlands in water pollution control* (ed. P.F. Cooper and B.C. Findlater), pp. 97-106. Oxford, UK: Pergamon Press.

Martin, C.D. and Moshiri, G.A. (1994) The use of in-series surface-flow wetlands for landfill leachate treatment. In *Proceedings of the Fourth IAWQ (International Association on Water Quality) Conference on Wetland Systems for Water Pollution Control, November 6 to 10, 1994, Guangzhou, China*, pp. 513-522. Guangzhou, P.R. China: Center for International Development and Research, South China Institute for Environmental Sciences.

Mellstrom, R.E. and Jager, R.A. (1994) Reedbed dewatering and treatment systems in New England. In *Proc. Water Environ. Fed. 67th Annual Conference and Exposition*, vol. 8, pp. 56-67. WEF, Washington, DC, USA.

Metcalf and Eddy, Inc. (1991) *Wastewater engineering, treatment, disposal, and reuse*, 3rd edn. (rev. G. Tchobanoglous and F.L. Burton). New York, USA: McGraw-Hill.

Mitsch, W. and Wu, X. (1995) *Olentangy River Wetland Research Park at the Ohio State University. Annual Report 1994.* Columbus, OH, USA: Ohio State University School of Natural Resources.

Mitsch, W.J. and Gosselink, J.G. (1993) *Wetlands*, 2nd edn. New York, USA: Van Nostrand Reinhold Company.

Mitsch, W.J., Wu, X. and Wang, N. (1995) *Modelling the Des Plaines experimental wetlands—An integrative approach to data management and ecosystem prediction.* Report to US EPA Region V, Contract 91044, Project 769641/725753.

Moore, J.A. and Skarda, S.M. (1992) Wetland treatment of pulp mill wastewater. *Wat. Sci. Technol.* **29** (4), 241-247.

Moore, J.A., Gamroth, M.J., Skarda, S.M. and Niswander, S.F. (1995) Treating dairy flush water in a constructed wetland. In *Proceedings of the Seventh International Symposium on Agricultural and Food Processing Wastes 18-20 June 1995, Chicago, Illinois*, pp. 74-86. ASAE, St Joseph, MI, USA.

Moorhead, K.K. and Reddy, K.R. (1988) Oxygen transport through selected aquatic macrophytes. *J. Envir. Qual.* **17**, 138-142.

Morris, M. and Herbert, R. (1997) The design and performance of a vertical flow reed bed for the treatment of high ammonia, low suspended solids organic effluents. *Wat. Sci. Technol.* **35** (5), 197-204.

Morrison, G.M.P., Revitt, D.M., Ellis, J.B., Balmer, P. and Stevenson, G. (1984) Heavy metal partitioning between the dissolved and suspended solids phases from stormwater runoff in residential areas. *Sci. Tot. Environ.* **33**, 287-296.

Mulamoottil, G., McBean, E. and Rovers, F.A. (eds.) (1998) Constructed wetlands for the treatment of landfill leachates. Boca Raton, FL, USA: Lewis Publishers.

NADB (North American Treatment Wetland Database) (1993) Electronic database created by R. Knight, R. Ruble, R. Kadlec, and S. Reed for the U.S. Environmental the Great Lake states, north central USA. *Nature* **358**, 139-141.

Netter, R., (1994) Flow charactersistics of planted soil filters. *Wat. Sci. Technol.* **29** (4), 37-44.

Ngo, V. (1987) Boosting pond performance with aquaculture. *Ops Forum* **4**, 20-24.

Nielsen, S.M. (1994) Biological sludge drying in reed bed systems - six years of operations experience. In *Proceedings of the Fourth International Conference on Wetland Systems for Water Pollution Control, Guangzhou, China.* ICWS '94 Secretariat, Guangzhou, China.

Niering, W.A. (1985) *Wetlands.* New York, USA: Knopf.

Noller, B.N., Woods, P.H. and Ross, B.J. (1994) Case studies of wetland filtration of mine waste water in constructed and naturally occurring systems in Northern Australia. *Wat. Sci. Technol.* **29** (4), 257-265.

O'Connor, D.J. and Dobbins, W.E. (1958) Mechanism of reaeration in natural streams. *Am. Soc. Civ. Engrs Trans.*, paper no. 2934, pp. 641-684.

Odum, H.T. (1983) *Systems ecology: an introduction.* New York, USA: John Wiley & Sons.

Patrick, W.H., Jr and Wyatt, R. (1964) Soil nitrogen loss as a result of alternate submergence and drying. *Proc. Soil Sci. Soc. Am.* **28**, 647-652.

Payne, W.J. (1981) *Denitrification.* New York, USA: John Wiley & Sons.

PBSJ (Post, Buckley, Schuh, and Jernigan) (1989) *Orlando Easterly Wetlands: monitoring report for 1988.* Report to City of Orlando, FL, USA.

Perdomo, S., Fujita, M. and Furukawa, K. (1996) Oxygen transport through *Pistia stratiotes* L. In *Proc. 5th Internat. Conf. Wetland Systems for Water Pollution Control*, pp. VIII/4-1–VIII/4-8. Universität für Bodenkultur Wien, Vienna, Austria.

Perfler, R. and Haberl, R. (1996) Efficiency of nutrient control in reed bed systems for domestic wastewater treatment. In *Nutrient cycling and retention in wetlands and their use for wastewater treatment* (ed. J. Vymazal), pp. 231–232. Ecology and Use of Wetlands, Prague and Botanical Institute Trebon.

Perfler, R., Haberl, R. and Laber, J. (1995) Abwasserreinigung in kleinen Verhältnissen. *Österr. Wasser- und AbfWirtsch.* **47**, 11–12.

Perfler, R., Haberl, R. and Laber, J. (1996) Einsatz und Betrieb von Pflanzenkläranlagen in der Praxis. *Wiener Mitt.* **130**, 307–337.

Perfler, R., Laber, J., Langergraber, G. Haberl, R. (1999) Constructed wetlands for rehabilitation and reuse of surface waters in tropical and subtropical areas - first results from small scale plots. *Wat. Sci. Technol.* (In the press.)

Pettecrew, E.L. and Kalff, J. (1992) Water flow and clay retention in submerged macrophyte bed. *Can. J. Fish. Aquat. Sci.* **49**, 2483–2489.

Pickett, J., McKellar, H. and Kelley, J. (1989) Plant community composition, leaf mortality, and aboveground production in a tidal freshwater marsh. In *Freshwater wetlands and wildlife* (Proceedings of a symposium held at Charleston, S.C.U.S. Dept. of Energy, CONF-8603101.) (ed. R.R. Sharitz and J.W. Gibbons), pp. 351-364. US DOE, Washington, DC, USA.

Platzer, C. (1996) Enhanced nitrogen elimination in subsurface flow artificial wetlands - a multi stage concept. In *Proc. 5th Internat. Conf. Wetland Systems for Water Pollution Control*, chapter I/7. Universität für Bodenkultur Wien, Vienna, Austria.

Platzer, C. (1998) Design recommendations for subsurface flow constructed wetlands for nitrification and denitrification. In *Proceedings of 6th International Conference on Wetland Systems for Water Pollution Control, Aguas de Sao Pedro, Brazil, 27 September to 2 October 1998* (ed. S.M. Tauk-Tornisielo), pp. 253–263. Universidade Estadual Paulista, São Paulo State, Brazil.

Pries, J.H. (1994) *Wastewater and stormwater applications of wetlands in Canada.* Issues paper no. 1994-1, North American Wetlands Conservation Council (Canada). Ottawa, Canada.

Reddy, K.R. and D'Angelo, E.M. (1994) Soil processes regulating water quality in wetlands. In *Global wetlands: Old World and new* (ed. W. Mitsch), pp. 309-324. Amsterdam, The Netherlands: Elsevier.

Reddy, K.R. and DeBusk, W.F. (1987) Nutrient storage capabilities of aquatic and wetland plants. In *Aquatic plants for water treatment and resource recovery* (ed K.R. Reddy and W.H. Smith), pp. 337-357. Orlando, FL, USA: Magnolia Publishing Inc.

Reddy, K.R. and Patrick, W.H. (1984) Nitrogen transformations and loss in flooded soils and sediments. *CRC Crit. Rev. Envir. Control* **13**, 273-309.

Reddy, K.R. and Smith, W.H. (eds.) (1987) *Aquatic plants for water treatment and resource recovery.* Orlando, FL, USA: Magnolia Publishing Inc.

Reddy, K.R., Khaleel, R., Overcash, M.R. and Westerman, P.W. (1979) A nonpoint source model for land areas receiving animal wastes. 1. Mineralization of organic nitrogen. *Trans. ASAE* **22**, 863-876.

Reed, S.C. (1985) Nitrogen removal in wastewater stabilization ponds. *J. Wat. Pollut. Control Fed.* **57** (1), 39-45.

Reed, S.C. and Brown, D.S. (1992) Performance of gravel bed wetlands in the United States. In *Wetland systems in water pollution control* (Proceedings of a conference, 30 November to 3 December 1992, Sydney, Australia) (ed. Organizing Committee of Wetland Systems in Water Pollution Control Conference, Conference Secretariat), pp. 16.1-16.8. Australian Water and Wastewater Association.

Reed, S.C., Crites, R.W. and Middlebrooks, E.J. (1995) *Natural systems for waste management and treatment*, 2nd edn. New York: McGraw-Hill.

Reed, S.C., Middlebrooks, E.J. and Crites, R.W. (1988) *Natural systems for waste management and treatment.* New York: McGraw-Hill.

Reid, G.K. (1987) *Pond life.* Racine, WI, USA: Western Publishing Co.

Richards, Moorehead & Laing Ltd (1992) *Constructed wetlands to ameliorate metal-rich mine waters. 1. Review of existing literature.* R&D note 102. National Rivers Authority, Bristol, UK.

Rivera, F., Warren, A., Ramirez, E., Decamp, O., Bonilla, P. Gallegos, E., Calderon, A. and Sanchez, J.T. (1994) Removal of pathogens from wastewaters by the Root Zone Method (RZM). In *Proceedings of the Fourth IAWQ Conference on Wetland Systems for Water Pollution Control, 6–10 November, 1994, Guangzhou, China* (ed. ICWS '94 Secretariat, Guangzhou, China), pp. 180–189. Guangzhou, P.R. China: Center for International Development and Research, South China Institute for Environmental Sciences.

Rivera, F., Warren, A., Ramirez, E., Decamp, O., Bonilla, P.,Gallegos, E., Calderon, A. and Sanchez, J.T. (1995) Removal of pathogens from wastewaters by the Root Zone Method (RZM). *Wat. Sci. Technol.* **32** (3), 211–218.

Rogozinski, L.R., Laubaucher, R.C. and Farmer, J.M. (1992) Groundwater treatment via constructed wetlands. Paper presented at the Petroleum Hydrocarbons Conference, Houston, TX, USA, 4–6 November.

Rovira, A.D. (1965) Interactions between plant roots and soil microorganisms. *A. Rev. Microbiol.* **19**, 241-266.

Rovira, A.D. (1969) Plant root exudates. *Bot. Rev.* **35**, 35–57.

Rushton, B.T., Miller, C., Hull, C. and Cunningham, J. (1997) *Three design alternatives for stormwater detention ponds.* Southwest Florida Water Management District, Brooksville, FL, USA. (109 pages + appendices.)

Sand-Jensen, K., Prahl, C. and Stokholm, H. (1982) Oxygen release from roots of submerged aquatic macrophytes. *Oikos* **38**, 349–354.

Sapkota, D.P. and Bavor, H.J. (1994) Gravel media filtration as a constructed wetland component for the reduction of suspended solids from maturation pond effluent. *Wat. Sci. Technol.* **29** (4), 55–66.

Sassaman, M.D. and Kauffman, T.R. (1992) Sludge dewatering and disposal utilising the reed system. In *Water Environment Federation 65th Annual Conference and Exposition, New Orleans, USA, September 1992*, vol. 9, pp. 147–152. WEF, Alexandria, VA, USA.

Scheuerman, P.R., Bitton, G. and Farrah, S.R. (1989) Fate of microbial indicators and viruses in a forested wetland. In *Constructed wetlands for wastewater treatment: municipal, industrial and agricultural* (ed. D.A. Hammer), pp. 657–663. Chelsea, MI, USA: Lewis Publishers.

Schierup, H.H., Brix, H. and Lorenzen, B. (1990*a*) Wastewater treatment in constructed reed beds in Denmark – state of the art. In *Proceedings of the International Conference on the Use of Constructed Wetlands in Water Pollution Control, Cambridge, UK, 24-28 September 1990* (ed. P.F. Cooper and B.C. Findlater), pp. 495-504. Oxford, UK: Pergamon Press.

Schierup, H.H., Brix, H. and Lorenzen, B. (1990*b*) *Spildevandsrensning i rodzoneanlæg.* Aarhus, Denmark: Botanical Institute, Aarhus University.

Schmidt, E.L. (1982) Nitrification in soils. In *Nitrogen in agricultural soil* (ed. F.J. Stevenson), Am. Soc. Agron., Madison, Wisconsin.

Schönerklee, M., Koch, F., Perfler, R., Haberl, R. and Laber, J. (1997) Tertiary treatment in a vertical flow reed bed system - a full scale pilot plant for 200–600 P.E. *Wat. Sci. Technol.* **35** (5), 223–230.

Schueler, T.R. (1992) *Design of stormwater wetland systems: guidelines for creating diverse and effective stormwater wetlands in the mid-Atlantic region.* Washington, DC, USA: Metropolitan Washington Council of Governments. (133 pages.)

Schwager, A. and Boller, M. (1997) Transport phenomena in intermittent filters. *Wat. Sci. Technol.* **35** (6), 13–20.

Schwegler, B.R. (1978) Effects of sewage effluent on algal dynamics of a Northern Michigan wetland. MS thesis, University of Michigan, Ann Arbor, MI, USA.

Seidel, K. (1964) Abbau von bacterium coli durch hohere Wasserpflanzen. *Naturwissenschaften* **51**, 395.

Seidel, K. (1966) Reinigung von Gewässern durch höhere Pflanzen. *Naturwissenschaften* **53**, 289–297.

Seidel, K. (1976) Macrophytes and water purification. In *Biological control of water pollution* (ed. J. Tourbier and R.W. Pierson, Jr), chapter 14. Philadelphia, PA, USA: University of Pennsylvania Press.

Shih, S.F., Federico, A.C., Milleson, J.F. and Rosen, M. (1979) Sampling programs for evaluating upland marsh to improve water quality. *Trans. Am. Soc. Agric. Engrs* **22**, 828–833.

Shutes, R.B.E., Ellis, J.B., Revitt, D.M. and Zhang, J. (1993) The use of *Typha lalifolia* for heavy metal pollution control in urban wetlands. In *Constructed wetlands for water quality improvement* (ed. G.A. Moshiri), pp. 407–464. Boca Raton, FL, USA: Lewis Publishers.

Sinicrope, T.L., Langis, R., Gersberg, R.M., Busanardo, M.J. and Zedler, J.B. (1992) Metal removal by wetland mesocosms subjected to different hydroperiods. *Ecol. Engng* **1** (4), 309–322.

Smith, I.D., Bis, G.N., Lemon, E.R. and Rozema, L.R. (1966) A thermal analysis of a sub-surface, vertical flow constructed wetland. In *Proc. 5th Internat. Conf. Wetland Systems for Water Pollution Control*, pp. I/1-1–I/1-8. Vienna, Austria: Universität für Bodenkultur Wien.

Somes, N.L.G., Breen, P.F. and Wong, T.H.F. (1996) Integrated hydrologic and botanical design of stormwater control wetlands. In *Proc. 5th Internat. Conf. Wetland Systems for Water Pollution Control*, pp. III/4-1–III/4-8. Vienna, Austria: Universität für Bodenkultur Wien.

Sorrell, B.K. and Armstrong, W. (1994) On the difficulties of measuring oxygen release by root systems of wetland plants. *J. Ecol.* **82**, 177–183.

Sorrell, B.K. and Boon, P.I. (1992) Biogeochemistry of billabong sediments. II. Seasonal variations in methane production. *Freshwat. Biol.* **27**, 435–445.

Spangler, F.L., Sloey, W.E. and Fetter, C.W., Jr (1976) *Wastewater treatment by natural and artificial marshes.* Report EPA-600/2-76-207 to the Robert S. Kerr Environmental Research Laboratory, Ada, OK, USA. (171 pages.)

Srinivasan, K.R. and Kadlec, R.H. (1995) *Wetland treatment of oil and gas well wastewaters.* Report to US Department of Energy on contract DE-AC22-92MT92010. (55 pages.)

Stairs, D.B. (1993) Flow characteristics of constructed wetlands: tracer studies of the hydraulic regime, MS Thesis, Oregon State University, Corvallis, OR, USA.

Stark, L.R., Wenerick, W.R., Williams, F.M. and Wuest, P.J. (1994) The effects of pH, flow rate, and carbon supplementation on manganese retention in mesocosm wetlands. *J. Envir. Qual.* **24**, 816–826.

Staubitz, W.W., Surface, J.M., Steenhuis, T.S., Peverly, J.H., Lavine, M.J., Weeks, N.C., Sanford, W.E. and Kopka, R.J. (1989) Potential use of constructed wetlands to treat landfill leachate. In *Constructed wetlands for wastewater treatment: municipal, industrial and agricultural* (ed. D.A. Hammer), pp. 735–742. Chelsea, MI, USA: Lewis Publishers.

Strecker, E.W., Kersnar, J.M., Driscoll, E.D. and Horner, R.R. (1992) *The use of wetlands for controlling stormwater pollution.* EPA/600. Washington D.C.: The Terrene Institute.

Sun, G., Gray, K.R. and Biddlestone, A.J. (1998) Treatment of agricultural wastewater in downflow reed beds: experimental trials and mathematical models. *J. Agric. Engng Res.* **69**, 63–71.

Surface, J.M., Peverly, J.H., Steenhuis, T.S. and Sanford, W.E. (1993) Effect of season, substrate composition, and plant growth on landfill-leachate treatment in a constructed wetland. In *Constructed wetlands for water quality improvement* (ed. G.A. Moshiri), pp. 461–472. Boca Raton, FL, USA: Lewis Publishers.

Tang, S.-Y. and Lu, X.-W. (1993) The use of *Eichhornia crassipes* to cleanse oil-refinery wastewater in China. *Ecol. Engng* **2**, 243–251.

Tanner, C.C. and Sukias, J.P. (1994) Accumulation of organic solids in gravel-bed constructed wetlands. In *Proceedings of the Fourth International Conference on Wetland Systems for Water Pollution Control, Guangzhou, China*, pp. 617–627. ICWS '94 Secretariat, Guangzhou, China.

Tettleton, R.P., Howell, F.G. and Reaves, R.P. (1993) Performance of a constructed marsh in the tertiary treatment of Bleach Kraft Pulp Mill effluent: results of a 2-year pilot project. In *Constructed wetlands for water quality improvement* (ed. G.A. Moshiri), pp. 437–440. Boca Raton, FL, USA: Lewis Publishers.

Thut, R.N. (1989) Utilization of artificial marshes for treatment of pulp mill effluents. In *Proceedings of the TAPPI Environmental Conference*, pp. 83–87. TAPPI, Washington, DC, USA.

Thut, R.N. (1990*a*) Treatment of pulp mill effluent by an artificial marsh – large scale pilot study. In *Proceedings of the TAPPI Environmental Conference*, pp. 121–126. TAPPI, Washington, DC, USA.

Thut, R.N. (1990*b*) Utilization of artificial marshes for treatment of pulp mill effluent. *TAPPI J.* **73**, 93–96.

Thut, R.N. (1993) Feasibility of treating pulp mill effluent with a constructed wetland. In *Constructed wetlands for water quality improvement* (ed. G.A. Moshiri), pp. 441–447. Boca Raton, FL, USA: Lewis Publishers.

TVA (1991) *General design, construction, and operation guidelines: constructed wetlands wastewater treatment systems for small users including individual residences.* TVA/WR/WQ—91/2. TVA, Chattanooga, TN, USA.

US Environmental Protection Agency (1988*a*) *Design manual. Constructed wetlands and aquatic plant systems for municipal wastewater treatment.* EPA/625/1-88/022. Cincinnati, OH, U.S.A.: Office of Research and Development, Center for Environmental Research Information.

US Environmental Protection Agency (1988*b*) *Report on the use of wetlands for municipal wastewater treatment and disposal.* EPA 430/09-88-005.

US Environmental Protection Agency (1989) *Nonpoint sources. Agenda for the future.* WH-556. Washington, D.C. USA: Office of Water.

US Environmental Protection Agency (1993*a*) *Constructed wetlands for wastewater treatment and wildlife habitat. 17 case studies.* EPA 832-R-93-005.

US Environmental Protection Agency (1993*b*) *Toxic substances spreadsheet.* Updated by Region IV Water Quality Standards Section. Section 304(a) of the Clean Water Act.

US Environmental Protection Agency (1993*c*) *Subsurface flow constructed wetlands for wastewater treatment: A technology assessment.* EPA Report 832-R-93-001. Washington, D.C. USA: Office of Water.

Valiela, I. (1984) *Marine ecological processes.* New York: Springer-Verlag.

van Oostrom, A.J. and Cooper, R.N. (1990) Meat processing effluent treatment in surface-flow and gravel-bed constructed wastewater wetlands. In *Constructed wetlands in water pollution control* (ed. P.F. Cooper and B.C. Findlater), pp. 321–332. Oxford, UK: Pergamon Press.

van Oostrom, A.J. (1995) Nitrogen removal in constructed wetlands treating nitrified meat processing effluent. *Wat. Sci. Technol.* **32** (3), 137–148.

Veber, K. (1978) Propagation, cultivation and exploitation of Common Reed in Czechoslovakia. In *Pond littoral ecosystems. Structure and functioning* (ed. D. Dykyjová and J. Kvet), pp. 416–425. Berlin, Germany: Springer-Verlag.

Vymazal, J. (1994) Constructed wetlands for wastewater treatment in the Czech Republic – state of the art. In *Proc. Conf. Wetlands Systems for Water Pollution Control (ICWS 94), Guangzhou, China*, pp. 129–137. ICWS '94 Secretariat, Guangzhou, China.

Vymazal, J. (1995) *Algae and element cycling in wetlands.* Boca Raton, Fl, USA: CRC Press/Lewis Publisher.

Vymazal, J. (1996*a*) The use of subsurface-flow contructed wetlands for wastewater treatment in the Czech Republic. *Ecol. Engng* **7**, 1–14.

Vymazal, J. (1996*b*) Subsurface horizontal-flow constructed wetlands for wastewater treatment—the Czech experience. In *Proc. Int. Workshop Nutrient Cyling and Retention in Wetlands and their Use for Wastewater Treatment* (ed. J. Vymazal), pp. 35–44. Prague, Czech

Republic: Ecology and Use of Wetlands/Trebon, Czech Republic: Institute of Botany.

Vymazal, J. (1997) Subsurface horizontal-flow constructed wetlands for wastewater treatment - the Czech experience. *Wetlands Ecol. Mgmt* 4, 199–206.

Vymazal, J. (1998*a*) Czech Republic. In *Constructed wetlands for wastewater treatment in Europe* (ed. J. Vymazal, H. Brix, P.F. Cooper, M.B. Green and R. Haberl), pp. 95–121. Leiden, The Netherlands: Backhuys Publishers.

Vymazal, J. (1998*b*) *Czech constructed wetlands database.* Prague, Czech Republic: Ecology and Use of Wetlands. [In Czech.]

Vymazal, J., Brix, H., Cooper, P.F., Green, M.B. and Haberl, R. (eds.) (1998) *Constructed wetlands for wastewater treatment in Europe.* Leiden, The Netherlands: Backhuys Publishers.

Vymazal, J., Dusek, J. and Kvet, J. (1999) Nutrient uptake and storage by plants in constructed wetlands with horizontal sub-surface flow: A comparative study. In *Nutrient cycling and retention in natural and constructed wetlands*, (ed. J. Vymazal). Leiden, The Netherlands: Backhuys Publishers. (In the press.)

Wallace, W. and Nicholas, D.J.D. (1969) The biochemistry of nitrifying microorganisms. *Biol. Rev.* **44**, 359–402.

Wass, R.D. and Fox, P. (1993) Constructed wetlands for nonpoint source control of wastewater from a vehicle maintenance yard. *Proc. Env. Fed. 66th Annual Conference, Anaheim, CA, USA*, vol. 9, pp. 9–20. WEF, Alexandria, VA, USA.

Watson, J.T., Reed, S.C., Kadlec, R.H., Knight, R.L. and Whitehouse, A.E. (1989) Performance expectations and loading rates for constructed wetlands. In *Constructed wetlands for wastewater treatment: municipal, industrial and agricultural* (ed. D.A. Hammer), pp. 319–351. Chelsea, MI, USA: Lewis Publishers.

WEF (Water Environment Federation) (1991) *Design of municipal wastewater treatment plants* (WEF Manual of Practice no. 8), vol. 2. WEF, Alexandria, VA, USA.

Weisner, S.E.B. and Ekstam, B. (1993) Influence of germination time on juvenile performance of *Phragmites australis* on temporarily exposed bottoms - implications for the colonization of lake beds. *Aquat. Bot.* **45**, 107–118.

Weisner, S.E.B., Graneli, W. and Ekstam, B. (1993) Influence of submergence on growth of seedlings of *Scirpus lacustris* and *Phragmites australis*. *Freshwat. Biol.* **29**, 371–375.

Wieder, R.K. (1989) A survey of constructed wetlands for acid coal mine drainage treatment in the eastern United States. *Wetlands* **9**, 299–315.

Wieder, R.K., Linton, M.N. and Heston, K.P. (1990) Laboratory mesocosm studies of Fe, Al, Mn, Ca, and Mg dynamics in wetlands exposed to synthetic acid coal mine drainage. *Wat. Air Soil Pollut.* **51**, 181–196.

Wolstenholme, R. and Bayes, C.D. (1990) An evaluation of nutrient removal by the Reed Bed Treatment System at Valleyfield, Fife, Scotland. In *Constructed wetlands in water pollution control* (ed. P.F. Cooper and B.C. Findlater), pp. 139–148. Oxford, UK: Pergamon Press.

Worrall, P. (1995) *Heathrow Airport Eastern balancing reservoir trial reedbed system interim report 3 (final) stage 1.* Stockport, Cheshire, UK: Penny Anderson Associates.

Worrall, P., Peberdy, K.J. and Millett, M.C. (1996) Constructed wetlands and nature conservation. In *Proc. 5th Internat. Conf. Wetland Systems for Water Pollution Control*, pp. V/1-1–V/1-9. Vienna, Austria: Universität für Bodenkultur Wien.

WPCF (Water Pollution Control Federation) (1990) *Natural systems for wastewater treatment* (Manual of Practice FD-16). WEF, Alexandria, VA, USA.

Ye, Z.H., Baker, A.J.M. and Wong, M.H. (1994) Heavy metal tolerance, uptake and accumulation in populations of *Typha latifolia* L. and *Phragmites australis* (Cav.) Trin. ex Steudel. In *Proceedings of 4th International Conference on Wetland Systems for Water Pollution Control, Guangzhou, China*, pp. 297–306. ICWS '94 Secretariat, Guangzhou, China.

Zhang, T., Ellis, J.B., Revitt, D.M. and Shutes, R.B. (1990) Metal uptake and associated pollution control by *Typha latifolia* in urban wetlands. In *Proceedings of the International Conference on the Use of Constructed Wetlands in Water Pollution Control, Cambridge, UK, 24–28 September 1990* (ed. P.F. Cooper and B.C. Findlater), pp. 451–459. Oxford, UK: Pergamon Press.

Index

A
acetic acid-forming bacteria, anaerobiosis 59
acid (coal mine) drainage 32–3, 76–7
adsorption processes
  metals 74–5
  nitrogen 66
aerobic degradation
  and BOD 58–9
  denitrifying bacteria 64
age of constructed wetlands 110
agricultural runoff 31–2
  Florida (Everglades), Nutrient Removal project 106, 124–5
  free water surface constructed wetlands, case studies 124–5
  nitrogen species 66
  nutrient–sediment control systems (NSCSs) 31
agricultural usage 30–1
algae
  characteristics 7–8
  suspended solids (SS) creation 25
aluminium, wetland treatment 33
ammonia volatilization 62
ammonification (mineralization) 5–6, 62–3
Amoco, petrochemical wastewater 34
anaerobiosis
  defined 3
  electron transfer 4–5
  heterotrophs 59
  nitrification 63–5
  processes, and BOD 59
animal feeding wastewaters 31, 32
  Oregon State University Dairy Farm Treatment Wetlands 121–4
animal processing waste, nitrogen species 66
animal–plant interactions 6, 9, 97
animals
  attached, terminology 8
  and transported solids (TSS) 55–6
antibiotics, release by *Scirpus* spp. (bulrush) 96
Arizona
  Show Low Wetland case study 118–19
  Tres Rios, nitrate removal 107
Arrhenius equation 53
asexual reproduction 8
Australia
  faecal coliforms (FC) 73
  pulp and paper mills 36
  SSF wetlands, TSS profile 57
Austria
  BOD removal and partial nitrification 135–6
  tertiary nitrification 133–4
autotrophs 50, 58
Azolla spp. (mosquito ferns) 19

B
bacteria luminescence test (Microtox) 35
basin sealing 86–7
  testing 103, 104
  see also liners
benthos, defined 7–8
bibliography 141–9
biochemical oxygen demand (BOD) 58–61
  and flow distance 53
  performance
    annual patterns 61
    surface flow 59–61
    transitions, system startup 106–8
    variability 61
  performance equations 51–3
  processes, aerobic/anaerobic degradation 58–9
  rate constants vs depth 80
  rate constants for reduction 60
  removal, and partial nitrification, vertical subsurface flow, case studies 134–7
biomass, roots 6
Bowater (pulp and paper mills) 37
British Water Research Centre (WRc), reed bed treatment systems 11
Brookhaven, first US constructed wetland treatment systems 11
bulrush *see Scirpus* spp.

C
cadmium, removal data 76, 127
California, Chevron Richmond Refinery Wetland 35, 119–20
California, Arcata
  BOD rate constants 60, 80
  faecal coliforms, free water surface (FWS) constructed wetlands 74
Camphill Village Trust, Oaklands Park, Gloucestershire (UK), BOD removal and partial nitrification 134–5
Canada, first systems 11
Carex spp. (sedges) 18
  hydroperiod and depth tolerance 92
case studies 117–39
  floating aquatic plant system 138–9
  free water surface constructed wetlands 117–27
  horizontal subsurface flow 128–32
    combined sewer overflow 131–2
    domestic secondary treatment 128–31
    domestic wastewater tertiary treatment 128
  integrated natural treatment system 137–8
  sludge lagoon supernatant 138–9
  soil-based reed beds 127–8
    domestic secondary treatment 127–8
  vertical subsurface flow 132–37
    BOD removal and partial nitrification 134–7
    tertiary nitrification 132–4
cation-binding sites 4
cattail *see Typha* spp.
Champion International (pulp and paper mills) 37
chemistry of water 77–8
Chevron Richmond Refinery Wetland, California 35, 119–20
China, petrochemical wastewater 35
chromium
  removal data 76, 127
  wetland treatment 33
climatic factors
  propagation 98–9
  see also precipitation
coal mining 32–3
combined sewer overflows (CSOs) 28
companion natural technologies
  lagoons 12–13
  land application 13–14
  natural treatment wetlands 14
  overland flow 13
  rapid infiltration 13
compartmentation, design of system 85–6
Connell, Washington, potato processing water treatment 137–8
constructed wetlands
  age 110
  defined 1
  design of system 79–90
  economics 109–16
    capital costs 109–13

constructed wetlands (*cont.*)
economics (*cont.*)
cost comparisons 112, 113
operating costs 113–15
historical development 11–12
hydrology 2–3
improvement of environment 10, 97
improvement of water quality, mechanisms 9–10
largest (Kis-Balaton, Hungary) 11
monitoring 114–15
minimum requirements 114
oxidation states 4
performance data, septic tank effluent 24
vs ponds, productivity 17
soils 6
structural components 1–2
system startup 103–8
technology description 9–11
water storage 3
see also design; free water surface (FWS); soils; subsurface flow (SSF) wetlands
copper
removal data 76, 127
wetland treatment 33
costs *see* economics of constructed wetlands
CO_2, reduction to methane 5
crop runoff *see* agricultural runoff
Cryptosporidium 72–3
cuttings, propagation 98
Czech Republic
HF SSF wetlands 54, 57, 60–1, 67
Prague, domestic secondary treatment 130–1

D

dairy wastewater 31, 121–2
Daphnia magna, toxicity test 35
Darcy's law 45
decomposition rates, macrophytes, parts 9
deep-water zones 86
denitrification 5–6, 18, 63–5
denitrifying bacteria 64
Denmark
performance equations, regression equations 57, 60, 67
soil-based reed beds, case study of domestic secondary treatment 127–8, 129
depth tolerance, and flooding duration (hydroperiod) 92
design of system 79–90
basin sealing 86–7
compartmentation 85–6
hydraulic sizing, subsurface horizontal/vertical flow wetlands 84–5
inlet/outlet structures 88–90
pond zones 86
sizing the system 79–85
substrate selection 87–8
subsurface flow systems 87–8
surface flow systems 87
system layout 85
tracer tests 48
treatment sizing 79–83
area setting 82–3
computations for each flow 79–80
computations for stormwater wetlands 83
hydraulic loading 81
parametric variability 80–1
stochastic variability 82
temperature 81–2
detention time 47, 58
detritus
components 3, 6, 21
decomposition rates 9
development 99–100
Dhulikhel hospital, Nepal, BOD removal and partial nitrification 136–7
domestic and municipal wastewaters 23–8
case studies
free water surface constructed wetlands 117–18
horizontal subsurface flow 128–31
vertical subsurface flow 132–7
combined sewer overflows (CSOs) 28
nitrogen species 66
tertiary and higher
SF/FWS 27–8
SSF 25–7
treatment
advanced treatment 23
natural vs conventional 12
preliminary treatment 23
pretreatment for irrigation use 14
primary treatment 23
secondary treatment 23–5
dosing events, and oxygen transfer 46

E

economics of constructed wetlands 109–16
capital costs 109–13
operating costs 113–15
present worth 115–16
Eichhornia crassipes (water hyacinth) 19
case study 138–9
nutrient uptake capacity 65, 95
electron transfer, anaerobiosis 4–5
Eleocharis spp. (spikerush) 18
hydroperiod and depth tolerance 92
Elodea canadensis (waterweed) 20
hydroperiod and depth tolerance 92
emergent macrophytes 17–18
energy balance
and available land area 109
and water temperature 48–9
environmental improvement, constructed wetlands 10, 97
epiphyte, defined 8
establishment of macrophytes 97–100
Europe, SSF wetlands in operation 23
European Design/Operations Guidelines for Reed Bed Treatment Systems (1990) 11–12
update (1996) 12
evapotranspiration 41, 43

F

faecal coliforms (FC)
free water surface (FWS) constructed wetlands 74
performance data, constructed wetlands 24, 72–3
see also pathogens
ferric iron, reduction to ferrous 5
field harvesting, propagules 97–8
fish, bottom disturbance 6
floating aquatic plant (FAP) treatment systems 18–19
case studies 138–9
floating mats 19–20
flood stress 101
flooding duration (hydroperiod), and depth tolerance 92
Florida, Tampa, Hidden River Corporate Office Park, stormwater 120–1
Florida (Everglades)
Nutrient Removal project 106, 124–5
petroleum storage, wetland treatment 35
wetland buffers, phosphorus 32
food wastes 33
free water surface (FWS) constructed wetlands
biological characteristics 10, 17
capital expenditure 10–11
case studies 117–27
agricultural runoff 124–5
animal wastewater 121–4
domestic 117–18
leachate 125–7
multiple benefit 118–19
petroleum 119–20
urban stormwater 120–1
faecal coliforms 74
historical development 11
hydraulic loading rates 10
inlets/outlets 88–90
planting density 91, 99
vegetation 91–2
with emergent macrophytes 17–18
with floating mats 19–20
with free-floating macrophytes 18–19
with rooted floating-leaved macrophytes 19
with submersed macrophytes 20
see also subsurface flow (SSF) wetlands
Freundlich equation 66

G

Georgia-Pacific, pulp and paper mills 37
Germany
first systems 11, 22
SSF wetlands in operation 23
Giardia 72–3
glossary of terms ix–xii
Glyceria maxima (sweet mannagrass) 10, 18

hydroperiod and depth tolerance 92
mats, floating 19–20
nutrient uptake capacity 65
granular medium filtration 55
gravel, costs 88, 109, 111

H
head loss 43–6
herbivory/grazing 101
heterotrophs 50, 58
Hidden River Corporate Office Park, Tampa, Florida 120–1
historical development of constructed wetlands 11–12, 21, 22
Hörbach, Upper Austria, BOD removal and partial nitrification 135–6
horizontal subsurface flow (SSF) wetlands 21–2
case studies 128–32
design of system
hydraulic sizing 84–5
substrate selection 88
water quality improvement
head loss 45–6
internal flow patterns 47
performance 67–8
see also reed bed treatment
Hornsby Bend Facility, Austin, Texas, floating aquatic plant system 138–9
humic substances, sorption properties 4
Hungary
largest constructed wetlands, (Kis-Balaton) 11, 106
petrochemicals, wastewater 35
hydraulic loading rate (HLR)
BOD data 81
defined 41–2
hydraulic profile, overland flow 44, 45
hydraulic retention time 12
see also soils
hydraulic sizing
horizontal subsurface flow (SSF) wetlands 84–5
vertical subsurface flow (SSF) wetlands 85
hydraulic variables
HEC2 code 45
SSF bed 45
Hydrocotyle umbellata (pennywort) 19
nutrient uptake capacity 65
hydrogen ions, water chemistry 78
hydrology, terminology 41–3
hydroperiod 2–3, 92
components 3

I
industrial wastewaters 32–7
coal mining 32–3
food wastes 33
metals, mining 33–4
petrochemicals 33–5
pulp and paper mills 36–7
infiltrating flows, water budget 42, 43
inlet/outlet structures, design 88–90
inspection and maintenance of macrophytes 100–1
integrated natural treatment systems 14–15, 137–8
International Conference on Constructed Wetlands 11
European Design/Operations Guidelines for Reed Bed Treatment 11–12
interpreting and predicting water quality improvement 41–54
dynamic effects 46–7
head loss 43–6
free water surface 43–5
horizontal subsurface 45–6
vertical subsurface 46
hydraulics 41
internal flow patterns 47–8
FWS wetlands 47
horizontal SSF wetlands 47
impact on performance and design 47–8
mineral background concentrations 50–1
performance 50–4
performance equations 51–4
stochastic variability 54
system hydrology 41–3
terminology 41
water budget 43
water temperature equations 53–4
thermal effects 48–50
energy balance and water temperature 48–9
snow and ice 49–50
iron
ferric, reduction to ferrous 5
removal data 76, 127
irrigation, wastewater pretreatment 14
Isanti–Chisago Leachate Treatment, Cambridge, Minnesota 125–7

J
Juncus spp. (rushes)
hydroperiod and depth tolerance 92
nutrient uptake capacity 65

L
lagoons 12–13, 25, 26
leachates 37–8
nitrogen species 66
land application, rapid infiltration systems 13–14
landfill leachate 37–8
layering, propagation 99
leachates
case studies 125–7
lagoons 37–8
lead, removal data 76, 127
Leek Wootton, Warwickshire (UK), domestic wastewater tertiary treatment 69, 128, 130
Lemna spp. (duckweed) 19
hydroperiod and depth tolerance 92
nutrient uptake capacity 65
light intensity, and *Phragmites australis* 95
Lighthorne Heath, Warwickshire (UK), combined sewer overflow 131–2
limestone drains, acid waters 32
liners
basin sealing 86–7, 103, 104
and gravel 88, 109, 111
synthetic 87
litter *see* detritus
Little Stretton, Leicestershire (UK), domestic secondary treatment 128–30
livestock feeding operations 31, 32, 121–4
long-term (low flow) systems, landfill leachates 38

M
macrophytes
aesthetic appearance 97
animal–plant interactions 6, 9, 97
annual vs perennial 8
capital costs 112
characteristics 8–9
colonization by volunteers 106
emergent 17–18
establishment 97–100
floating mats 19–20
free-floating 18–19
functions 7, 93–7
effects on soil hydraulic conductivity 94–5
nutrient uptake 95–6
physical effects 94
surface for microbial growth 95
summary 94
herbivory 101
hydroperiod and depth tolerance 92
inspection and maintenance 100–1
list 92
monitoring 101–2
nutrient uptake capacity 65
pathogens 101
plant–animal interactions 6
recycling 9
reproduction 8
rooted floating-leaved 19
roots 6, 8, 94–7
submersed macrophytes 20
system startup, ecological transitions 104–6
temperature effects 101
types (emergent, floating and submerged) 8, 17–20
weeds 101
see also named species
maintenance of macrophytes 100–1
mammals, bottom disturbance 6
manganese
reduction 5, 6
removal data 76, 127
Manning's equation, turbulent flow 44
Marathon, petrochemical wastewater 36
matrix adsorption processes, nitrogen 66
Max Planck Institute Process (MPIP), RZM 11, 22
meadow/marsh/pond (MMP) system 11
see also water meadows
meat processing waste 33
nitrogen species 66

mechanisms and results for water quality improvement 55–78
ancillary water chemistry 77–8
hydrogen ions 78
oxygen 77–8
performance
biochemical oxygen demand 59–61
metals 76–7
nitrogen 66–9
pathogens 72–4
phosphorus 70–2
suspended solids 56–8
processes
biochemical oxygen demand 58–9
metals 74–6
nitrogen 61–6
pathogens 72
phosphorus 69–70
suspended solids 55–6
Medmenham, Buckinghamshire (UK), tertiary nitrification 132–3
mercury, removal data 76, 127
metals 74–7
buildup in constructed wetlands 4
coal mining 32–3
metal mining 33–4
performance
mine drainage 76–7
surface flow 76
processes 75
adsorption and cation exchange 74–5
filtration 75
microbially mediated processes 75
plant uptake 75–6
removal data 76, 127
stormwater 29
methane, formation 6, 59
Michigan, Vermontville, domestic FWS scheme 117–18
microbes
growth on macrophyte surface 95
see also pathogens
microbial processes 5–6
denitrification 5–6, 18, 63–5
metals 75
nitrogen tranformation 5–6
Middleton, Shropshire (UK), domestic secondary treatment 130
mine drainage, metals 32–3, 76–7
mineral solids accumulation and blockage 112
mineralization rates 5–6, 62–3
Minnesota, Cambridge, Isanti–Chisago Leachate Treatment 125–7
Minnesota barrel test, basin testing 103, 104
monitoring
constructed wetlands 114–15
macrophytes 101–2
municipal waste *see* domestic and municipal wastewaters
Myriophyllum spp. (water milfoil) 20

N

Najas spp. 20
naphthenic acids (NAs) 35
natural gas, produced water 35–6
natural treatment systems, companion 12–15, 137–8
Nelumbo spp. (lotus) 19
Nepal, Dhulikhel hospital, BOD removal and partial nitrification 136–7
net primary productivity 8
Netherlands, combined sewer overflows (CSOs) 28
Nevada, Incline Village system, capital costs 110
nickel, removal data 76, 127
nitrate
reduction 4–5, 6
removal patterns 107
nitrifiers, nitrification 5–6, 18, 63–5
nitrogen 61–9
composition
outlet concentrations FWS wetlands 67, 68, 69
various wastewaters 66
control in constructed wetland, 'water meadows' 33
mass loading rates, influents/effluents 26
performance
annual patterns 69
horizontal subsurface flow 67–8
surface flow 66–7
transitions, system startup 107
variability 69
vertical subsurface flow 68–9
processes
ammonia volatilization 62
ammonification (mineralization) 62–3
matrix adsorption 66
nitrification/denitrification 5–6, 18, 63–5
plant uptake 65–6
nitrogen cycle 63
nitrogenous oxygen demand (NOD) 77
nominal detention time, defined 42
North American Treatment Wetland Database (NADB) 51
regression correlation 56–7
Nuphar spp. (cowlily) 19
Nutrient Removal project, Florida (Everglades) 106, 124–5
nutrient uptake capacity, macrophytes 65, 95–6
nutrient–sediment control systems (NSCSs) 15, 31
Nymphaea spp. (water lily) 19

O

oil and grease treatment 35
oil refining *see* petrochemicals
oil-sand processing 35
Ontario, Canada, Listowel
BOD concentrations 53, 60, 62
faecal coliforms 74
P concentrations 72
operating costs, economics of constructed wetlands 113–15
Oregon State University Dairy Farm Treatment Wetlands 121–4
organic matter
accretion/accumulation 3, 6, 99–100
degradation in rhizosphere 9, 21
outlet structures, design 90
overland flow treatment systems 13
oxygen
ancillary water chemistry 77–8
BOD loading rates 12–13
BOD treatment 10
diffusion, flooded soils 3–4
dissolved, inlet/outlet 78
nitrogenous oxygen demand (NOD) 77
release by roots 77, 96–7
transfer, and dosing events 46
see also biochemical oxygen demand
oxygen deficiency
HF systems 21, 22
peat accumulation 3

P

paper and pulp mills 36–7
pathogens 72–4, 101
annual patterns 73
subsurface flow 73
surface flow 72–3
variability 73–4
peat accumulation, oxygen deficiency 3
performance equations
basic equations 51–2
first order equations 52–3
regression equations 52
periphyton, defined 7–8
person equivalent (PE), volume and loading of pollutants 82–3
petrochemicals
Chevron Richmond Refinery Wetland, California 35, 119–20
free water surface constructed wetlands, case studies 119–20
NPDES permit 34
oil-sand processing 35
polishing wastewater 35
produced water 35–6
refinery effluents 33–4
spills and washing 35
pH, buffer capacity 78
Phalaris arundinacea (reed canarygrass) 10, 18
hydroperiod and depth tolerance 92
nutrient uptake capacity 65
phosphorus
antecedent loading 108
ligand exchange reactions leading to removal 22
mass balance model 107
mass loading rates, influents/effluents 26
performance
annual patterns 71
subsurface flow 71
surface flow 70–1
transitions, system startup 107–8
variability 71–2
pretreatment, FWS vs SSF wetlands 28

processes 69–70
sorption/burial/precipitation 6, 17, 18, 69–70
start-up phenomena 107–8
wetland buffers, Everglades 32
wetlands vs chemical treatment, cost comparisons 116
Phragmites australis (common reed) 10, 18
hydroperiod and depth tolerance 92
incident light intensity 95
mats, floating 19–20
metals, removal data 76, 127
nutrient uptake capacity 65
oxygen release 96–7
roots/rhizomes
biomass 57
depth distribution 100
development 100
sludge consolidation 38, 39
special features 92–3
water temperature effects 95
wind velocity 95
phytoplankton, defined 7–8
Pistia stratiotes (water lettuce) 19
planting density
free water surface (FWS) wetlands 91, 99
subsurface flow (SSF) wetlands 93, 99
planting depth, distribution of roots/rhizomes 100
plants *see* macrophytes
polishing
petrochemicals wastewater 35
ultra-polishing by vegetation 91
pollutants
loading rate (LR_i) 51
NPS *see* agricultural runoff
performance equations 51–3
removal efficiency (RED) 51–2, 76
stormwater concentrations 29
and water mass balance 43
see also named elements and pollutants
polycyclic aromatic hydrocarbons (PAHs) 35
pond systems (lagoons) 12–13
pond zones, design 86
Pope & Talbot (pulp and paper mills) 37
potato processing waste 33
Connell, Washington 137–8
integrated natural treatment system 15
nitrogen species 66
Prague, Kolodeje, domestic secondary treatment 130–1
precipitation
rain events 43, 46, 47
snow and ice, thermal effects 49–50
see also stormwater
productivity
constructed wetlands vs ponds 17
net primary 8
tropical marshes 9
tundra 8
propagation
climatic factors 98–9
cuttings 98
layering 99
soil moisture 99
soil preparation 99
propagules
field harvesting 97–8
seeds and seedlings 97–8
sources 97
Pseudomonas 72–3
pulp and paper mills 36–7

R
rain *see* precipitation
rapid infiltration systems 13
redox potential 4
reed
canarygrass *see Phalaris arundinacea*
common *see Phragmites australis*
reedmace *see Typha* spp.
reed bed treatment systems
British Water Research Centre (WRc) 11
sludge consolidation 38, 39
USA *see* vegetated submersed bed (VSB)
reproduction, asexual vs sexual 8
rhizosphere (root zone) 6
wastewater cleaning 21
riparian buffer strips 32
root zone method (RZM), origins 11, 21
roots/rhizomes
biomass 6
cuttings, propagation 98
depth distribution 100
development 100
nutrient uptake 95–6
oxygen release 96–7
winter 77
rhizosphere 6, 21
rush *see Eleocharis* spp. (spikerush); *Juncus* spp.; *Scirpus* spp. (bulrush)

S
Salmonella 72–3
Scandinavia, 'water meadows' 32
Schoenoplectus see Scirpus spp. (bulrush)
Scirpus spp. (bulrush) 10, 18
antibiotic release 96
hydroperiod and depth tolerance 92
nutrient uptake capacity 65
special features 93
SS detention 58
seeds and seedlings 97–8
septic tank effluent, CW performance data 24
Severn Trent Water
database 51
regression correlation 56–7
tertiary treatment 25–7
Show Low Wetland, Arizona 118–19
silver, removal data 76
sizing *see* design of system
sludge consolidation 38–9
sludge lagoon supernatant, floating aquatic plant system 138–9
soil-based reed beds, case study 127–8
soils 3–6
chemical environment 4–5
formation 3–4
hydraulic conductivity, macrophyte effects 94–5
hydraulic loading rates 10, 12
inundation changes 104–5
layering 105
microbial processes 5–6
moisture control 99
plant-animal interactions 6, 9
preparation, for propagation 99
previous upland 104–5
previous wetland 105
SF/SFF wetlands 6, 104–5
topsoil depth 99
sorption 4
phosphorus, expanded clay 6
Spartina cynosuroides (cordgrass) 36
Sphagnum spp. 32
hydroperiod and depth tolerance 92
Stone Container (pulp and paper mills) 37
stormwater
computations for wetlands 83
concentrations of pollutants 29
detention ponds 29
free water surface constructed wetlands, case studies 120–1
Hidden River Corporate Office Park, Tampa, Florida 120–1
treatment systems 14–15, 28, 29–30
layout 30
see also precipitation
Strengberg, Lower Austria, tertiary nitrification 133–4
submersed macrophytes 20
substrate selection, design of system 87–8
subsurface flow (SSF) wetlands
bed depth 10
capital expenditure 10–11
historical development 11–12
horizontal vs vertical flow systems 21–2
see also reed bed treatment
hydraulic loading rates 10
multicell configuration 24, 25
planting density 93, 99
sediments 6
vegetation 92–3
see also horizontal flow; vertical flow
sugar production 33
sulphate, reduction to sulphide 6, 75
sulphide, energy source for chemoautotrophs 6, 75
surface flow (SF) wetlands *see* free water surface (FWS) wetlands
suspended solids 55–6
budget 56
detention time 47, 58
performance
annual patterns 58
subsurface flow 56–8
surface flow 56
variability 58
processes
subsurface wetlands 55–6
surface-flow wetlands 55
transportable (TSS) 55–6

sweet mannagrass *see Glyceria maxima*
system startup 103–8
  antecedent conditions 103
  basin testing 103, 104
  ecological transitions 103–6
    soils and vegetation 104–6
  performance transitions 106–8

T
temperature *see* water temperature
Tenneco, natural gas wastewater 34
Texas, Austin, floating aquatic plant system 138–9
thermal effects, water quality improvement 48–50
Thiobacillus ferrooxidans 75
tidal inundation 46
tracer tests, design of system 48
tropical marshes, productivity 9
tundra, productivity 8
Typha spp. (cattail) 10, 18, 19, 36, 93
  hydroperiod and depth tolerance 92
  mats, floating 19–20
  metals, removal data 76
  nutrient uptake capacity 65
  special features 93

U
UK
  first systems 11
  SSF wetlands in operation 23
urban *see* domestic and municipal wastewaters
urban stormwater *see* stormwater treatment systems
USA
  EPA Nationwide Urban Runoff Project 29–30
  first systems 11
  SSF wetlands in operation 24
    peformance data 27
USDA Natural Resources Conservation Service, nutrient–sediment control systems (NSCSs) 15, 31

V
vegetable processing 33
  see also potato processing waste
vegetation *see* macrophytes
Vermontville, Michigan, domestic FWS scheme 117–18
vertical subsurface flow (SSF) wetlands 22
  BOD removal and partial nitrification 134–7
  case studies 132–7
  design of system
    hydraulic sizing 85
    substrate selection 88
  interpreting and predicting water quality improvement, head loss 46
  tertiary nitrification 132–34
void fraction, defined 42

W
wading birds, bottom disturbance 6
wastewaters, nitrogen species composition 66
water budget 2–3, 42, 43
components 4
water depth 42
water hyacinth *see Eichhornia crassipes*
'water meadows'
  nitrogen control wetland 33
  Scandinavia 32
water quality improvement *see* interpreting and predicting; mechanisms and results; pollutants
water regime 2–3
Water Research centre (WRc), Medmenham, Buckinghamshire (UK), tertiary nitrification 132–3
Water Services Association Reed Bed Treatment Systems Coordinating Group (1985) 11
water stress 101
water temperature
  coefficients 81–2
    BOD models 62
  effects of common reed 95
  effects on macrophytes 101
  and energy balance 48–9
  equations
    system hydrology 53–4
    water quality improvement 53–4
  snow and ice, thermal effects 49–50, 50
weeds 101
wetland plants *see* macrophytes
wetlands *see* constructed wetlands; free water surface (FWS) wetlands; subsurface flow (SSF) wetlands
Weyerhaeuser (pulp and paper mills) 36–7
wildlife
  animal–plant interactions 6, 9, 97
  exposure to poor quality waters/habitat 28
  see also case studies
wind velocity, *Phragmites australis* 95
Wolfern/farmhouse Schillhuber, Upper Austria, BOD removal and partial nitrification 135–6

Y
Yersinia 72–3

Z
zinc
  removal data 76, 127
  wetland treatment 33

IWA Publishing's authorised EU representative for General Product Safety Regulations is Diane D'Arras, 15 rue Duret, 75116 Paris, France, e-mail: safety@iwap.co.uk.

Printed and bound by CPI Group (UK) Ltd, Croydon, CR0 4YY
12/05/2026
02108893-0017